Antioxidant Capacity of Anthocyanins and other Vegetal Pigments: Modern Assisted Extraction Methods and Analysis

Antioxidant Capacity of Anthocyanins and other Vegetal Pigments: Modern Assisted Extraction Methods and Analysis

Editors

Agustín G. Asuero
Noelia Tena

MDPI • Basel • Beijing • Wuhan • Barcelona • Belgrade • Manchester • Tokyo • Cluj • Tianjin

Editors
Agustín G. Asuero
Departamento de Química
Analítica
Facultad de Farmacia
Universidad de Sevilla
Sevilla
Spain

Noelia Tena
Departamento de Química
Analítica
Facultad de Farmacia
Universidad de Sevilla
Sevilla
Spain

Editorial Office
MDPI
St. Alban-Anlage 66
4052 Basel, Switzerland

This is a reprint of articles from the Special Issue published online in the open access journal *Antioxidants* (ISSN 2076-3921) (available at: www.mdpi.com/journal/antioxidants/special_issues/ Antioxidant_Pigments_Extraction).

For citation purposes, cite each article independently as indicated on the article page online and as indicated below:

LastName, A.A.; LastName, B.B.; LastName, C.C. Article Title. *Journal Name* **Year**, *Volume Number*, Page Range.

ISBN 978-3-0365-5254-5 (Hbk)
ISBN 978-3-0365-5253-8 (PDF)

© 2022 by the authors. Articles in this book are Open Access and distributed under the Creative Commons Attribution (CC BY) license, which allows users to download, copy and build upon published articles, as long as the author and publisher are properly credited, which ensures maximum dissemination and a wider impact of our publications.

The book as a whole is distributed by MDPI under the terms and conditions of the Creative Commons license CC BY-NC-ND.

Contents

About the Editors vii

Preface to "Antioxidant Capacity of Anthocyanins and other Vegetal Pigments: Modern Assisted Extraction Methods and Analysis" ix

Noelia Tena and Agustin G. Asuero
Antioxidant Capacity of Anthocyanins and Other Vegetal Pigments: Modern Assisted Extraction Methods and Analysis
Reprinted from: *Antioxidants* 2022, 11, 1256, doi:10.3390/antiox11071256 1

Dalia Urbonaviciene, Ramune Bobinaite, Pranas Viskelis, Ceslovas Bobinas, Aistis Petruskevicius and Linards Klavins et al.
Geographic Variability of Biologically Active Compounds, Antioxidant Activity and Physico-Chemical Properties in Wild Bilberries (*Vaccinium myrtillus* L.)
Reprinted from: *Antioxidants* 2022, 11, 588, doi:10.3390/antiox11030588 5

Lourdes Carmona, Maria Sulli, Gianfranco Diretto, Berta Alquézar, Mónica Alves and Leandro Peña
Improvement of Antioxidant Properties in Fruit from Two Blood and Blond Orange Cultivars by Postharvest Storage at Low Temperature
Reprinted from: *Antioxidants* 2022, 11, 547, doi:10.3390/antiox11030547 19

Michal Fanyuk, Manish Kumar Patel, Rinat Ovadia, Dalia Maurer, Oleg Feygenberg and Michal Oren-Shamir et al.
Preharvest Application of Phenylalanine Induces Red Color in Mango and Apple Fruit's Skin
Reprinted from: *Antioxidants* 2022, 11, 491, doi:10.3390/antiox11030491 35

Rima Urbstaite, Lina Raudone and Valdimaras Janulis
Phytogenotypic Anthocyanin Profiles and Antioxidant Activity Variation in Fruit Samples of the American Cranberry (*Vaccinium macrocarpon* Aiton)
Reprinted from: *Antioxidants* 2022, 11, 250, doi:10.3390/antiox11020250 49

Ana V. González-de-Peredo, Mercedes Vázquez-Espinosa, Estrella Espada-Bellido, Marta Ferreiro-González, Ceferino Carrera and Gerardo F. Barbero et al.
Development of Optimized Ultrasound-Assisted Extraction Methods for the Recovery of Total Phenolic Compounds and Anthocyanins from Onion Bulbs
Reprinted from: *Antioxidants* 2021, 10, 1755, doi:10.3390/antiox10111755 65

Julio Salazar-Bermeo, Bryan Moreno-Chamba, María Concepción Martínez-Madrid, Domingo Saura, Manuel Valero and Nuria Martí
Potential of Persimmon Dietary Fiber Obtained from Byproducts as Antioxidant, Prebiotic and Modulating Agent of the Intestinal Epithelial Barrier Function
Reprinted from: *Antioxidants* 2021, 10, 1668, doi:10.3390/antiox10111668 87

Ceferino Carrera, María José Aliaño-González, Monika Valaityte, Marta Ferreiro-González, Gerardo F. Barbero and Miguel Palma
A Novel Ultrasound-Assisted Extraction Method for the Analysis of Anthocyanins in Potatoes (*Solanum tuberosum* L.)
Reprinted from: *Antioxidants* 2021, 10, 1375, doi:10.3390/antiox10091375 111

Tamara Ortiz, Federico Argüelles-Arias, Belén Begines, Josefa-María García-Montes, Alejandra Pereira and Montserrat Victoriano et al.
Native Chilean Berries Preservation and In Vitro Studies of a Polyphenol Highly Antioxidant Extract from Maqui as a Potential Agent against Inflammatory Diseases
Reprinted from: *Antioxidants* **2021**, *10*, 843, doi:10.3390/antiox10060843 . . . . **127**

Noelia Tena and Agustin G. Asuero
Up-To-Date Analysis of the Extraction Methods for Anthocyanins: Principles of the Techniques, Optimization, Technical Progress, and Industrial Application
Reprinted from: *Antioxidants* **2022**, *11*, 286, doi:10.3390/antiox11020286 . . . . **145**

Bianca Enaru, Georgiana Drețcanu, Teodora Daria Pop, Andreea Stănilă and Zorița Diaconeasa
Anthocyanins: Factors Affecting Their Stability and Degradation
Reprinted from: *Antioxidants* **2021**, *10*, 1967, doi:10.3390/antiox10121967 . . . . **185**

Antonio Morata, Carlos Escott, Iris Loira, Carmen López, Felipe Palomero and Carmen González
Emerging Non-Thermal Technologies for the Extraction of Grape Anthocyanins
Reprinted from: *Antioxidants* **2021**, *10*, 1863, doi:10.3390/antiox10121863 . . . . **209**

María Roca and Antonio Pérez-Gálvez
Metabolomics of Chlorophylls and Carotenoids: Analytical Methods and Metabolome-Based Studies
Reprinted from: *Antioxidants* **2021**, *10*, 1622, doi:10.3390/antiox10101622 . . . . **225**

About the Editors

Agustín G. Asuero

Agustin G. Asuero received his Ph.D. in Chemistry from the University of Seville, Spain in 1976. Professor of Analytical Chemistry in 1996. Professor Emeritus of the University of Seville since 2021. He is interested in parameter estimation methods (solution chemistry), qualimetrics, chemometrics, natural antioxidants (anthocyanins), history of science, and in the methodological aspects of analytical chemistry. He is a fellow of the United Kingdom Royal Society of Chemistry (FRSC) and a member of the Spanish Royal Society of Chemistry (RSEQ). He is a full member of the Iberoamerican Academy of Pharmacy and (elected) full member of the Royal National Academy of Pharmacy (Spain), and correspondent of the Royal Catalonian Academy of Pharmacy and of the Peruvian Academy of Pharmacy. University of Seville "Fame Award" for Scientific Research in Health Sciences. He has published over 250 research, educational and divulgation papers, 3 books and 20 book chapters. He has been Dean of the Faculty of Pharmacy and Vice Chancellor of the University of Seville.

Noelia Tena

Noelia Tena is Assistant Professor of Analytical Chemistry at the University of Seville. She holds a degree in Chemical Sciences (2005), a Master's Degree in High Specialisation in Fats (2006) and a PhD from the University of Seville in 2010. She is currently part of the editorial team of the journal *Antioxidants* as thematic advisor. Tena's research focuses on the analysis of the characterisation and sensory quality of food matrices. Her research line, "food safety and quality", is aimed at developing analytical methods to detect, quantify and prevent possible food fraud while guaranteeing food integrity. She currently focuses her interest on the impact of functional foods on health, in particular on the characterisation of these compounds in different matrices, the study of the antioxidant activity of some chemical compounds such as anthocyanin and polyphenols, and their role in human health and disease.

Preface to "Antioxidant Capacity of Anthocyanins and other Vegetal Pigments: Modern Assisted Extraction Methods and Analysis"

"Hence we must believe that all the Sciences are so inter-connected, that it is much easier to study the all together than to isolate one from the others."René Descartes (1596–1650), in Rules for the Direction of the Mind (1701).

The investigation of phenolic compounds from the chemical, biochemical and biological points of view during the last few decades has focused intensely on the study of anthocyanins. Interest in understanding the medicinal, therapeutic, and nutritional values of these naturally present phytochemical compounds has increased notably, with the number of papers published in this regard experiencing a significant increase. It is important to highlight the interdisciplinary nature of the research that is currently being carried out in this prolific area of work.

The highly bioactive anthocyanin compounds present in all plant tissues including leaves, stems, flowers, roots and fruits, e.g., berries, impart the colors orange, red, purple, and blue to many fruits, vegetables, grains, flowers, and plants. Anthocyanins are part of a group of secondary metabolites known as flavonoids, a subclass of the polyphenol family, being common components of the human diet. The identification and quantification of anthocyanins in natural products, as well as the elucidation of their effects in vivo and in vitro, requires the development of analytical techniques, and constitutes a challenge.

Anthocyanins also show an antioxidant activity largely dependent on their chemical structure. Numerous epidemiological studies have shown the beneficial effects of a diet rich in fruits and vegetables on human health, and in the prevention of various disorders related to oxidative stress, such as cancers and cardiovascular diseases. On the other hand, extracts rich in anthocyanins are highly valued in the food industry, given their coloring properties, as an alternative to the use of dyes and synthetic lakes. This is because various adverse effects on human health are attributed to synthetic dyes.

Anthocyanins are found in solution as a mixture of different secondary structures: a flavylium ion, a quinoid base, a carbinol base, and a chalcone pseudobase. Various stabilization mechanisms occur through self-association, and intermolecular and intramolecular copigmentation of anthocyanins, leading to the formation of tertiary structures. Anthocyanins show enormous potential for the food, pharmaceutical and cosmetic industries, although their use is limited by their relative instability (the Achilles heel) and low extraction percentages. Much progress has been made in the pretreatment of samples, having also developed some novel extraction techniques, as well as in stabilization techniques.

Thus, this reprint shows studies (Urbonaviciene et al.; Carmona et al.; Fanyuk et al. and Urbstaite et al.) applied to exotic or wild fruits, in which the aim is to elucidate how different external factors, such as climatic and/or geographical factors, affect the production, antioxidant capacity and physicochemical properties of anthocyanins and polyphenols. Moreover, other studies (González-de-Peredo et al.; Salazar-Bermeo et al.; Carrera et al. and Ortiz et al.) show the feasibility of applying certain extraction methods to obtain anthocyanins from different wild fruits or by-products. Finally, other authors (Tena et al.; Enaru et al.; Morata et al. and Roca et al.) review the state of the art of extraction methods for these precious compounds, not only polyphenols and anthocyanins, but also carotenes and chlorophyll pigments. These studies provide, among other things, interesting information on the experimental parameters to be applied for each type of extraction, including material preparation, extraction procedures, different recommendations and information on recovery rates. A critical discussion on the effects of biotic and abiotic stressors on living organisms, in which chlorophylls and carotenoids are involved, was also included in the last revision of this reprint.

In short, the possibilities of these vegetable chameleons, as Mikhail Tsvet (father of chromatography) baptized them, are enormous, and some interesting applications can be seen in this reprint published today by *MDPI*, which covers a fairly wide space, honoring the interdisciplinary nature of this polyhedric subject.

Last but not least, we would like to thank the authors for their brilliant contributions, which have made this reprint possible. We would also like to thank *Antioxidants* for trusting us as editors with the publication of this reprint. In addition, we cannot fail to mention the reviewers for the important work they have done, thus ensuring that the results shown in this reprint have sufficient rigor and scientific quality, and for which we also show our gratitude.

Agustín G. Asuero and Noelia Tena
Editors

 *antioxidants*

Editorial

Antioxidant Capacity of Anthocyanins and Other Vegetal Pigments: Modern Assisted Extraction Methods and Analysis

Noelia Tena and Agustin G. Asuero *

Departamento de Química Analítica, Facultad de Farmacia, Universidad de Sevilla, Prof. García González 2, 41012 Sevilla, Spain; ntena@us.es
* Correspondence: asuero@us.es

Citation: Tena, N.; Asuero, A.G. Antioxidant Capacity of Anthocyanins and Other Vegetal Pigments: Modern Assisted Extraction Methods and Analysis. *Antioxidants* **2022**, *11*, 1256. https://doi.org/10.3390/antiox11071256

Received: 22 June 2022
Accepted: 23 June 2022
Published: 26 June 2022

Publisher's Note: MDPI stays neutral with regard to jurisdictional claims in published maps and institutional affiliations.

Copyright: © 2022 by the authors. Licensee MDPI, Basel, Switzerland. This article is an open access article distributed under the terms and conditions of the Creative Commons Attribution (CC BY) license (https://creativecommons.org/licenses/by/4.0/).

Anthocyanins [1,2], chlorophylls, and carotenoids [3] are pigments responsible for the colour of many fruits, flowers, and plant tissues. In particular, anthocyanins show potential utility as natural dyes and for having a beneficial impact on health (as demonstrated by many epidemiological studies), as in addition to being safe and innocuous molecules, they have antioxidant properties and exert an effect on the gut microbiome. This special issue of Antioxidants contains twelve contributions—eight research articles and four reviews—including recent advances in the field. The interdisciplinary nature of the subject and the breadth of the content presented by the authors make this special issue very interesting and comprehensive. Modern methods of analysis of anthocyanins, their geographical variability, the improvement of their antioxidant properties, the valorisation of by-products, stability studies, and the metabolomics of chlorophylls and carotenoids are the subject of research or review. "Today, science has few boundaries and collaboration is the name of the game" [4].

Urbonaviciene et al. [5] carried out an evaluation of the antioxidant capacity and physicochemical properties of biologically active compounds (total polyphenol and total anthocyanin content) of wild blueberries by developing chemometric tools that make it possible to relate authenticity and quality control to geographical origin. This is the first study of its kind carried out on wild blueberries from several northern European countries.

The production of anthocyanins from blood oranges requires low temperature conditions, thus being a useful postharvest strategy to be applied in hot climates. Peña et al. [6] found a different response in the case of pear and Moorish oranges at the biochemical and molecular level, the changes being more prominent in the latter case. Blood orange has found use in traditional Asian medicine due to the vital bioactivity of the polyphenols it contains.

The red skin colour of some fruits, such as mango and apple, is vital for marketing and consumer acceptance. Alkan et al. [7] sought to determine whether external pre-harvest treatment of apples and mangoes with phenylalanine can promote red skin colouring of the fruit, and experimentally proved the hypothesis to be true, especially when combined with exposure to sunlight. The level of anthocyanin content of the treated peel of Cripps pink or May Kent apples, as determined by HPLC, increased in both cases.

Janulis et al. [8] performed a qualitative and quantitative analysis of anthocyanin and anthocyanidin composition in a variety of cultivars and genetic clones of American blueberries. The novelty is the growth under Lithuanian climatic conditions. Chemometric tools, such as hierarchical cluster analysis and principal component analysis, indicate that the Woodman cultivar is different from other cranberry cultivars, as its samples contain twice the average total amounts of anthocyanins. A correlation was observed between the total anthocyanin content and the anti-radical and reducing activity of the in vitro extracts.

Allium cepa L. has a wide abundance worldwide, its versatility being a feature in culinary uses with the bulbs also showing many interesting medicinal uses, due to their high content of bioactive compounds. Barbero et al. [9] developed assisted extraction methods for the phenolic and anthocyanin compounds present, making use of a Box–Behnken design for their optimisation. Both methods show high repeatability and intermediate precision, short extraction times, and good recoveries.

Food by-products with high content of dietary fibre and free and bound bioactive compounds are usually discarded. Persimmon by-products are an interesting source of fibre and bioactive compounds, as demonstrated by Valero et al. [10]. The effects of solvent extraction of persimmon dietary fibre by-products after in vitro gastrointestinal digestion and probiotic bacterial fermentation on techno- and physico-functional properties were evaluated.

The purple potato variety is not well known although it is as rich in nutrients, amino acids, and starches as other potato varieties. In addition, it has a high anthocyanin content and its consumption is attractive in relation to human health. Barbero et al. [11] developed a methodology based on ultrasound-assisted extraction to achieve a higher anthocyanin yield. The method has been applied to successfully extract and quantify anthocyanins found in Vitelotte, Double Fun, Highland, and Violet Queen potatoes.

Native Chilean berries (rich in total polyphenols and anthocyanins) were studied by Alcudia et al. [12], and a large-scale extract of maqui berries was tested on intestinal epithelial and immune cells and shown to have potential as a nutraceutical agent with health benefits for the treatment of inflammatory bowel disease (IBD). Total polyphenol content (Folin–Ciocullteau) and antioxidant capacity (DPPH, FRAP, and ORAC) were estimated, and the anthocyanin profile was assessed by ultra-high-performance liquid chromatography (UHPL-MS/MS).

Non-conventional extraction techniques meet the requirements of the food industry in terms of legal aspects, waste policy, safety, and environmental protection. However, the selection of a particular process is not an easy task and multiple factors are involved in planning the choice of the most suitable one. Tena et al. [13] provided an overview of recent applications in the field of anthocyanins extracted from different natural matrices, both by conventional and non-conventional techniques. Aspects such as the principles of the techniques involved, optimisation, technical progress, and industrial applications were considered and some useful recommendations were made.

The Achilles heel of anthocyanins is their lack of stability, which is affected by a number of factors such as pH, light, co-pigmentation, sulphites, ascorbic acid, oxygen, and enzymes. Diaconeasa et al. [14] reviewed all these factors affecting anthocyanin stability and degradation, assessing the impact of each parameter in order to minimise negative behaviour and consequently enhance the beneficial health effects.

The unstable nature of anthocyanins, which are affected as we have seen by changes in pH, oxidation, or high temperatures, requires the application of gentle non-thermal technologies for their extraction. Morata et al. [15] reviewed the characteristics, advantages, and disadvantages in the extraction of anthocyanins from grapes by applying non-thermal technologies such as Hugh hydrostatic pressure (HHP), ultra-high pressure homogenisation (UHPH), pulsed electric fields (PEF), ultrasound (US), irradiation, and pulsed light (PL). These techniques significantly increase extraction capacity while reducing extraction times and maintaining antioxidant capacity.

Chlorophylls and carotenoids are two families of antioxidants that include a large and complex number of compounds, present in daily food intake, with added value ingredients and functional properties. Their extraction and analysis require more powerful, precise, and accurate methods at hand, as well as a better understanding of the technical and biological context. Roca and Pérez-Gálvez [16] reviewed recent advances in the metabolomics of chlorophylls and carotenoids (pigmentomics), including material preparation and extraction procedures, and the use of instrumental techniques, e.g., spectroscopic and spectrometric (mass spectrometry to pigment metabolomics). The review also covered a critical account of studies showing the effects of biotic and abiotic stressors on living organisms, in which chlorophylls and carotenoids are involved.

Many thanks to the authors for their brilliant contributions, which have made this special issue possible. Many thanks also to Antioxidants for having us as guest editors for this special issue. Finally, we cannot leave out the reviewers for the important work they have done, which is worthy of mention and praise, and for which we also show our gratitude.

Funding: This research received no external funding.

Conflicts of Interest: The authors declare no conflict of interest.

References

1. Martin Bueno, J.; Sáez-Plaza, P.; Ramos-Escudero, F.; Jiménez, A.M.; Fett, R.; Asuero, A.G. Analysis and Antioxidant Capacity of Anthocyanin Pigments. Part II. Chemical Structure, Colour and Intake of Anthocyanins. *Crit. Rev. Anal. Chem.* **2012**, *42*, 126–151. [CrossRef]
2. Tena, N.; Martin, J.; Asuero, A.G. State of the Art of Anthocyanins: Antioxidant Activity, Sources, Bioavailability, and Therapeutic Effect in Human Health. *Antioxidants* **2020**, *9*, 451. [CrossRef] [PubMed]
3. Pérez-Gálvez, A.; Viera, I.; Roca, M. Carotenoids and Chlorophylls as Antioxidants. *Antioxidants* **2020**, *9*, 505. [CrossRef] [PubMed]
4. Jalavi-Heravi, M.; Arrastia, M.; Gómez, F.A. How can Chemometrics Improve Microfluid Research? *Anal. Chem.* **2015**, *87*, 3544. [CrossRef] [PubMed]
5. Urbanoviciene, D.; Bobinaite, R.; Viskelis, P.; Bobinas, C.; Petruskevicius, A.; Klavins, L.; Viskeli, J. Geographic Variability of Biologically Active Compounds, Antioxidant Activity and Physico-Chemical Properties in Wild Bilberries (*Vaccinium myrtillus* L.). *Antioxidants* **2022**, *11*, 588. [CrossRef] [PubMed]
6. Carmona, L.; Sulli, M.; Diretto, G.; Alquézar, B.; Alves, M.; Peña, L. Improvement of Antioxidant Properties in Fruit from Two Blood and Blond Orange Cultivars by Postharvest Storage at Low Temperature. *Antioxidants* **2022**, *11*, 547. [CrossRef] [PubMed]
7. Fanyuk, M.; Patel, M.K.; Ovadia, R.; Maurer, D.; Feygenberg, O.; Oren-Shamir, M.; Alkan, N. Preharvest Application of Phenylalanine Induces Red Color in Mango and Apple Fruit's Skin. *Antioxidants* **2022**, *11*, 491. [CrossRef] [PubMed]
8. Urbstaite, R.; Raudone, L.; Janulis, V. Phytogenotypic anthocyanin profiles and antioxidant activity variation in Fruit Samples of the American Cranberry (Vaccinium Macrocarpon Aiton). *Antioxidants* **2022**, *11*, 250. [CrossRef] [PubMed]
9. González-de-Peredo, A.V.; Vázquez-Espinosa, M.; Espada-Bellido, E.; Ferreiro-González, M.; Carrera, C.; Barbero, G.F.; Palma, M. Development of Optimized Ultrasound-Assisted Extraction Methods for the Recovery of Total Phenolic Compounds and Anthocyanins from Onion Bulbs. *Antioxidants* **2021**, *10*, 1755. [CrossRef] [PubMed]
10. Salazar-Bermeo, J.; Moreno-Chamba, B.; Martínez-Madrid, M.C.; Saura, D.; Valero, M.; Martí, N. Potential of Persimmon Dietary Fiber Obtained from Byproducts as Antioxidant, Prebiotic and Modulating Agent of the Intestinal Epithelial Barrier Function. *Antioxidants* **2021**, *10*, 1668. [CrossRef] [PubMed]
11. Carrera, C.; Aliaño-González, M.J.; Valaityte, M.; Ferreiro-González, M.; Barbero, G.F.; Palma, M. A Novel Ultrasound-Assisted Extraction Method for the Analysis of Anthocyanins in Potatoes (*Solanum tuberosum* L.). *Antioxidants* **2021**, *10*, 1375. [CrossRef] [PubMed]
12. Ortiz, T.; Argüelles-Arias, F.; Begines, B.; García-Montes, J.M.; Pereira, A.; Victoriano, M.; Vázquez-Román, V.; Bernal, J.L.P.; Callejón, R.M.; De-Miguel, M.; et al. Native Chilean Berries Preservation and In Vitro Studies of a Polyphenol Highly Antioxidant Extract from Maqui as a Potential Agent against Inflammatory Diseases. *Antioxidants* **2021**, *10*, 843. [CrossRef] [PubMed]
13. Tena, N.; Asuero, A.G. Up-To-Date Analysis of the Extraction Methods for Anthocyanins: Principles of the Techniques, Optimization, Technical Progress, and Industrial Application. *Antioxidants* **2022**, *11*, 286. [CrossRef] [PubMed]
14. Enaru, B.; Dretcanu, G.; Pop, T.D.; Stanila, A.; Diaconeasa, Z. Anthocyanins: Factors affecting their stability and degradation. *Antioxidants* **2021**, *10*, 1967. [CrossRef] [PubMed]
15. Morata, A.; Escott, C.; Loira, I.; López, C.; Palomero, F.; González, C. Emerging Non-Thermal Technologies for the Extraction of Grape Anthocyanins. *Antioxidants* **2021**, *10*, 1863. [CrossRef] [PubMed]
16. Roca, M.; Pérez-Gálvez, A. Metabolomics and Chlorophylls and Carotenoids: Analytical Methods and Metabolome-Based Studies. *Antioxidants* **2021**, *10*, 1622. [CrossRef] [PubMed]

Article

Geographic Variability of Biologically Active Compounds, Antioxidant Activity and Physico-Chemical Properties in Wild Bilberries (*Vaccinium myrtillus* L.)

Dalia Urbonaviciene [1,*], Ramune Bobinaite [1], Pranas Viskelis [1], Ceslovas Bobinas [1], Aistis Petruskevicius [1], Linards Klavins [2] and Jonas Viskelis [1]

[1] Lithuanian Research Centre for Agriculture and Forestry, Institute of Horticulture, 54333 Babtai, Lithuania; ramune.bobinaite@gmail.com (R.B.); pranas.viskelis@lammc.lt (P.V.); ceslovas.bobinas@lammc.lt (C.B.); aistis.petruskevicius@lammc.lt (A.P.); jonas.viskelis@lammc.lt (J.V.)

[2] Department of Environmental Science, University of Latvia, 1004 Riga, Latvia; linards.klavins@lu.lv

* Correspondence: d.urbonav@gmail.com; Tel.: +370-683-08757

Abstract: The aim of this study was to characterize the variation in biologically active compounds, antioxidant activity and physico-chemical properties in naturally grown bilberries gathered from different sites in Northern Europe. The variability in the biologically active compounds, antioxidant capacity and physico-chemical properties, as well as the development of tools for the authenticity and quality control of wild bilberries (*V. myrtillus* L.) in different geographical locations was evaluated. The berries of bilberries were handpicked during the summers of 2019 and 2020 during the time periods when they are typically harvested for commercial purposes in Northern Europe (Norway (NOR), Finland (FIN), Latvia (LVA) and Lithuania (LTU)). Berries from locations in NOR were distinguished by their higher mean TPC (791 mg/100 g FW, average), whereas the mean TPC of samples from the most southern country, LTU, was the lowest (587 mg/100 g FW). The TPC of bilberries ranged from 452 to 902 mg/100 g FW. The TAC values of investigated bilberry samples varied from 233 to 476 mg/100 g FW. A high positive correlation was found between TPC and antioxidant activity of the bilberry samples (R = 0.88 and 0.91 (FRAP and ABTS assays, respectively)), whereas the correlation between TAC and antioxidant activity was lower (R = 0.65 and 0.60). There were variations in the TPC and TAC values of investigated berries, suggesting that genotype also affects the TPC and TAC in berries. In 2020, the pH values and TSS contents of berries were significantly lower than in 2019. To the best of our knowledge, this is the first comprehensive reported evaluation of the biologically active compounds in wild bilberries from different Northern European countries using one laboratory-validated method.

Keywords: bilberry; natural habitats; polyphenol; anthocyanins; antioxidant activity; different geographical locations

1. Introduction

Authenticity and traceability are very important criterions of product quality. In order to assure quality, the chemometric approach can be efficiently used for berries and their products. Phenolic compounds are the major group of phytochemicals found in berries. They possess one or more aromatic rings with hydroxyl groups and their structures may range from that of a simple phenolic molecule to that of a complex high-molecular mass polymer [1]. The content of phenolic compounds in berries is determined by many factors, such as genotype, growing conditions, ripeness, storage time and other conditions [2–5]. Researchers have shown that the fruits of berry plants that grow in a cold climate with a short vegetation season and without fertilizers are marked by higher contents of phenolic compounds than the same varieties that grow in a milder climate [6,7]. Phenolic compounds can be used for the assessment of authenticity since genotype has a profound impact on the concentration and qualitative composition of these compounds in berries [2,8,9].

Anthocyanin pigments represent a very important group of phenolic substances because they determine the color of both berry fruits and their products [10]. Anthocyanins are glycosides of anthocyanidins of which the most important in the plants are: cyanidin, delphinidin, pelargonidin, peonidin, malvidin and petunidin. The distribution of these anthocyanidins in fruits and vegetables is 50%, 12%, 12%, 12%, 7% and 7%, respectively [11]. Glucose and rhamnose are the more common sugar moieties attached to aglycone (anthocyanidin), but galactose, arabinose, xylose, rutinose, sambubiose and other sugars are also frequently found [12]. Anthocyanins are usually present in colored flavylium cation form but may also be in uncolored form depending on the pH. This characteristic is an important issue when analyzing anthocyanins [13].

The berry genotype predetermines its typical profile of anthocyanins [10,14], thus the anthocyanin profile can be used to identify berry variety [10] for product quality control, label claim verification and raw material source identification [15].

The official method of anthocyanin pigment analysis is the spectrophotometric pH differential method. It is a rapid and simple quantitative method. The pH differential method is not useful for qualitative analysis of anthocyanins [16], but on the other hand, the spectrophotometric method for the quantification of total anthocyanins is widely used for the standardization of natural raw materials [17]. Other validated methods for identification and quantification of anthocyanins include high-pressure liquid chromatography (HPLC) [18,19]. In general, liquid chromatography remains the most common technique used for identification and characterization of anthocyanins in fruit extracts, whereas its combination with electrospray ionization mass spectrometry (ESI-MS) [15] or quadrupole/time-of-flight mass spectrometry (QTOF-MS) gives better quantitation of anthocyanins in more complex samples [19,20], because retention times, UV–vis spectra, mass/charge ratios as well as the type of anthocyanidin and the nature of sugar substitution can be easily recorded and used together to solve anthocyanin structures [21].

Different laboratories use different analysis parameters and conditions in order to optimize separation, thus it is difficult to point to one single (standard) HPLC procedure for determination of anthocyanins [19], but the columns chosen for the determination of anthocyanins are almost exclusively reverse phase columns composed of a C18 stationary phase. The solvent system (mobile phase) typically consists of an aqueous phase and an organic phase (mainly methanol or acetonitrile) with the acidic modifier (mainly formic or acetic acid) [19,22]. Acidic modifiers ensure that anthocyanins predominantly exist in their colored (flavylium cationic) form, having a maximum absorbance in the visible region around 520 nm [23].

The acid hydrolysis HPLC method simplifies the anthocyanin profile of berry samples by converting anthocyanins to anthocyanidin aglycones that can be separated and quantified using external standards. However, in this case part of the information concerning the precise composition and quantity of individual anthocyanins present is lost [24,25]. Furthermore, acid hydrolysis as part of the sample preparation tends to overestimate the anthocyanin content, since proanthocyanidins may be de-polymerized and converted into anthocyanidins [26].

Bilberry (*Vaccinium myrtillus* L.) is one of the richest sources of anthocyanin pigments among berry fruits [27,28]. These colored compounds accumulate not only in the skin of bilberries but also throughout the fruit flesh [25]. Bilberries are characterized by 15 anthocyanins. These include delphinidin, cyanidin, petunidin, peonidin and malvidin, glycosidically linked to glucose, arabinose or galactose [24,27,29]. As mentioned previously, the HPLC profiles of anthocyanins and anthocyanidins present in berries can be used as fingerprints for the evaluation of the authenticity of raw materials [30] as well as finished food products [31].

The aim of this study was to characterize the variation in biologically active compounds, antioxidant activity and physico-chemical properties in naturally grown bilberries gathered from different sites in Northern Europe. The variability of the biologically active compounds (polyphenol (TPC) and total anthocyanin (TAC) content), antioxidant capacity (FRAP; ABTS RSA) and physico-chemical properties (total soluble solids (TSS), pH), as well as the development of tools for the authenticity and quality control of wild bilberries (*V. myrtillus* L.) in different geographical locations was evaluated.

2. Materials and Methods

2.1. Extraction

For determination of anthocyanins, total phenols and antioxidant activity in 50 g of defrosted wild blueberries (*V. myrtillus* L.) were homogenized using Polytron (PT 1200E) (Kinematica, Luzern, Switzerland) for 5 min, then 5.000 g of the homogenized sample was extracted with 40 mL of acidified (0.5% HCl) aqueous ethanol solution (70% v/v). After a 24 h extraction at 25 °C with constant shaking at 160 rpm (Sklo Union LT, Teplice, Czech Republic), samples were filtered with a filter paper (Watman no.1) in a Buchner funnel and stored at 4 ± 1 °C until analyzed.

2.2. Analysis

2.2.1. Determination of Total Phenolic Content (TPC)

The total phenolic contents of berry extracts were determined using the Folin–Ciocalteu method as previously described by Bobinaite et al. [32]. Briefly, the test tubes were filled with 1.0 mL of appropriately diluted extract and mixed with 5.0 mL of Folin–Ciocalteu's phenol reagent diluted in distilled water (1/10, v/v) and 4.0 mL of Na_2CO_3 (7.5%). The absorbance of the test solution was read at 765 nm after 60 min of incubation in the darkness using a Genesys-10 UV/Vis spectrophotometer (Thermo Spectronic, Rochester, NY, USA). Gallic acid was used as the standard for the calibration curve and results are expressed in mg of gallic acid equivalents in 100 g of berries (FW) (Table 1).

Table 1. Summary of analysis methods.

Analysis Method	Assay	Unit
Polyphenol content (TPC)	Folin–Ciocalteu	mg GAE/100 g FW
Total anthocyanin content (TAC)	Liquid chromatography	mg C_3G/100 g FW
Antioxidant capacity	FRAP, ABTS	µmol TE/g FW
Total soluble solids (TSS)	Refractometric	Brix°
pH	Potentiometric	-

2.2.2. HPLC Analysis of Total Anthocyanin Content (TAC)

Anthocyanins were analysed using a Waters 2695 series HPLC system, equipped with a Waters 2998 photo diode array detector (DAD) (Waters Corporation, Milford, MA, USA). Analytical separation was carried out using a LiChroCART Purospher® STAR RP-18 endcapped column (250 × 4.6 mm, 5 µm particle size) with a guard column, Purospher STAR RP 18e 4.0 × 4.0 mm 5 µm (Merck KgaA, Darmstadt, Germany), using a slightly modified procedure of Lätti et al. [27]. The temperature of the column oven was set at 25 °C. The mobile phase consisted of aqueous 10% formic acid (eluent A) and ACN–MeOH (85:15, v/v) (eluent B). The gradient program was as follows: 0–2 min, 4–6% eluent B; 2–4 min, 6–8% eluent B; 4–12 min, 8–9% eluent B; 12–46 min, 9–11% eluent B; 46–48 min, 11–24% eluent B; 48–52 min, 24–34% eluent B; 52–59 min, 34–80% eluent B; 59–61 min, 80–20% eluent B; 61–65 min, 4% eluent B. The injection volume was 10 µL.

Anthocyanins were detected at the wavelength of 520 nm. DAD data were recorded from 200 to 600 nm. Anthocyanins in bilberry extracts were identified according to the HPLC retention times (RT) and UV absorbance maximum, in comparison with commercial standards or with literature data [27]. The chromatogram is added to supplementary materials (Figure S1).

Commercial standard (cyanidin-3-glucoside) was dissolved in solvent B (10%) and solvent A (90%) to generate a seven-point external standard calibration curve (with a concentration range from 1 to 100 mg/L) whose linearity was acceptable ($R^2 = 0.999$).

The total content of anthocyanins in the extracts was determined by the sum of the amounts of the individually quantified compounds as equivalents of cyanidin-3-glucoside (C_3G) per 100 g of FW of berry (Table 1).

2.3. Determination of Ferric-Reducing Antioxidant Power (FRAP)

A FRAP assay was performed according to the method of Benzie and Strain [33], with slight modifications [4].

For the FRAP assay, 0.3 M of sodium acetate buffer (pH 3.6) was prepared by dissolving 3.1 g of sodium acetate and 16 mL of acetic acid in 1000 mL of distilled water; a 10 mM TPTZ solution was prepared by dissolving 0.031 g of TPTZ in 10 mL of 40 mM HCl; and a 20 mM ferric solution was prepared by dissolving 0.054 g of $FeCl_3 \cdot 6H_2O$ in 10 mL of distilled water. Working FRAP reagent was prepared by freshly mixing acetate buffer, TPTZ and ferric solutions at a ratio of 10:1:1.

For the analysis, 2 mL of freshly prepared FRAP working solution and 20 µL of diluted extract were mixed and incubated for 30 min at ambient temperature. The change in absorbance due to the reduction of ferric-tripyridyltriazine (Fe III-TPTZ) complex by the antioxidants present in the samples was monitored at 593 nm using a Genesys-10 UV/Vis (Thermo Spectronic, Rochester, NY, USA) spectrophotometer. The absorptions of blank samples (by applying the same analysis conditions) were tested each time before and after analysis.

Trolox was used as the standard and the antioxidant activity is expressed as µmol of trolox equivalents (µmol TE) per g of berries (FW) (Table 1).

2.4. Determination of ABTS Radical Scavenging Activity (ABTS RSA)

The RSA of extracts was also measured by $ABTS^{\bullet+}$ radical cation assay [34]. ABTS solution (2 mM) was prepared by dissolving 2,2′-azinobis(3-ethylbenzothiazoline-6-sulfonic acid) diammonium salt in 50 mL of phosphate-buffered saline (PBS) obtained by dissolving 8.18 g NaCl, 0.27 g KH_2PO_4, 1.42 g Na_2HPO_4 and 0.15 g KCl in 1 L of pure water. The pH of the prepared solution was adjusted to 7.4 using NaOH. Then $K_2S_2O_8$ solution (70 mM) was prepared in pure water.

The working solution ($ABTS^{\bullet+}$ radical cation) was produced by reacting 50 mL of ABTS solution with 200 µL of $K_2S_2O_8$ solution and allowing the mixture to stand in the dark at room temperature for 15–16 h before use.

For the assessment of antiradical activity of the extracts, 2 mL of $ABTS^{\bullet+}$ solution was mixed with 20 µL extract in a 1 cm path length cuvette. The reaction mixture was kept at room temperature in the dark for 30 min and the absorbance was read at 734 nm.

Trolox was used as the standard and ABTS RSA is expressed as µmol of trolox equivalents (µmol TE) per g of berries (FW) (Table 1).

2.5. Determination of Dry Matter (DM), Total Soluble Solids (TSS) and pH

Dry matter content was determined after forced air convention drying at 105 °C to a constant weight (Table 1).

The total soluble solids were determined using a digital refractometer (ATAGO PR-32, Atago Co., Ltd., Tokyo, Japan) (Table 1).

The pH was measured using an inoLab Level 1 pH meter with a SenTix 81 (WTW) electrode (Table 1).

2.6. Statistical Analysis

All the experiments were carried out in triplicate. The mean values and standard deviations of the experimental data were calculated using SPSS 20 software (SPSS Inc., Chicago, IL, USA). One-way analysis of variance (ANOVA) along with the post hoc Tukey's HSD test were employed for statistical analysis. Differences were considered to be significant at $p < 0.05$.

3. Results

3.1. Sample Locations

The ripe berries of bilberry (*V. myrtillus* L.) were handpicked during the summers of 2019 and 2020 during the time periods when they are typically harvested for commercial purposes in Norway, Finland, Latvia and Lithuania in three different locations (Table 2).

Table 2. Locations of bilberry sample collection in Norway (NOR), Finland (FIN), Latvia (LVA) and Lithuania (LTU).

Sample No. (Collection Location)	Country Code	Coordinates	
		Latitude	Longitude
B1	NOR	69°41.66521'	18°59.46854'
B2	NOR	69°45.07693'	19°01.54336'
B3	NOR	69°40.25058'	18°37.08972'
B4	FIN	64°51.69040'	26°42.26600'
B5	FIN	64°59.17020'	25°54.21950'
B6	FIN	65°13.75280'	25°33.59240'
B7	LVA	57°08.55168'	21°51.95172'
B8	LVA	57°09.05976'	21°51.08952'
B9	LVA	57°08.79108'	21°52.32660'
B10	LTU	54°07.37150'	24°43.01752'
B11	LTU	54°43.29541'	23°30.53200'
B12	LTU	55°04.48091'	22°28.23829'

The berry samples were cooled immediately to below 10 °C then frozen and stored at −55 °C until use.

3.2. The pH and Total Soluble Solids (TSS) Content

The pH measures the acidity, and the total soluble solids (TSS) shows a high positive correlation with sugar content of fruits and berries [35]. The organoleptic quality and storage life of berries is related to their TSS content and acidity [36]. The pH values of investigated bilberries varied from 2.94 to 3.47 (Figure 1). In 2019, lower mean pH values were found in berries from LVA and NOR (3.30 and 3.32, respectively), whereas in 2020, the lowest mean pH was measured in bilberries from LTU (2.95). In 2020, the pH of bilberries from all countries was significantly lower than in 2019 (Figure 1).

Previously, Giovanelli and Buratti (2009) [37] reported that the pH of Italian bilberries ranged from 3.13 to 3.22. Turkben et al. (2008) [38] reported pH values between 2.77 and 2.95 among wild bilberries from western Turkey. These results are in accordance with the pH values estimated in our study.

The content of TSS in bilberries varied from 9.4 to 15.8 Brix° (Figure 2). In 2019, berries from NOR and FIN had higher mean TSS content (12.6 and 13.0 Brix°, respectively) than berries from LVA and LTU, whereas in 2020, the mean TSS content of berries from all countries was similar (Figure 2). In 2020, TSS values of the bilberry samples, with the exception of the ones collected in LTU location B10, were significantly lower than the respective values determined in 2019.

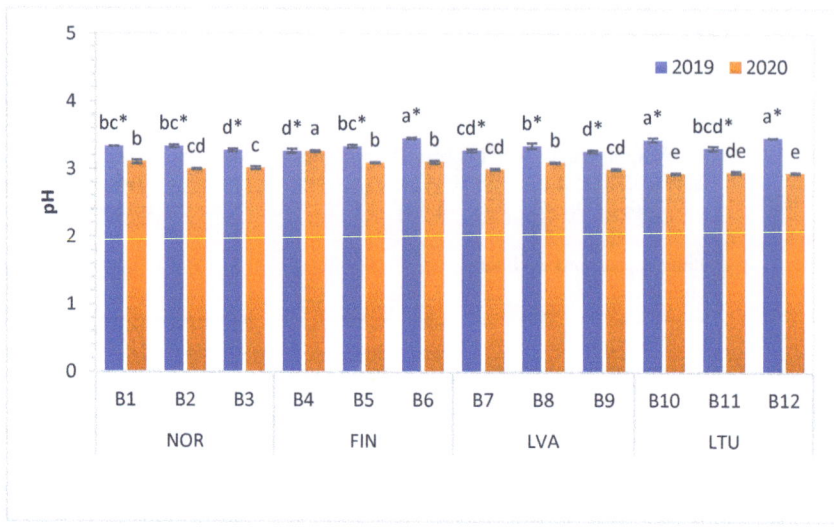

Figure 1. pH values of bilberries. Values are presented as means ± standard deviations. Different letters within the same column indicate significant differences between the collection locations (B1–B12) ($p < 0.05$). Significant differences between 2019 and 2020 are indicated by asterisks (*) ($p < 0.05$). Locations of bilberry sample collection include Norway (NOR), Finland (FIN), Latvia (LVA) and Lithuania (LTU).

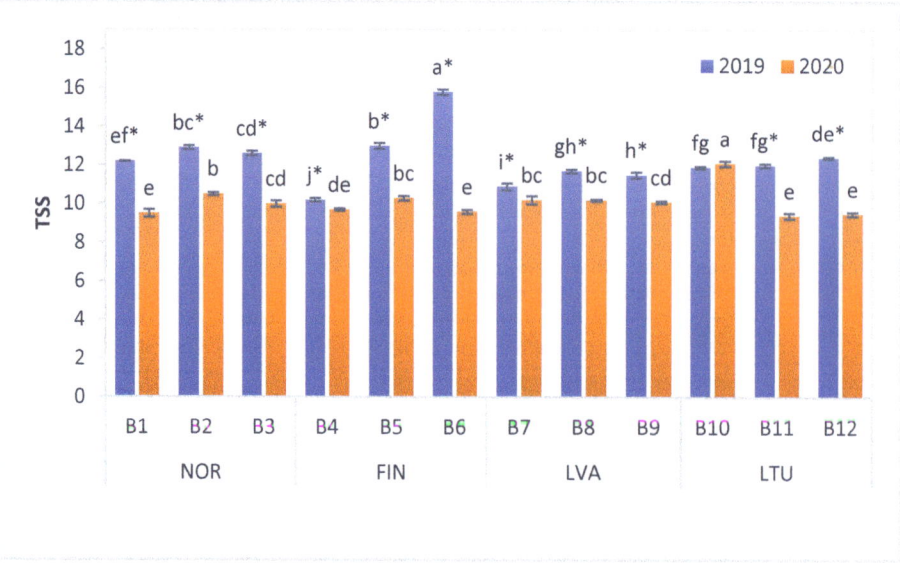

Figure 2. TSS values of bilberries. Values are presented as means ± standard deviations. Different letters within the same column indicate significant differences between the collection locations (B1–B12) ($p < 0.05$). Significant differences between 2019 and 2020 are indicated by asterisks (*) ($p < 0.05$). Locations of bilberry sample collection include Norway (NOR), Finland (FIN), Latvia (LVA) and Lithuania (LTU).

TSS values of investigated bilberries were in accordance with previously reported findings. Turkben et al. (2008) [38] reported that the TSS content in *V. myrtillus* berries from Turkey ranged from 9.0 to 11.0%. The TSS content in wild bilberries from Italy varied from 10.8 to 11.1% [37], whereas the TSS content in bilberries from Romania— from 9.2 to 13.7% [39]. TSS content in Polish bilberries was 13.0% [40]. Different growth and environment conditions such as temperature, day length and light intensity possibly influenced the TSS of the berries [15].

3.3. Total Phenolic Content (TPC)

The total phenolic content (TPC) of the tested bilberry samples ranged from 452 (LTU B12 in 2020) to 902 mg/100 g FW (NOR B2 in 2019) (Figure 3). The highest average TPC value was found in the berries collected in Norway (791 mg/100 g FW in 2019 and 660 mg/100 g FW in 2020). For both years (in 2019 and 2020), the lowest mean TPC value was found in berry samples from Lithuania (587 mg/100 g FW in 2019 and 546 mg/100 g in 2020).

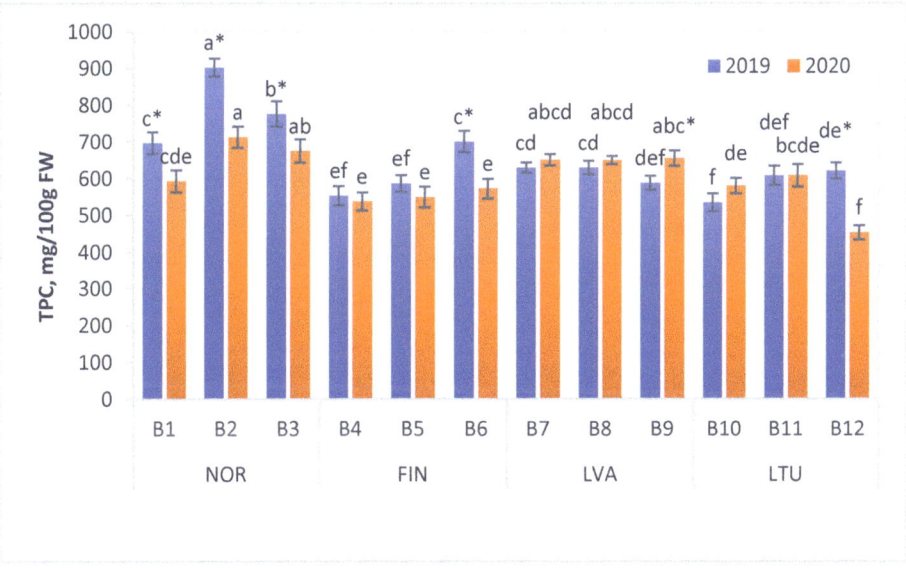

Figure 3. Total phenolic content (TPC) of bilberries mg/100 g FW. Values are presented as means ± standard deviations. Different letters within the same column indicate significant differences between the collection locations (B1–B12) ($p < 0.05$). Significant differences between 2019 and 2020 are indicated by asterisks (*) ($p < 0.05$). Locations of bilberry sample collection include Norway (NOR), Finland (FIN), Latvia (LVA) and Lithuania (LTU).

Total phenolic contents of investigated samples were in accordance with previously reported findings. For instance, the TPC value of bilberries collected from a natural population in Macedonia was 706 mg/100 g FW [41] and in bilberries from Serbia— 890 mg/100 g FW [42]. However, Milivojević et al. [43] determined a somewhat lower TPC of bilberries collected in Serbia (387 mg/100 g FW). The TPC of bilberries from the forest of Poland was reported to be 640 mg/100 g FW [39]. The TPC values of bilberries from natural populations in Norway were reported to range between 512 and 674 mg/100 g FW [44,45].

It has previously been shown that both the growing conditions and the genetic origin of the wild bilberries affect the contents of phenolic compounds [46,47]. The latitude-related factor was reported as having a high influence on the quality and quantity of phenolic compounds in bilberries, suggesting that higher phenolic contents may be supported by northern latitudes, altitude and sunny weather [48]. In previous studies, higher contents of phenolic compounds and anthocyanins were detected in bilberry clones originating from higher latitudes [27,46,49]. Interestingly, our data also show that the mean TPC of berries from Norway (the most northern country covered in the study) was the highest, whereas the mean TPC of samples from the most southern country (Lithuania) was the lowest (Figure 3). Furthermore, bilberry samples from Norway collected in the northernmost location (B2) had the highest TPC value, whereas the TPC values of the samples collected in the southernmost location (B1) in 2019 and 2020 were 23 and 17% lower, respectively. Similarly, among samples collected in Finland, the highest TPC value (700 mg/100 g FW in 2019 and 572 mg/100 g FW in 2020) was determined in berries from the northernmost location (B6) (Figure 3). In 2019, the same trend could also be observed for the bilberry samples from Lithuania, where TPC values also slightly increased with higher latitudes. On the other hand, for Lithuanian samples, this trend was not observed in 2020. The samples collected in Latvia had very similar TPC values, most likely due to the proximity of the sample collection sites. Our results also indicate that there were significant yearly variations in the TPC values of berries (Figure 3), suggesting that although genotype affects the TPC in bilberries, its final content also depends on weather conditions.

3.4. Total Anthocyanin Content (TAC)

Bilberry is one of the richest sources of anthocyanins, which have multiple biological activities [11]. The mean TAC values of investigated bilberry samples were 401.9 and 327.5 mg/100 g FW in 2019 and 2020, respectively. It is worth noting that within the same year, the mean TAC values of the berry samples from different countries did not differ significantly, with one exception—Lithuanian bilberries in 2019, which had a significantly lower mean TAC (Figure 4). The highest TAC values were from two berry samples collected in 2019 in the northernmost locations (B2 and B6) in Norway and Finland (475.4 and 454.6 mg/100 g FW, respectively).

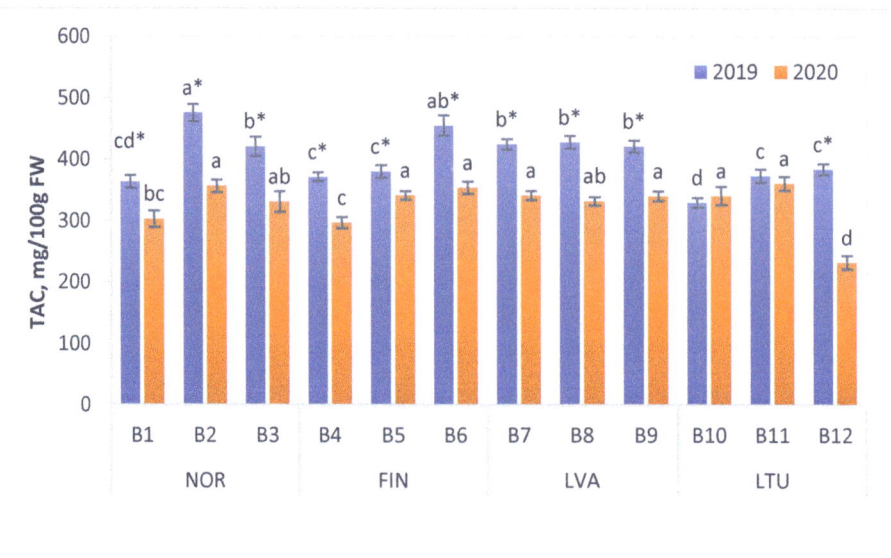

Figure 4. Total anthocyanin content (TAC) of bilberries mg/100 g FW. Values are presented as means ± standard deviations. Different letters within the same column indicate significant differences between the collection locations (B1–12) ($p < 0.05$). Significant differences between 2019 and 2020 are indicated by asterisks (*) ($p < 0.05$). Locations of bilberry samples collection include Norway (NOR), Finland (FIN), Latvia (LVA) and Lithuania (LTU).

TAC values of investigated bilberry samples were in accordance with previously reported findings. For instance, Skrede and colleagues [44] reported that the concentration of anthocyanins in bilberry samples ranged from 429 to 627 mg/100 g FW [44]. The TAC of bilberries from Macedonia was 507 mg/100 g FW [41]. Rohloff et al. [45] reported somewhat lower amounts of total anthocyanins in Norwegian bilberries (from 330 to 449 mg/100 g FW) [45], whereas the TAC of Finnish bilberries from 20 different populations varied from 350 to 525 mg/100 g FW [27].

It has been shown that genotype and environment interaction affect accumulation of anthocyanins in bilberries [7,45,47]. The increasing trend in anthocyanin content has been repeatedly observed in bilberries toward northern latitudes of Europe [27,49]. However, when effects of different environmental factors on berry chemical composition were studied in eight forest fields of bilberry in the north, the middle and the south of Norway, previous findings concerning latitudinal effects on anthocyanin concentration were not confirmed [45]. The authors concluded that environmental impacts probably confounded the genetic (population) effects [45].

With regard to increased TAC towards northern latitudes, the trend was not clear in the present study. In 2019, the highest TAC values were from two samples collected in the northernmost locations (B2 and B6), however, in 2020, the berries from locations B2 and B6 had high, but not the highest TAC values (Figure 4).

In 2020, TAC of the bilberry samples, with the exception of the ones collected in LTU location B10, were 3 to 39% lower than the respective values determined in 2019, which suggest significant influence of environmental factors on the accumulation of anthocyanins.

3.5. Antioxidant Activity (AA)

The antioxidant activity of berry samples was evaluated using FRAP and ABTS assays. In the ABTS assay, antioxidants suppress the generation of a blue-green ABTS radical cation by electron donation radical scavenging, whereas in the FRAP assay, there are no free radicals involved, but the reduction of ferric-to-ferrous iron is monitored.

In 2019, the FRAP of bilberry samples ranged from 36.0 (LTU B10) to 57.7 µmol TE/g FW (NOR B2) and in 2020, from 35.1 (LTU B12) to 49.1 µmol TE/g FW (NOR B2) (Figure 5). For both years of investigation, the highest mean FRAP values were from berry samples collected in Norway (50.6 µmol TE/g and 46.6 µmol TE/g FW in 2019 and 2020, respectively), followed by samples collected in Latvia (45.0 µmol TE/g and 46.3 µmol TE/g FW in 2019 and 2020, respectively). Bilberry samples from Lithuania had the lowest mean FRAP values (41.2 µmol TE/g and 40.2 µmol TE/g FW in 2019 and 2020, respectively). The FRAP results obtained in our study are close to those previously reported (53 and 57 µmol TE/g FW) in *V. myrtillus* fruits [50].

Berries showed higher antioxidant activity in the ABTS reaction system (Figure 6). The ABTS RSA of bilberries ranged from 60.9 (LTU B12 in 2020) to 106.0 µmol TE/g FW (NOR B2 in 2020). In 2019, the highest mean ABTS RSA was from berry samples collected in Norway (95.1 µmol TE/g FW), followed by samples collected in Finland (81.3 µmol TE/g FW), whereas in 2020, the highest mean ABTS RSA was from berry samples from Latvia (89.9 µmol TE/g FW) followed by samples from Norway (83.3 µmol TE/g FW) (Figure 6).

It was observed that bilberries from different geographical locations vary significantly in quantity of antioxidant activity. It is stated that plants are exposed to natural climatic stress due to environmental differences. This study's results show that in the case of wild bilberries, the natural variation in biologically active compounds and antioxidant activity has shown an authenticity of the berries from northern countries in comparison to more southern natural habits.

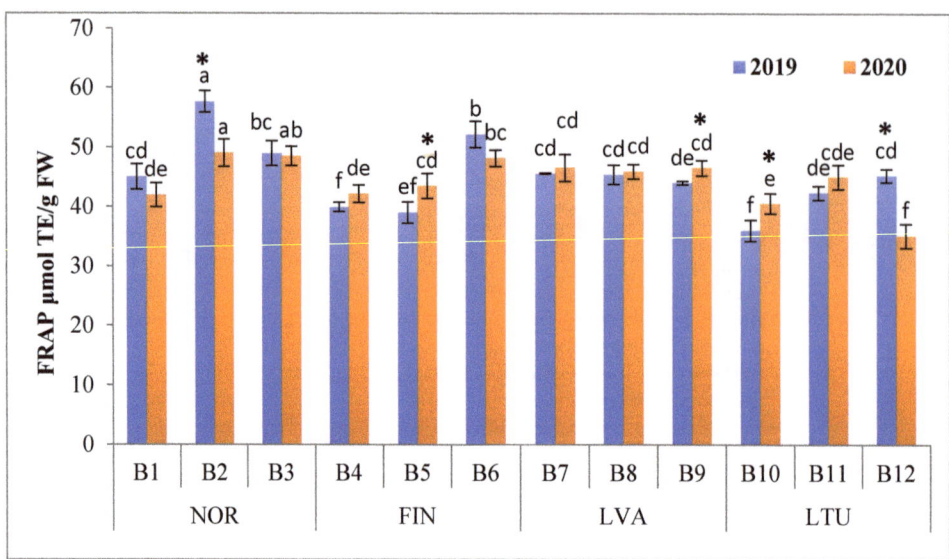

Figure 5. Ferric-reducing antioxidant power (FRAP) of bilberries (µmol TE/g FW). Different letters above the same color bars indicate significant differences between the mean values ($p < 0.05$). Significant differences between 2019 and 2020 are indicated by asterisks (*) ($p < 0.05$). Locations of bilberry samples collection include Norway (NOR), Finland (FIN), Latvia (LVA) and Lithuania (LTU).

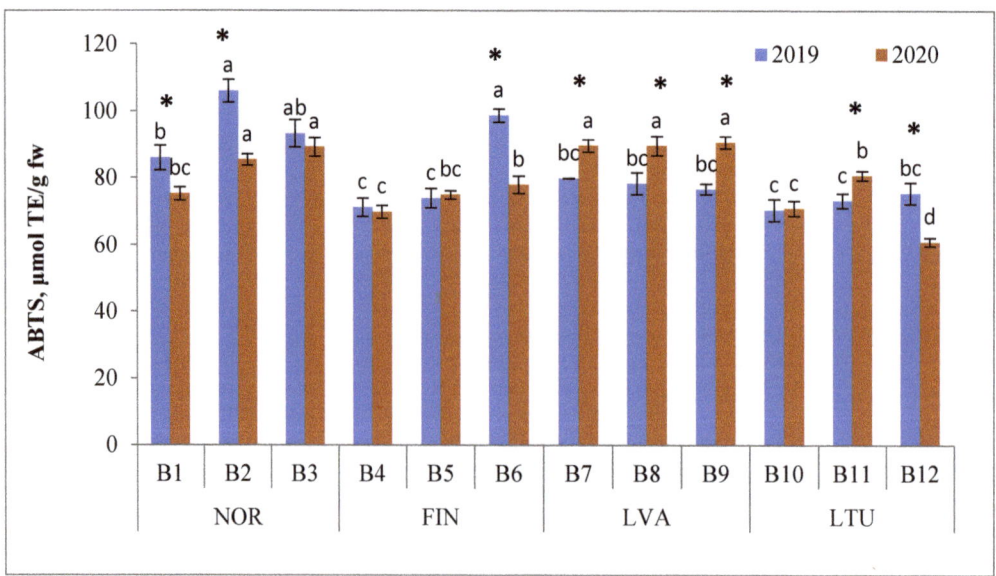

Figure 6. ABTS radical scavenging activity (RSA) of bilberries (µmol TE/g FW). Different letters above the same color bars indicate significant differences between the mean values ($p < 0.05$). Significant differences between 2019 and 2020 are indicated by asterisks (*) ($p < 0.05$). Locations of bilberry samples collection include Norway (NOR), Finland (FIN), Latvia (LVA) and Lithuania (LTU).

In agreement with previous findings [37,51] a high positive correlation was found between TPC and antioxidant activity of the bilberry samples (R = 0.88 and 0.91 as determined by the FRAP and ABTS assays, respectively), whereas the correlation between TAC and antioxidant activity was somewhat lower (R = 0.65 and 0.60 as determined by the FRAP and ABTS assays, respectively).

Previously, Giovanelli and Buratti [37] investigated cultivated blueberries (*Vaccinium corymbosum*) and wild bilberries (*Vaccinium myrtillus*) and reported that the antioxidant capacity of berries strongly correlated with the content of total anthocyanins and total phenolics. However, contrary to our findings, a higher correlation coefficient was found between the antioxidant capacity and total anthocyanin content (R = 0.93) than between the antioxidant capacity and total phenolic content (R = 0.89) of berries [37]. Uleberg et al. [46] also reported quite strong correlations between anthocyanins, total phenolics and antioxidant activity of bilberries.

4. Conclusions

With the increasing globalization of the food supply chain, food quality and safety have become of increased concern for consumers, food producers and governments. Although there is no reconciled definition of food fraud, it is generally accepted that food fraud is committed intentionally for financial gain through consumers. Nowadays, the most common food categories susceptible to any type of food fraud are the so-called premium food products, including non-timber forest products such as bilberries. The biochemical composition and nutritional value of berries are particularly relevant in the production of various pharmaceutical products and food supplements. In addition, there are natural or unavoidable flaws which may be caused by climatic or annual conditions. Nevertheless, berry authenticity research is important to ensure that the berries offered for sale are of the nature, substance and quality expected by the purchaser. Bilberries have different amounts of bioactive substances (anthocyanins, phenolics, etc.), depending on the place of growth and climatic conditions, so buyers are interested in knowing where those berries came from, as berries from certain regions or countries will also fetch a higher price due to their unique qualities. Spectrophotometric methods are fast and easy and therefore highly suitable to be used in the comparison of berry quality, especially for commercial purposes.

This study provided information concerning polyphenol (TPC) and total anthocyanin (TAC) content, antioxidant capacity (FRAP; ABTS RSA) and physico-chemical properties (total soluble solids (TSS), pH) of wild bilberry (*V. myrtillus* L.) populations from the Northern Europe (Norway (NOR), Finland (FIN), Latvia (LVA) and Lithuania (LTU)), including three locations in each country. Variations in biologically active compounds, antioxidant activity and physico-chemical properties in wild bilberries (*Vaccinium myrtillus* L.) have not been previously systematically investigated. Some studies have been conducted in order to evaluate the extent of the variation in the anthocyanins of bilberries or blueberries in one country, but there is little information in the results indicating geomorphological influence on the biochemical and physico-chemical compositions of bilberries depending on their place of origin. These may be used as discriminating criteria for distinguishing between bilberries from different geographical origins.

In conclusion, our study results show that the pH values of investigated bilberries varied from 2.94 to 3.47. These results are in accordance with the pH values estimated from literature data (2.77–3.22) [36,37]. The TSS content of bilberries varied within a similar range—from 9.4 to 15.8 Brix°. In 2020, the pH values and TSS content of berries were significantly lower than in 2019, showing that different growth and environment conditions such as temperature, day length and light intensity possibly influence the TSS of the berries.

The TPC of bilberries ranged from 452 to 902 mg/100 g FW. The mean TPC of bilberries from Norway (the most northern country covered in the study) was the highest, whereas the mean TPC of samples from the most southern country (Lithuania) was the lowest. The TAC values of investigated bilberry samples varied from 232.7 to 475.5 mg/100 g FW and were somewhat lower in 2020 than in 2019, which could be due to several factors, including genetic variation and environmental conditions.

The ABTS RSA of bilberries ranged from 60.9 (LTU B12 in 2020) to 106.0 µmol TE/g FW (NOR B2 in 2020). In 2019, the highest mean ABTS RSA had berry samples collected in Norway (95.1 µmol TE/g FW), followed by samples collected in Finland (81.3 µmol TE/g FW), whereas in 2020, the highest mean ABTS RSA was from berry samples from Latvia (89.9 µmol TE/g FW) followed by samples from Norway (83.3 µmol TE/g FW). A high positive correlation was found between TPC and antioxidant activity of the bilberry samples ($R = 0.88$ and 0.91 as determined by the FRAP and ABTS assays, respectively), whereas the correlation between TAC and antioxidant activity was lower ($R = 0.65$ and 0.60 as determined by the FRAP and ABTS assays, respectively).

In recent years, a growing demand for healthy food has been noted in the market. Consumers are primarily interested in food which is appealing, helps in preventing various diseases and contains high levels of promoted bioactive compounds. Our results could therefore be very useful for nutritional studies on anthocyanin- and phenolic compound-rich food and berry authenticity research. Results indicated a geomorphological influence on the biochemical and physico-chemical compositions of bilberries depending on their place of origin. These may be used as discriminating criteria for distinguishing between bilberries from different geographical origins.

Supplementary Materials: The following supporting information can be downloaded online: https://www.mdpi.com/article/10.3390/antiox11030588/s1, Figure S1: HPLC-PDA chromatogram ($\lambda = 520$ nm) of the wild bilberry B2 (*Vaccinium myrtillus* L.) samples extracts of 2019 and 2020.

Author Contributions: Conceptualization, D.U., R.B., P.V., C.B., L.K. and J.V.; writing—original draft preparation, D.U., R.B., P.V., C.B., A.P. and P.V.; writing—review and editing, D.U., R.B., P.V., L.K. and J.V.; visualization, D.U. and J.V.; supervision, D.U., R.B. and J.V.; project administration, D.U., R.B., P.V. and J.V.; funding acquisition, D.U., R.B., P.V., C.B. and L.K. All authors have read and agreed to the published version of the manuscript.

Funding: This study was financed by the Lithuanian Research Centre for Agriculture and Forestry and attributed to the long-term research program, "Horticulture: agrobiological foundations and technologies".

Institutional Review Board Statement: Not applicable.

Informed Consent Statement: Not applicable.

Data Availability Statement: Data is contained within the article and Supplementary Materials.

Conflicts of Interest: The authors declare no conflict of interest.

References

1. Nile, S.H.; Park, S.W. Edible berries: Bioactive components and their effect on human health. *Nutrition* **2014**, *30*, 134–144. [CrossRef] [PubMed]
2. Josuttis, M.; Verrall, S.; Stewart, D.; Krüger, E.; McDougall, G.J. Genetic and environmental effects on tannin composition in strawberry (*Fragaria* × *ananassa*) cultivars grown in different European locations. *J. Agric. Food Chem.* **2013**, *61*, 790–800. [CrossRef] [PubMed]
3. Skrovankova, S.; Sumczynski, D.; Mlcek, J.; Jurikova, T.; Sochor, J. Bioactive compounds and antioxidant activity in different types of berries. *Int. J. Mol. Sci.* **2015**, *16*, 24673–24706. [CrossRef] [PubMed]
4. Bobinaitė, R.; Pataro, G.; Lamanauskas, N.; Šatkauskas, S.; Viškelis, P.; Ferrari, G. Application of pulsed electric field in the production of juice and extraction of bioactive compounds from blueberry fruits and their by-products. *J. Food Sci. Technol.* **2015**, *52*, 5898–5905. [CrossRef] [PubMed]
5. Raudonė, L.; Liaudanskas, M.; Vilkickytė, G.; Kviklys, D.; Žvikas, V.; Viškelis, J.; Viškelis, P. Phenolic profiles, antioxidant activity and phenotypic characterization of *Lonicera caerulea* L. berries, cultivated in Lithuania. *Antioxidants* **2021**, *10*, 115. [CrossRef] [PubMed]
6. Szajdek, A.; Borowska, E.J. Bioactive compounds and health-promoting properties of berry fruits: A review. *Plant Foods Hum. Nutr.* **2008**, *63*, 147–156. [CrossRef] [PubMed]
7. Zoratti, L.; Jaakola, L.; Häggman, H.; Giongo, L. Anthocyanin profile in berries of wild and cultivated *Vaccinium* spp. along altitudinal gradients in the Alps. *J. Agric. Food Chem.* **2015**, *63*, 8641–8650. [CrossRef] [PubMed]

8. Plessi, M.; Bertelli, D.; Albasini, A. Distribution of metals and phenolic compounds as a criterion to evaluate variety of berries and related jams. *Food Chem.* **2007**, *100*, 419–427. [CrossRef]
9. Zhang, J.; Yu, Q.; Cheng, H.; Ge, Y.; Liu, H.; Ye, X.; Chen, Y. Metabolomic approach for the authentication of berry fruit juice by liquid chromatography quadrupole time-of-flight mass spectrometry coupled to chemometrics. *J. Agric. Food Chem.* **2018**, *66*, 8199–8208. [CrossRef] [PubMed]
10. Veberic, R.; Slatnar, A.; Bizjak, J.; Stampar, F.; Mikulic-Petkovsek, M. Anthocyanin composition of different wild and cultivated berry species. *LWT-Food Sci. Technol.* **2015**, *60*, 509–517. [CrossRef]
11. Khoo, H.E.; Azlan, A.; Tang, S.T.; Lim, S.M. Anthocyanidins and anthocyanins: Colored pigments as food, pharmaceutical ingredients, and the potential health benefits. *Food Nutr. Res.* **2017**, *61*, 1361779. [CrossRef]
12. He, J.; Giusti, M.M. Anthocyanins: Natural colorants with health-promoting properties. *Annu. Rev. Food Sci. Technol.* **2010**, *1*, 168–187. [CrossRef]
13. Pascual-Teresa, S.; Sanchez-Ballesta, M.T. Anthocyanins: From plant to health. *Phytochem. Rev.* **2007**, *7*, 281–299. [CrossRef]
14. Wu, X.; Prior, R.L. Systematic identification and characterization of anthocyanins by HPLC-ESI-MS/MS in common foods in the United States: Fruits and berries. *J. Agric. Food Chem.* **2005**, *53*, 2589–2599. [CrossRef]
15. Primetta, A.K.; Jaakola, L.; Ayaz, F.A.; Inceer, H.; Riihinen, K.R. Anthocyanin fingerprinting for authenticity studies of bilberry (*Vaccinium myrtillus* L.). *Food Control* **2013**, *30*, 662–667. [CrossRef]
16. Lee, J.; Durst, R.; Wrolstad, R. AOAC 2005.02: Total Monomeric Anthocyanin Pigment Content of Fruit Juices, Beverages, Natural Colorants, and Wines—pH Differential Method. In *Official Methods of Analysis of AOAC International*; AOAC: Washington, DC, USA, 2005; pp. 37–39.
17. Council of Europe. *European Pharmacopoeia*. Strasbourg: Council of Europe, 4th ed.; p. 1099. Available online: https://www.edqm.eu/en (accessed on 10 January 2022).
18. United States Pharmacopeia (USP). *31–NF 26 Second Supplement; Powdered Bilberry Extract (Identification and Assay)*; United States Pharmacopeia (USP): Rockville, MD, USA, 2008; p. 685.
19. Ongkowijoyo, P.; Luna-Vital, D.A.; de Mejia, E.G. Extraction techniques and analysis of anthocyanins from food sources by mass spectrometry: An update. *Food Chem.* **2018**, *250*, 113–126. [CrossRef] [PubMed]
20. Karaaslan, N.M.; Yaman, M. Determination of anthocyanins in cherry cranberry by high-performance liquid chromatography-electrospray ionization-mass spectrometry. *Eur. Food Res. Technol.* **2016**, *242*, 127–135. [CrossRef]
21. Kähkönen, M.P.; Heinämäki, J.; Ollilainen, V.; Heinonen, M. Berry anthocyanins: Isolation, identification and antioxidant activities. *J. Sci. Food Agric.* **2003**, *83*, 1403–1411. [CrossRef]
22. Valls, J.; Millán, S.; Martí, M.P.; Borràs, E.; Arola, L. Advanced separation methods of food anthocyanins, isoflavones and flavanols. *J. Chromatogr. A* **2009**, *1216*, 7143–7172. [CrossRef] [PubMed]
23. Willemse, C.M.; Stander, M.A.; Tredoux, A.G.J.; Villiers, A. Comprehensive two-dimensional liquid chromatography analysis of anthocyanins. *J. Chromatogr. A* **2014**, *1359*, 189–201. [CrossRef] [PubMed]
24. Zhang, Z.; Kou, X.; Fugal, K.; McLaughlin, J. Comparison of HPLC methods for determination of anthocyanins and anthocyanidins in bilberry extracts. *J. Agric. Food Chem.* **2004**, *52*, 688–691. [CrossRef] [PubMed]
25. Burdulis, D.; Sarkinas, A.; Jasutiene, I.; Stackevicené, E.; Nikolajevas, L.; Janulis, V. Comparative study of anthocyanin composition, antimicrobial and antioxidant activity in bilberry (*Vaccinium myrtillus* L.) and blueberry (*Vaccinium corymbosum* L.) fruits. *Acta Pol. Pharm.* **2009**, *66*, 399–408. [PubMed]
26. Cassinese, C.; Combarieu, E.D.; Falzoni, M.; Fuzzati, N.; Pace, R.; Sardone, N. New liquid chromatography method with ultraviolet detection for analysis of anthocyanins and anthocyanidins in *Vaccinium myrtillus* fruit dry extracts and commercial preparations. *J. AOAC Int.* **2007**, *90*, 911–919. [CrossRef] [PubMed]
27. Lätti, A.K.; Riihinen, K.R.; Kainulainen, P.S. Analysis of anthocyanin variation in wild populations of bilberry (*Vaccinium myrtillus* L.) in Finland. *J. Agric. Food Chem.* **2008**, *56*, 190–196. [CrossRef] [PubMed]
28. Može, Š.; Polak, T.; Gašperlin, L.; Koron, D.; Vanzo, A.; Ulrih, N.P.; Abram, V. Phenolics in Slovenian bilberries (*Vaccinium myrtillus* L.) and blueberries (*Vaccinium corymbosum* L.). *J. Agric. Food Chem.* **2011**, *59*, 6998–7004. [CrossRef] [PubMed]
29. Pataro, G.; Bobinaité, R.; Bobinas, Č.; Šatkauskas, S.; Raudonis, R.; Visockis, M.; Ferrari, G.; Viškelis, P. Improving the extraction of juice and anthocyanins from blueberry fruits and their by-products by application of pulsed electric fields. *Food Bioprocess Technol.* **2017**, *10*, 1595–1605. [CrossRef]
30. Koswig, S. Determination of foreign fruit types and fruit varieties—Analyses, evaluation and practical problems. *Fruit Processing*. **2006**, *6*, 401–412.
31. Filip, M.; Vlassa, M.; Copaciu, F.; Coman, V. Identification of anthocyanins and anthocyanidins from berry fruits by chromatographic and spectroscopic techniques to establish the juice authenticity from market. *JPC-J. Planar Chromatogr.-Mod. TLC* **2012**, *25*, 534–541. [CrossRef]
32. Bobinaité, R.; Viškelis, P.; Venskutonis, P.R. Variation of total phenolics, anthocyanins, ellagic acid and radical scavenging capacity in various raspberry (*Rubus* spp.) cultivars. *Food Chem.* **2012**, *132*, 1495–1501. [CrossRef] [PubMed]
33. Benzie, I.F.; Strain, J.J. The ferric reducing ability of plasma (FRAP) as a measure of "antioxidant power": The FRAP assay. *Anal. Biochem.* **1996**, *239*, 70–76. [CrossRef] [PubMed]
34. Re, R.; Pellegrini, N.; Proteggente, A.; Pannala, A.; Yang, M.; Rice-Evans, C. Antioxidant activity applying an improved ABTS radical cation decolorization assay. *Free. Radic. Biol. Med.* **1999**, *26*, 1231–1237. [CrossRef]

35. Viljakainen, S.; Visti, A.; Laakso, S. Concentrations of organic acids and soluble sugars in juices from Nordic berries. *Acta Agric. Scand. Sect. B-Soil Plant Sci.* **2002**, *52*, 101–109. [CrossRef]
36. Retamales, J.B.; Hancock, J.F. *Blueberries. Crop Production Science in Horticulture Series*; CAB International: Wallingford, UK, 2012; pp. 1–336.
37. Giovanelli, G.; Buratti, S. Comparison of polyphenolic composition and antioxidant activity of wild Italian blueberries and some cultivated varieties. *Food Chem.* **2009**, *112*, 903–908. [CrossRef]
38. Turkben, C.; Barut, E.; Incedayi, B. Investigations on population of blueberry (*Vaccinium myrtillus* L.) in Uludag (Mount Olympus) in Bursa, Turkey. *Akdeniz Üniversitesi Ziraat Fakültesi Derg.* **2008**, *21*, 41–44.
39. Oancea, S.; Moiseenco, F.; Traldi, P. Total phenolics and anthocyanin profiles of Romanian wild and cultivated blueberries by direct infusion ESI-IT-MS/MS. *Rom. Biotechnol. Lett.* **2013**, *18*, 8351.
40. Ochmian, I.; Oszmianski, J.; Skupien, K. Chemical composition, phenolics, and firmness of small black fruits. *J. Appl. Bot. Food Qual.* **2009**, *83*, 64–69.
41. Stanoeva, J.P.; Stefova, M.; Andonovska, K.B.; Vankova, A.; Stafilov, T. Phenolics and mineral content in bilberry and bog bilberry from Macedonia. *Int. J. Food Prop.* **2017**, *20*, S863–S883. [CrossRef]
42. Šavikin, K.; Zdunić, G.; Janković, T.; Tasić, S.; Menković, N.; Stević, T.; Đorđević, B. Phenolic content and radical scavenging capacity of berries and related jams from certificated area in Serbia. *Plant Foods Hum. Nutr.* **2009**, *64*, 212–217. [CrossRef]
43. Milivojević, J.; Rakonjac, V.; Pristov, J.B.; Maksimović, V. Classification and fingerprinting of different berries based on biochemical profiling and antioxidant capacity. *Pesqui. Agropecuária Bras.* **2013**, *48*, 1285–1294. [CrossRef]
44. Skrede, G.; Martinsen, B.K.; Wold, A.B.; Birkeland, S.E.; Aaby, K. Variation in quality parameters between and within 14 Nordic tree fruit and berry species. *Acta Agric. Scand. Sect. B-Soil Plant Sci.* **2012**, *62*, 193–208. [CrossRef]
45. Rohloff, J.; Uleberg, E.; Nes, A.; Krogstad, T.; Nestby, R.; Martinussen, I. Nutritional composition of bilberries (*Vaccinium myrtillus* L.) from forest fields in Norway–Effects of geographic origin, climate, fertilization and soil properties. *J. Appl. Bot. Food Qual.* **2015**, *88*, 274–287. [CrossRef]
46. Uleberg, E.; Rohloff, J.; Jaakola, L.; Trôst, K.; Junttila, O.; Häggman, H.; Martinussen, I. Effects of temperature and photoperiod on yield and chemical composition of northern and southern clones of bilberry (*Vaccinium myrtillus* L.). *J. Agric. Food Chem.* **2012**, *60*, 10406–10414. [CrossRef] [PubMed]
47. Mikulic-Petkovsek, M.; Schmitzer, V.; Slatnar, A.; Stampar, F.; Veberic, R. A comparison of fruit quality parameters of wild bilberry (*Vaccinium myrtillus* L.) growing at different locations. *J. Sci. Food Agric.* **2015**, *95*, 776–785. [CrossRef] [PubMed]
48. Ștefănescu, B.E.; Călinoiu, L.F.; Ranga, F.; Fetea, F.; Mocan, A.; Vodnar, D.C.; Crișan, G. Chemical composition and biological activities of the nord-west romanian wild bilberry (*Vaccinium myrtillus* L.) and lingonberry (*Vaccinium vitis-idaea* L.) leaves. *Antioxidants* **2020**, *9*, 495. [CrossRef] [PubMed]
49. Åkerström, A.; Jaakola, L.; Bång, U.; Jaderlund, A. Effects of latitude-related factors and geographical origin on anthocyanidin concentrations in fruits of *Vaccinium myrtillus* L. (bilberries). *J. Agric. Food Chem.* **2010**, *58*, 11939–11945. [CrossRef] [PubMed]
50. Nestby, R.; Percival, D.; Martinussen, I.; Opstad, N.; Rohloff, J. The European blueberry (*Vaccinium myrtillus* L.) and the potential for cultivation. *Eur. J. Plant Sci. Biotechnol.* **2011**, *5*, 5–16.
51. Dincheva, I.; Badjakov, I. Assesment of the Anthocyanin Variation in Bulgarian Bilberry (*Vaccinium Myrtillus* L.) and Lingonberry (*Vaccinium Vitis-Idaea* L.). *Int. J. Med. Pharm. Sci. (IJMPS)* **2016**, *6*, 39–50.

Article

Improvement of Antioxidant Properties in Fruit from Two Blood and Blond Orange Cultivars by Postharvest Storage at Low Temperature

Lourdes Carmona [1,2], Maria Sulli [3], Gianfranco Diretto [3], Berta Alquézar [1,2], Mónica Alves [2,4] and Leandro Peña [1,2,*]

[1] Instituto de Biología Molecular y Celular de Plantas, Consejo Superior de Investigaciones Científicas, Universidad Politécnica de Valencia, CP 46022 Valencia, Spain; lcarmona@ibmcp.upv.es (L.C.); beralgar@ibmcp.upv.es (B.A.)
[2] Fundo de Defesa da Citricultura (Fundecitrus), Sao Paulo 14807-040, Brazil; monicanelialves@gmail.com
[3] Agenzia Nazionale per le Nuove Tecnologie, l'Energia e lo Sviluppo Economico Sostenibile, Centro Ricerche Casaccia, Via Anguillarese, 301, Santa Maria di Galeria, 00123 Rome, Italy; maria.sulli@enea.it (M.S.); gianfranco.diretto@enea.it (G.D.)
[4] Faculdade de Ciências Agrárias e Veterinárias (FCAV), Universidade Estadual Paulista (UNESP), Jaboticabal 14884-900, Brazil
* Correspondence: lpenya@ibmcp.upv.es

Abstract: Numerous studies have revealed the remarkable health-promoting activities of citrus fruits, all of them related to the accumulation of bioactive compounds, including vitamins and phytonutrients. Anthocyanins are characteristic flavonoids present in blood orange, which require low-temperature for their production. Storage at low-temperature of blood oranges has been proven to be a feasible postharvest strategy to increase anthocyanins in those countries with warm climates. To our knowledge, no studies comparing the effect of postharvest storage effect on phenylpropanoid accumulation in cultivars with and without anthocyanins production have been published. We have investigated the effect of postharvest cold storage in flavonoid accumulation in juice from *Citrus sinensis* L. Osbeck in two different oranges: Pera, a blond cultivar, and Moro, a blood one. Our findings indicate a different response to low-temperature of fruit from both cultivars at biochemical and molecular levels. Little changes were observed in Pera before and after storage, while a higher production of phenylpropanoids (3.3-fold higher) and flavonoids (1.4-fold higher), including a rise in anthocyanins from 1.3 ± 0.7 mg/L to 60.0 ± 9.4 mg/L was observed in Moro concurrent with an upregulation of the biosynthetic genes across the biosynthetic pathway. We show that postharvest storage enhances not only anthocyanins but also other flavonoids accumulation in blood oranges (but not in blond ones), further stimulating the interest in blood orange types in antioxidant-rich diets.

Keywords: antioxidants; blood oranges; flavonoids; anthocyanins

1. Introduction

Nutraceuticals are phytochemical compounds found in vegetables and fruits, which are getting high consideration for their health-promoting effects when consumed with a certain frequency. Both fruits and vegetables are rich sources of polyphenols, including flavonoids, involved in reducing inflammation and oxidative stress related with chronic diseases, such as those derived from cardiovascular risks, and different types of cancer or diabetes [1–3]. Flavonoids, biosynthesised from the phenylpropanoid pathway (Figure 1), constitute the largest class of nutraceuticals in our diet [4]. Depending on their structure, flavonoids can be grouped into six main categories: flavones, flavonols, isoflavones, flavanones, flavanols, and anthocyanins. Among them, flavanones and anthocyanins present a higher antioxidant activity [5].

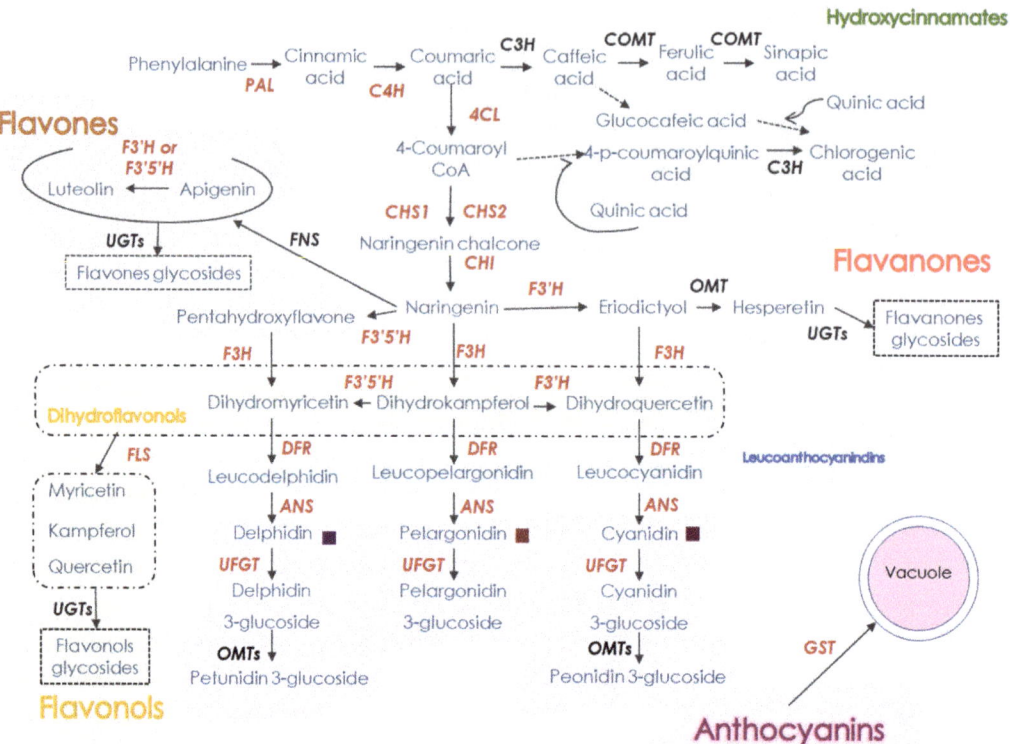

Figure 1. A schematic representation of the phenylpropanoid pathway. Red-bold labelled genes are those studied in this work. Gene names are abbreviated as follows: *PAL, phenylalanine ammonia-lyase; C3H, p-coumarate 3-hydroxylase, C4H, cinnamate 4-hydroxylase; 4CL, 4-hydroxy-cynnamoyl CoA ligase; CHS, chalcone synthase; CHI, chalcone isomerase; COMT, caffeic acid 3-O-methyltransferase; FNS, flavone synthase; F3H, flavanone 3-hydroxylase; F3'H, flavonoid 3'-hydroxylase; F3'5'H, flavonoid 3'5'-hydroxylase; GST, glutathione-S- transferase; OMTs, O-methyltransferases; FLS, flavonol synthase; DFR, dihydroflavonol 4- reductase; ANS, anthocyanidin synthase; UFGT,* uridine *diphosphate-glucose:flavonoid 3-O-glucosyltransferase* and *UGTs, O-methyltransferase.*

Citrus fruits are excellent sources of nutrients due to their abundance in vitamin C, sugars, dietary fibre, minerals and phytochemicals, including flavonoids [6–12]. The vital bioactivities of these secondary metabolites have made blood oranges (BO) to be used as traditional medicine in different Asian countries [13,14]. Flavones are found mainly in their peels, while flavanones are present in both peel and juice of oranges, mandarins, lemons and grapefruits [15]. Regarding anthocyanins, they are only accumulated in peel and juice of the BO, providing not only vivid colours but also a higher antioxidant activity to their pulp and juices [16–18]. In fact, it has been proposed that dietary anthocyanins are more effective antioxidants than vitamins E and C [19]. Other beneficial effects of anthocyanins include their anticancer activities, antiviral properties and protective effects against various metabolic, degenerative and cardiovascular diseases as well as eyesight or inhibiting viral replication [20–22].

Anthocyanin production in BO is very dependent on cold-temperature, making their quality at harvest to fluctuate between geographical locations and seasons. In specific cold regions of China, Spain and Italy, BO develop optimal colour, while countries with tropical climates such as Sao Paulo (Brazil) or Florida (USA) yield BO with a very low pigmentation [23]. The number of hours of exposition to low-temperature has been proposed as a crucial factor to get a strong purple/red coloration in the fruit [24]. Storage at either 4 or 8 °C has been reported to promote BO characteristic coloration because of the activation of anthocyanin biosynthetic genes [25–29], and by the increase in proteins related to anthocyanin biosynthesis, energy input, and other metabolic pathways associated with defence, oxidative and stress responses [30]. Storage at 9 °C in comparison to 4 °C has been demonstrated to be more effective for enrichment of anthocyanin production in BO [31] and an additional stress treatment (curing) further promotes anthocyanin production, increasing additionally flavonoids accumulation [32]. Low temperature conservation offers also the advantage of being the most frequently used technique to extend the postharvest life and preserve the quality of citrus fruits [33]. Additionally, in order to control pests, several countries require storage at low temperature for citrus fruits exportation to other markets [34].

The consumption of BO may be an important contribution to healthy diets [18,22]. Thus, juices and beverages flaunting red/purple orange in their composition increase their market value, owing to the established health-promoting potential of purple/red orange bioactive components [35]. Besides, increasing flavonoids accumulation in blond oranges by postharvest treatment can also improve the health-promoting properties of derived products or even of fruit consumption. To our knowledge, there is little information comparing the postharvest storage effect on phenylpropanoids accumulation in blond and BO. In this work, we investigated low-temperature storage effect on flavonoids and other phenylpropanoids in two selected cultivars of *Citrus sinensis* (L.) Osbeck, Pera (blond) and Moro (blood). Pera blond sweet orange is the most important citrus cultivar in Brazil (2nd worldwide citrus producer), constituting more than a third of the commercial acreage in São Paulo State [36]. Regarding BOs, Moro, characterised by yielding deeply purple-coloured fruits [17], is the most widely grown cultivar in Europe for food processing and other industrial applications, and the most common BO cultivar grown in the United States [36].

2. Materials and Methods

2.1. Plant Materials and Storage Conditions

The effect of postharvest storage on fruit colour was determined in two different cultivars of *Citrus sinensis* L. Osbeck, Moro (blood type) and Pera (blond type) mature fruits (8 months after the bloom). Fruits were harvested from adult trees grown (13 years old) and grafted on citrumelo Swingle under standard cultivation in a commercial orchard in Maringá (21°45′53″ S; 48°28′21.15″ O, 540 m), Gavião Peixoto-Sao Paulo State (Brazil). The local climate is Cwa type (mountain subtropical), characterised by dry winter (<1230 mm total rainfall of the year) and a normal average air temperature of ≥ 17 °C and ≥ 28 °C in the coldest and in the warmest month, respectively [37]. Maringá soil type is classified as *latossolo vermelho* (Red-oxisol), dystrophic A moderate type, and soft-moderately wavy relief [38]. A total of 80 fruits were harvest from 15 trees planted at 7.0 m × 3.0 m spacing. Fruits were uniform in size and colour, as well as free of damage or external defects. Harvested fruits were stored for 45 days at 9 °C, and 90–95% relative humidity in constant darkness. At the time of harvest (zero) and after 15, 30 and 45 days of storage, pulp and juice samples were taken and stored at −80 °C until analysis. Three replicate samples of 5 fruits each per storage time were used. Pulp was separated with a scalpel, and immediately frozen in liquid nitrogen and ground into a fine powder. Juice was extracted with a domestic squeezer (Braun GmbH, Germany), filtered through a metal sieve with a pore size of 0.8 mm and frozen in liquid nitrogen.

2.2. Determination of Internal Maturity Index, pH and Total Anthocyanin Quantification

Titration with phenolphthalein was used to determine the juice acidity, and data were expressed as mg citric acid per 100 mL [31]. Briefly, 5 mL of centrifuged juice were diluted to 45 mL with water (Sigma) and supplemented with 5 drops of phenolphthalein. Acidity was evaluated by titration with NaOH (0.1 N) and expressed as mg of citric acid per 100 mL. The determination of soluble solids content (°Brix) was estimated by refractometer, using an Atago® refractometer (Tokyo, Japan) as described by Carmona et al. [31]. Maturity index was calculated and expressed as the ratio of °Brix/acidity. The pH value was measured with a pH-meter Gehaka® (Sao Paulo, Brasil) [31].

2.3. Flavonoids, Anthocyanins and Phenylpropanoid Extraction and Identification

All compounds were extracted using 2 mg of Moro (blood type) or Pera (blond type) mature freeze-dried fruits pulp re-suspended in 20 mL of water (LC-MS Grade, LiChrosolv®, Merck, Darmstadt, Germany). Flavonoids and other phenylpropanoids were extracted as previously described by Carmona et al. [32], with the following modifications: 0.5 mL of each re-suspended pulp solution was shaken for 1 h at 20 Hz using a Mixer Mill 300 (Qiagen, Hilden, Germany) with 1.5 mL of methanol containing 0.3% formic acid, plus 2 µg/L of formononetin as the internal standard, and centrifuged at 15,000× g for 20 min (15 °C). The supernatants (0.4 mL) were filtered into vials for LC/MS analysis (Mini-UniPrep® syringeless filters with 0.2 µm pore size PTFE membrane, Whatman®, Maidstone, UK). LC/MS analysis was performed using an HPLC system equipped with a photodiode array detector (Dionex, ThermoFisher Scientific, Sunnyvale, CA, USA) coupled to a quadrupole-Orbitrap Q-exactive system (ThermoFisher Scientific, Sunnyvale, CA, USA). HPLC analysis was performed using a C18 Luna Column (Phenomenex, Aschaffenburg, Germany) (150 × 2.0 mm; 3 µm). Total of 5 µL of each extract were injected at a flow of 0.25 mL/min. Total run time was 32 min using an elution system running at 0.250 mL/min and consisting of (A) water (0.1% formic acid) and (B) acetonitrile: H_2O 90:10 (0.1% formic acid). Gradient was 0 to 0.5 min 95/5%-A/B, 24 min 25/75%-A/B, and 26 min 5/95%-A/B. MS analysis of flavonoids and other phenylpropanoids was carried out with a heated electrospray ionization (HESI) source operating in positive and negative ion mode. Mass spectrometer parameters were as follows: sheath and aux gas flow rate set at 40 and 10 units, respectively; capillary temperature was at 250 °C, discharge current was set at 3.5 µA and S-lens RF level at 50. The acquisition was carried out with m/z 110–1600 Full MS scan range, and the following parameters: resolution 70,000, microscan 1, AGC target 1e6, and maximum injection time 50. Data were analysed using the Xcalibur 4.4 software (ThermoFisher Scientific, Sunnyvale, CA, USA). Metabolites were identified as M+H and M-H adducts, based on their accurate masses (m/z) and MS fragmentation, using both in house database and public sources (e.g., KEGG, ChemSpider, PubChem, MetaCyc, Metlin, Phenol-Explorer). Relative abundances of the investigated flavonoids and other phenylpropanoids were calculated as fold average and the standard deviation of integrated areas under the m/z peak of the adduct of each compound and the internal standard peak area (Fold/ISTD), calculated with the Xcalibur 4.4 software (ThermoFisher Scientific, Sunnyvale, CA, USA).

Anthocyanins were analysed in samples obtained by mixing 500 µL of each re-suspended pulp solution with 500 µL 85:15 methanol:HCl (1 N) containing 0.3% formic acid and 2 µg/L of formononetin (Sigma-Aldrich, San Luis, MO, USA) as internal standard [39]. Samples were shaken for 12 h and subsequently centrifuged for 10 min at 15,300× g at 25 °C. Supernatants (0.25 mL) were transferred to new Eppendorf vials, dried by Speedvac concentrator, resuspended in 0.1 mL 75% methanol (plus 0.1% formic acid) and centrifuged (10 min at 15,300× g at 25 °C). Total of 70 µL of supernatant was transferred to HPLC vials for LC/MS analysis, and 10 µL of extract was injected to the HPLC-PDA/MS. The method used for separation performed with a C18 Luna column (Phenomenex, Aschaffenburg, Germany) (150 × 2.0 mm; 3 µm) was as previously described [39] and PDA detection was performed by an online Accela Surveyor photodiode array detector (PDA; ThermoFisher Scientific, Sunnyvale, CA, USA), acquiring continuously from 200 to 600 nm. Mass spectrometry analysis was performed using a quadrupole-Orbitrap Q-exactive system (ThermoFisher scientific, Sunnyvale, CA, USA) and ionization was carried out with a heated electrospray ionization (HESI) source operating in positive ion mode. The MS parameters used are as follows: nitrogen was used as sheath and auxiliary gas (45 and 15 units, respectively), capillary and vaporizer temperatures 30 °C and 270 °C, respectively, discharge current 4.0 KV, probe heater temperature at 370 °C, S-lens RF level at 50 V. The acquisition was carried out in the 110–1600 *m/z* scan range, resolution 70,000, microscan 1, AGC target 1e6, and maximum injection time 50 [39]. In detail, MS analysis was performed using a first full scan with data-dependent MS/MS fragmentation in order to identify the anthocyanins in pulp extracts. Subsequently, a single ion monitoring (SIM) with targeted MS/MS fragmentation was applied to identify anthocyanins for which dd-MS/MS fragmentation was not successful, and to further validate the tentative identifications. Anthocyanins were analysed using Xcalibur 3.1 software (ThermoFisher Scientific, Sunnyvale, CA, USA) and identified as M+ adducts, based on their accurate masses (m/z) and MS fragmentation, compared with in house database and public sources (e.g., KEGG, MetaCyc, ChemSpider, PubChem, Metlin, Phenol-Explorer), as well as with comigration with available authentic standards (cyanidin 3-glucoside, peonidin 3-glucoside and dephinidin 3-glucoside) (Extrasynthese, Genay, France). Absolute amounts were measured as previously described [39] using the two most abundant fragments per compound, and data were normalised based on integrated peak areas of external calibration curves of previously described standards [39]. LOD (limit of detection) was estimated from signal-to-noise ratio (S/N) as described [40] and defined as signal intensity corresponding to three times of that noise, while LOQ (limit of quantification) was nine times of that noise. All data are presented as means and standard deviation of at least three independent biological replicates. All the chemicals and solvents used during both the procedures were of LC/MS grade.

2.4. Quantitative RT-PCR Analysis

Plant material used for flavonoids and anthocyanins analysis was the same as used for total RNA isolation. Total RNA extraction, DNase treatment, cDNA synthesis and quantitative real-time PCR (qPCR) and relative gene expression were performed as described previously by Carmona et al. [27]. Briefly, qPCR was achieved with a StepOne Plus Real Time PCR System (Applied Biosystem, Waltham, MA, USA) and analysed using StepOne Software version 2.3 (Thermo Fisher, Valence, Spain). RT-PCR was carried out with 50 ng of total cDNA adding 6 µL of SYBR Green PCR Master Mix (Applied Biosystems, Waltham, MA, USA) and 0.3 µM of gene specific primers in a total volume of 12 µL. The RT-PCR procedure consisted of 95 °C 10 min followed by 40 cycles at 95 °C 15 s and 60 °C 40 s. Primers sequences for analysing *phenylalanine ammonia-lyase* (*PAL*), *cinnamate 4-hydroxylase* (*C4H*), *4-hydroxy-cynnamoyl CoA ligase* (*4CL*), *chalcone synthases 1* and *2* (*CHSs*), *chalcone isomerase* (*CHI*), *flavonoid 3-hydroxylase* (*F3H*), *flavonoid 3′5′-hydroxylase* (*F3′5′H*), *flavonol synthase* (*FLS*), *dihydroflavonol 4-reductase* (*DFR*), *anthocyanidin synthase* (*ANS*), *uridine diphos-phate-glucose:flavonoid 3-O-glucosyltransferase* (*UFGT*) and *glutathione-S-transferase* (*GST*) genes are described in Table S1. The relative expression between cold-treated and control samples (zero time of orange fruits) was determined by the method described by Livak et al. [41]. Values are presented as the mean of at least three independent analyses. Statistical analyses were performed using ANOVA.

3. Results

3.1. Pulp and Juice Appearance and Quality Parameters in Moro and Pera Oranges after Storage

The effect of postharvest storage on visual aspect, maturity index (MI), total flavonoid and anthocyanin contents were assessed in pulp and juice from Moro (blood type) and Pera (blond type) mature oranges subjected to postharvest storage at 9 °C (Figure 2 and Table 1). The visual aspect of pulp and juice was evaluated at the onset, at 30 and 45 days. No changes in colour were detected in pulp and juice from Pera during all the storage period, while a notable enhancement in the red/purple coloration was observed in those from Moro (Figure 2).

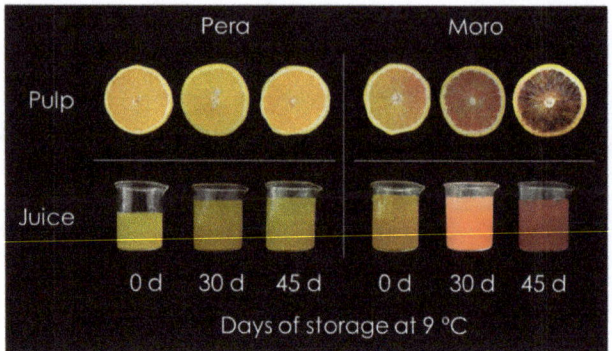

Figure 2. Internal oranges appearance and colour of pulp (**up**) and juices (**down**) of Pera (**left**) and Moro (**right**) oranges from Sao Paulo (Brazil), during storage at 9 °C for 0, 30 and 45 days.

Table 1. Maturity index, pH, flavonoids and anthocyanin content of Pera and Moro pulp during storage for 0, 30 and 45 days. Statistical analyses were performed using ANOVA and different letters indicate statistically significant different values ($p \leq 0.01$) for a given time.

	Pera			Moro		
	0 Days	30 Days	45 Days	0 Days	30 Days	45 Days
Maturity index (MI)	7.9 ± 0.7 [a]	7.9 ± 0.4 [a]	7.9 ± 0.6 [a]	9.5 ± 0.7 [a]	9.2 ± 0.2 [a]	9.4 ± 0.1 [a]
pH	3.8 ± 0.1 [a]	3.8 ± 0.1 [a]	3.9 ± 0.0 [a]	3.6 ± 0.2 [a]	3.8 ± 0.1 [a]	3.7 ± 0.0 [a]
Total flavonoids	22.4 ± 1.5 [a]	22.1 ± 1.5 [a]	22.4 ± 1.4 [a]	31.6 ± 3.3 [b]	28.2 ± 2.8 [b]	45.4 ± 2.8 [c]
Total anthocyanin content (mg/L)	-	-	-	1.3 ± 0.1 [a]	43.4 ± 4.2 [b]	60.0 ± 3.5 [c]

No differences were found in MI and pH along the storage period in any of the two fruit types investigated (Table 1). Total flavonoids content was different between both types, being 1.4-fold higher in Moro than in Pera at the onset of the experiment (Table 1). Moreover, no enhanced accumulation was detected in the blond cultivar during the storage period, while an increment of 1.4-fold was observed in Moro at the end of the storage. No anthocyanin presence was detected in Pera orange, while Moro fruit displayed a noticeable presence and considerable enhancement of anthocyanins from 1.3 ± 0.7 mg/L to 60.0 ± 9.4 mg/L under storage conditions.

3.2. Accumulation of Hydroxycinnamates and Flavonoids (Non-Anthocyanins) in Moro and Pera Oranges during Postharvest Storage

Contents of hydroxycinammic acids (HA) and main flavonoids were assessed in the pulp of Pera and Moro fruit during 9 °C post-harvest storage (Figure 3 and Table S2). The profile of the eleven HA identified showed significant differences between both cultivars (Table S2). In general, the BO presented a higher HA content at the onset of the experiment, and a much higher accumulation along all the postharvest experimental period. For instance, the content of coumaric acid, precursor of both HA and flavonoids, was 4.9- and 13.6-fold higher in Moro than in Pera at the onset of the experiment and after 45 days, respectively (Figure 3 and Table S2).

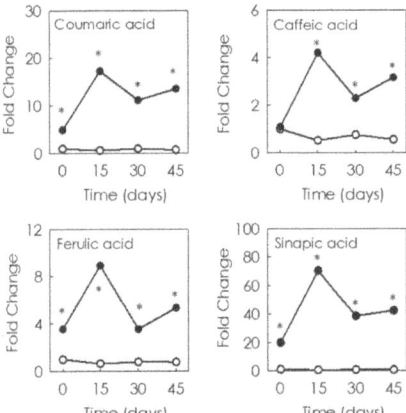

Figure 3. Fold change of the 4 main hydroxycinnamic acids identified in the pulp of Moro (●) and Pera (○) oranges during storage for 0, 15, 30 and 45 days. Data are expressed as the mean fold change ± SD of each sample as compared to the control Pera fruits sample (at harvest time). Asterisk indicates statistically significant different values ($p \leq 0.01$) for each given time point when comparing cultivars.

A total of 76 flavonoids, including flavanones, flavonols and flavones were identified and measured during the storage period in juice from both fruit types (Figure 4 and Table S2). Among them, the flavonol class was the main group constituting 38.2% of the total, followed by flavones (35.5%) and flavanones (26.3%). Additionally, polymethoxylated derivatives of each class were identified, with polimethoxyflavones (PMFs) being the most represented. The initial flavonoids profile, their content and the accumulation patterns along storage displayed drastic differences between Moro and Pera. Eight flavonoid compounds present in Moro were not detected in Pera fruit either at the onset or after postharvest storage, such as isosakuranetin or the flavonols kaemferol and myricetin, while only two flavonoids (poncirin and natsudaidain) were not present at the beginning in Moro, but were found later and progressively increased with the storage (0.19 and 122.2-fold, respectively) (Table S2). Regarding the initial content of the individual flavonoids, 25 flavonoids showed low (<0.6-fold) accumulation in Moro than in Pera, as the flavanones naringenin or eriocitrin (0.60 an 0.12-fold, respectively). Conversely, 24 flavonoid compounds exhibited a higher content in Moro than in Pera at the onset, which varied between the 2.21-fold increase of the chrysoriol-8-C-glucoside (scoparin) and 202.4-fold enrichment of dihydrokaempherol (Table S2). Considering the content after the storage period, most of the identified flavonoids (54%) were more than two-fold enhanced in Moro compared to Pera, being among them the main precursors naringenin chalcone (206.4-fold) and the dihydroflavonoids dihydrokaemferol and dihydroquercetin (1602.8 and 239.8-fold, respectively). However, 19.7% of the identified flavonoids displayed less than 0.70-fold accumulation in Moro along storage, such as eriocitrin (0.43-fold) or methoxykaempferol-3-O-neohesperidoside (0.07-fold) (Figure 4 and Table S2). Individual flavonoids followed also a different accumulation profile during the storage at 9 °C depending on the cultivar. Whereas in Pera most of the identified flavonoids barely changed with the storage, the individual profile varied in Moro depending on each metabolite (Figure 5 and Table S2). In general, main citrus flavanones (isosakuranetin derivatives), flavones (apigenin, luteolin and their derivatives), and flavonols (quercetin, kampferol and their derivatives) increased along storage (1.2 up to 361.6-fold) in Moro fruit (Figure 5). Although other compounds such as 3,3',4',5,6,7,8-heptamethoxyflavone, nobiletin or sinensetin showed a decreasing profile (1.4, 2.2 and 2.9-fold, respectively), their content was still higher in Moro than in Pera fruit along the storage period (Figure 5 and Table S2).

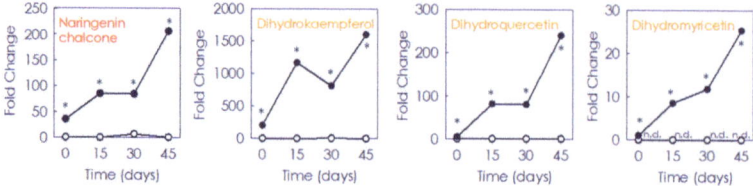

Figure 4. Fold change of naringenin chalcone (flavonoids precursor, in red colour) and the dihydroflavonoids (anthocyanins precursors, in orange colour) dihydrokampherol, dihydroquecetin and dihydromyricetin in the pulp of Moro (•) and Pera (○) oranges during storage for 0, 15, 30 and 45 days. Data are expressed as the mean fold change ± SD of each sample compared to the control Pera fruits sample (at harvest time). Asterisk indicates statistically significant different values ($p \leq 0.01$) for each given time.

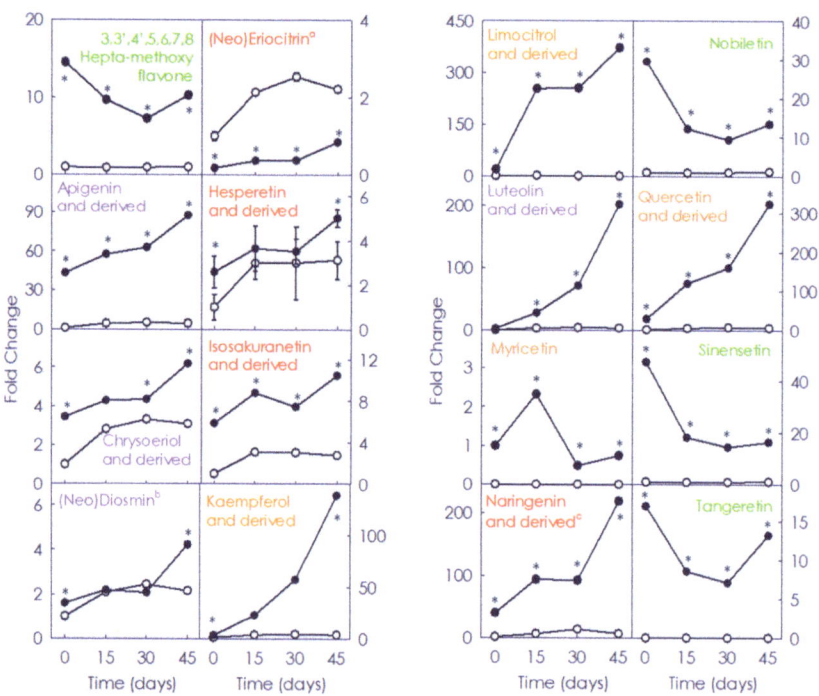

Figure 5. Fold change of representative flavonoids identified in the pulp of Moro (•) and Pera (○) oranges during storage for 0, 15, 30 and 45 days. Compounds belonging to flavanones class are represented in red colour, flavonols class in orange and the flavones in purple and light green (polymethoxyflavones subgroup). [a]. Sum of eriocitrin and neoeriocitrin. [b]. Sum of diosmin and neodiosmin. [c]. Including the naringenin chalcone. Data are expressed as the mean fold change ± SD of each sample compared to the control Pera fruits sample (at harvest time). Asterisk indicates statistically significant different values ($p \leq 0.01$) for each given time.

3.3. Accumulation of Anthocyanins in Moro Orange during Postharvest Storage

Variations in anthocyanins composition and contents in Moro orange during postharvest storage were evaluated for 45 days (Figure 6). No anthocyanins were detected in Pera, while a total of 11 anthocyanins were identified in Moro pulp. At the onset of the study, cyanidin 3-O-glucoside (C3-glu) and cyanidin 3-(6″-malonyl)-glucoside (C3-(6M)-glu) were the most abundant anthocyanins in Moro, representing 61.1% and 24.9% of total anthocyanins, respectively. Storage at 9 °C promoted a progressive increase in anthocyanins accumulation, with C3-glu and C3-(6M)-glu remaining as the most abundant ones, representing 27.7% and 49.4% of total anthocyanins at day 45, respectively (Figure 6). Altogether, cyanidin 3-rhamnoside (C3-rha), delphinidin 3-(6″ malonyl)-glucoside (D3-(6M)-glu) and delphinidin 3-glucoside (D3-glu) accounted for 20.8% of total anthocyanins by day 45, in comparison with 9.4% at the onset (Figure 6). At 45 days of storage, the main anthocyanins experimented showed an increase of 5.2-, 4.5-, 3.6- and 3.5-fold for C3-(6M)-glu, C3-rha, Peo3-(6M)-glu and D3-(6M)-glu, respectively. Other minor compounds also increased at the end of the storage period as it was observed for pelargonidin 3-glucoside and cyanidin 3-(ferulyl)glucoside (C3-Fe-glu) (5.3-fold) and cyanidin 3-O-sophoride (C3-sph) (2.3-fold). Other anthocyanin pigments that were not found initially, such as petunidin 3,5-glucoside (Pe3,5-glu) and petunidin 3-(6″-malonyl)-glucoside (Pet3-(6M)-glu), were detected at the end of storage (0.14 ± 0.00 and 0.12 ± 0.00 mg/mL, respectively).

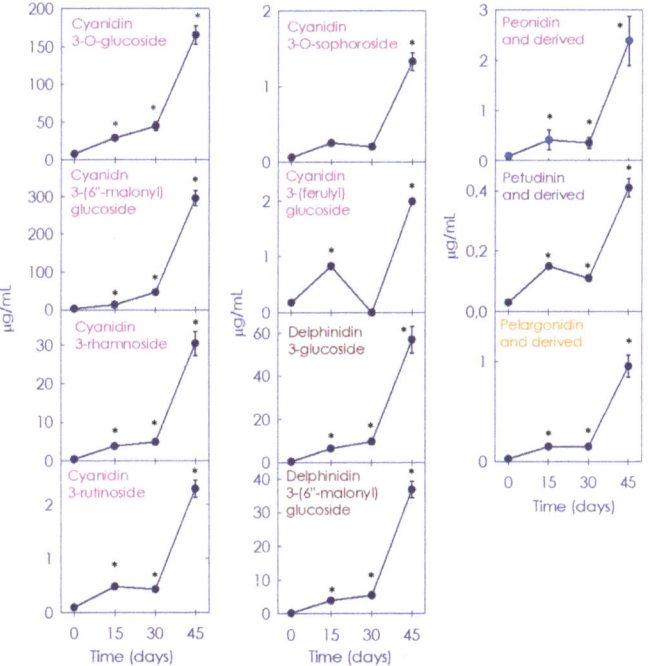

Figure 6. Composition of anthocyanins in the pulp of Moro blood orange during storage for 0, 15, 30 and 45 days. Data are expressed as the mean ± SD of each sample as compared to the control sample (Moro fruit at harvest time). Asterisk indicates statistically significant different values ($p \leq 0.01$) for each given time.

3.4. Gene Expression Ratio of Phenylpropanoid Biosynthetic Genes between Pera and Moro Orange Pulp during Postharvest Storage

Transcript accumulation levels between Moro and Pera during postharvest storage were determined and compared as the ratio between relative quantification of Moro vs. Pera at the onset and each postharvest time (Figure 7). A total of 13 genes were evaluated: *PAL*, *C4H* and *4CL* involved in the initial steps of the general phenylpropanoid pathway, seven genes involved in the initial steps of flavonoids biosynthesis (*CHSs*, *CHI*, *F3H*, *F3′H*, *F3′5′H*), three structural anthocyanin biosynthesis genes (*DFR*, *ANS* and *UFGT*) and one gene involved in the transport of the purple/red pigments to vacuoles (*GST*). In general, expression profile revealed that all genes presented a higher ratio at the onset and during the storage period in Moro vs. Pera, with the only exception of *FLS*. At the onset, *F3′5′H* showed up to 168-fold higher expression in Moro than in Pera, and upstream genes *CHS1* and *CHS2* were 20 and 28-fold higher, respectively. Anthocyanin structural gene expression presented also a positive ratio in Moro vs. Pera at the onset of storage, between 20 and 122-fold. During the storage period, the highest induction was shown for *CHS2* and $F3'5'H$ with a final ratio of 1915-fold and 1330-fold increase, respectively (Figure 7).

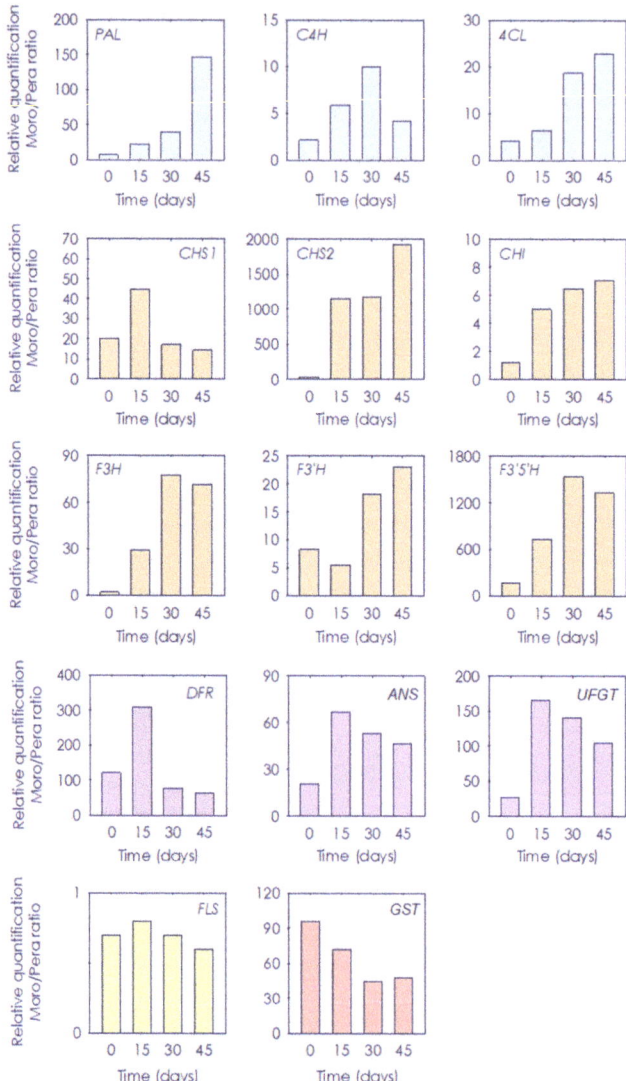

Figure 7. Ratio of the relative expression of *phenylalanine ammonia-lyase* (PAL), *cinnamate 4-hydroxylase* (C4H), *4-hydroxy-cynnamoyl CoA ligase* (4CL), *chalcone synthases 1* and *2* (CHSs), *chalcone isomerase* (CHI), *flavonoid 3-hydroxylase* (F3H), *flavonoid 3'5'-hydroxylase* (F3'5'H), *flavonol synthase* (FLS), *dihydroflavonol 4-reductase* (DFR), *anthocyanidin synthase* (ANS), *uridine diphos-phate-glucose:flavonoid 3-O-glucosyltransferase* (UFGT) and *glutathione-S-transferase* (GST) in the pulp of Moro vs. Pera oranges during storage for 0, 15, 30 and 45 days. Blue, and orange bars indicate genes involved in the general phenypropanoids and flavonoids, respectively. Yellow and purple bars indicate genes involved in the flavonols and anthocyanins biosynthesis, respectively. Red bars belong to those genes involved in the anthocyanin transport.

4. Discussion

Among fruits, citrus pulp, juice and by-products constitute one of the most important sources of phytonutrients, especially in countries where these fruits are extensively produced. In the last years, the beneficial health-promoting effects claimed for flavonoids has stimulated the interest for investigating these phytochemical compounds in citrus fruits [6,42–46]. Postharvest storage of the fruit at low temperature, combined or not with a stress (i.e., curing) treatment, has been proven to induce anthocyanins and other flavonoids accumulation in BO via stimulation of their biosynthesis [26,28,31]. However, to our knowledge, information regarding changes in the accumulation of specific flavonoid contents in blond and BO under postharvest storage conditions is limited. In this work, the effect of storage on flavonoids and other phenylpropanoids accumulation was investigated and compared in fruits from two important orange cultivars for juice production, Pera (blond) and Moro (blood).

Different responses to the storage at low-temperature were noticed in Pera and Moro fruits. While no visual changes were observed in the blond fruit pulp and juice, Moro fruit displayed a conspicuous enhancement of its purple/red coloration in the pulp and juice, as expected (Figure 2) [31,32]. Juice pH, which can influence anthocyanin by changing their coloration [47], did not vary along the storage period in our study (Table 1). Instead, the progressive darkening of BO pulp and juice could be associated with the increase of anthocyanins content along storage (Figure 2 and Table 1) [31]: from a low content (1.3 ± 0.7 mg/L) at harvest time, it increased 46-fold along storage period due to the cold induction effect (Table 1) [25,26,31,32]. Red/purple colour enhancement in BO during cold storage is related to the greater increase of all individual anthocyanins (Figure 6 and Table 1) [26,28,31,32]. C3-glu and C3-(6M)-glu have been described as the main anthocyanins in BO juice, together with D3-glu, Peo3-(6M)-glu and cyanidin 3-(6″ dioxalyl)-glucoside (C6D-glu) [48]. Accordingly, the two main anthocyanins found in Moro orange were also C3-glu and C3-(6M)-glu, followed by D3-glu and D3-(6M)-glu (Figure 6). In contrast, although no C6D-glu was detected, other anthocyanins such as C3-rha, C3-Fe-glu and C3-sph were accumulated in response to storage at low-temperature in Moro orange (Figure 6) [32]. Due to their electron-donating properties, anthocyanins are potent antioxidants [49]. In the case of BO, the antioxidant activity of anthocyanins was favourable for human health, with impact on some diseases, derived from their anti-inflammatory, anticancer, and antidiabetic properties, due to the prevention of oxidation and free-radical chain reactions [50]. The absence of anthocyanins accumulation in blond orange fruit has been widely documented [23], and is mainly due to the lack of expression in key positive transcriptional factors required for the production of these compounds. Similarly, anthocyanins were not detected in Pera fruit (Table 1).

The beneficial effect of low temperature storage on anthocyanin enhancement has been shown in fruits from other BO cultivars, and was related to a strong boost in the induction and expression of the initial genes of the phenylpropanoids biosynthetic pathway concomitant with the induction of anthocyanin structural genes, leading also to the higher accumulation of other flavonoids [31,32]. Regarding flavonoids, their content and composition in citrus fruits vary among species, cultivars and fruit organs [13,17,51–53]. Moreover, stressful temperatures alter the general phenylpropanoid metabolism in citrus fruits [54,55]. In agreement with that, very different accumulation patterns of the individual flavonoid and HA metabolites were observed in fruit from the blood and blond cultivars, the BO pulp being richer at the onset and during postharvest storage (Figures 2 and 3 and Table S2). A higher abundance of HA has been reported in BO fruit compared to that of blond cultivars, being Moro one of the BO types accumulating higher amounts of total HA [17,25]. Additionally, the evaluation of flavonoid profiles revealed a higher initial content and a progressive increment along the storage of the main flavonoids (non-anthocyanins) and anthocyanins only in BO fruit (Figures 4–6 and Table S2). Interestingly, storage promoted the enhanced accumulation of precursor substrates in BO, as an effect of a major request of precursor for flavonols and anthocyanins to respond to temperature stress [25,26,31,32]. Concordantly, storage at low temperature induced an increased accumulation of the flavonoid (naringenin chalcone) and anthocyanin (dihydroflavonols) main precursor substrates in Moro in comparison with Pera fruit, concurrent with a higher induction of early phenylpropanoid genes expression (Figure 7) [28,31,32]. Taken together, these results support the higher substrate availability for downstream production of phenylpropanoids in Moro and a different regulation from the initial steps in Moro vs. Pera oranges (Figures 2–4 and Table S2) [32].

Flavonoids are potent inducers of antioxidant defence mechanisms in animal cells, by stimulation of different enzymes activity such as glutathione peroxidase, catalase, and superoxide dismutases, or by inhibition of the accumulation of other enzymes such as xanthine oxidase and also the lipid peroxidation as well as protecting other biomolecules, such as DNA from oxidation [43]. In citrus fruits, flavanones have been described as the predominant class of flavonoids, being flavonols the less representative group [51,56]. In our study, flavonoid profiles revealed that the flavonols and flavones classes were the main ones, being flavanones the third (Table S2). These differences might be determined by the different cultivars used [17,52,53], as is supported by the less flavanones accumulation in fruit from blond cultivars stored at low temperature when compared with that of BO cultivars (Table 1 and Table S2) [25]. In our study, although flavanones class was not so representative in the edible portion of the fruit at the onset, storage promoted the increase in the accumulation of hesperidin, hesperitin, naringenin, narirutin or isosakuranetin (and derivatives) (Figure 5 and Table S2). Flavanones from citrus fruits have been awarded important biological activities, as helping in cardiovascular and cancer risk prevention and avoiding the onset of oxidative stress involved in inflammatory damage due to their antioxidant potential [57,58]. The main studies of citrus flavanones antioxidant and anti-inflammatory properties focused on hesperidin and its aglycones (hesperitin), and narirutin [59–63]. In the case of isosakuranetin anti-oxidative activities, they have been related with potential free radical scavenging mechanisms [64]. Numerous studies describing the effects of naringenin on human health reported increasing antioxidant defences, scavenging reactive oxygen species, antiviral responses or exerting anti-atherogenic and anti-inflammatory effects [65]. Naringenin is predominantly found in the edible citrus part and, although is poorly absorbed by oral ingestion, a positive orange juice prebiotic effect due to its bioavailability has been shown [66]. In relation to flavones, compounds grouped into this class also showed a better enrichment in Moro than in Pera fruit under postharvest storage. Among them, luteolin, apigenin and nobiletin (including some glycosylated forms) were the most enhanced by 9 °C storage (Table S2). Citrus flavones have been proposed as the most suitable and capable compounds in terms of antioxidant and anti-inflammatory activities due to their substitutions groups [67]. In the case of nobiletin, it has been indicated to prevent obesity, hepatic steatosis, dyslipidemia, and insulin resistance [42]. Luteolin effects on activation of antioxidant enzymes involved in cancer prevention as well as cardio-protective effects have been reported, as well as properties in inhibiting the onset and development of inflammatory diseases as asthma [68–71]. Multiple other beneficial bioactivities of apigenin have been proposed on different types of cancer or interactions on gut microbiota [72]. Finally, compounds belonging to the flavonols class were the main groups of flavonoids identified and accumulated in Moro fruit stored at low temperature, mainly limocitrol and quercetin (and their derivatives) (Table S2). Recently, the high intake of flavonol in the diet has been associated to a reduction in the risk of developing Alzheimer dementia [73]. Among the flavonols, limocitrol presented the highest increment observed in our study (Table S2). This compound has been reported as one of the main flavonoids in finger citron (*Citrus medica* cv. sarcodactylis) and their strong antioxidation and antiaging activities have been indicated in both in vitro and in vivo studies [74]. Regarding quercetin and its derivatives, the second most induced flavonol by storage at 9 °C, it is considered as one of the most relevant antioxidant metabolites due to its chemical structure. Its involvement in lipid peroxidation prevention and tocopherol regeneration has been described, as well as in ischemia injury reduction by the induction of nitric oxide synthase [43]. Quercetin has anticarcinogenic and anti-inflammatory properties with antioxidant and free radical scavenging effects. Moreover, other antimicrobial, antiviral, and biological effects, which include anti-inflammatory activity has been attributed to it [75,76].

5. Conclusions

BO is an excellent source of natural antioxidant and bioactive compounds, promoting the interest of consumers and researchers in the recent years [6,7]. Many in vivo studies associate the beneficial health-effects of BO juice consumption in reduction of inflammatory processes related to its remarkable antioxidant power [18]. Protection against oxidative stress of flavonoids by induction of reactive oxygen and nitrogen species have been described to play a role as markers of different degenerative diseases [14,18,22]. All these protective bioactivities are likely due to the marked presence of phenolic acids, flavonoids and other phytochemicals in BO [35], although the contribution of each phytochemical in such antioxidant properties requires further research. In this study, storage at low temperature induced a great enrichment on anthocyanins and flavonoids accumulation levels in Moro orange, suggesting that all these compounds could be contributing to their higher antioxidant capacity. However, we did not observe a similar effect of cold storage in Pera fruit. Taken together, we show here that through a regular postharvest practise the content of not only anthocyanins, but also specific health-related flavonoids is enhanced in Moro blood orange pulp and juice (but not in the Pera blond orange counterpart), reinforcing the interest of blood orange to improve natural antioxidant diets. This work should further analyse flavonoids composition and content in other blood and blond orange cultivars to assess whether the drastic increases observed in flavonoids accumulation in Moro fruit during cold storage may be extended to blood-orange types or it is independent of anthocyanins production.

Supplementary Materials: The following supporting information can be downloaded at: https://www.mdpi.com/article/10.3390/antiox11030547/s1. Table S1: Primer sequences used for the quantitative RT-PCR analyses. Table S2: Fold changes of flavonoids identified in the pulp of Pera and Moro orange during storage at 9 °C for 0, 15, 30 and 45 days. Data are expressed as the mean fold change ± SD of each sample as compared to the control sample (harvest time of Pera fruit).

Author Contributions: Conceptualisation: L.C. methodology: L.C., B.A. and M.A. formal analysis: L.C., M.S. and G.D. investigation: L.P. data curation: L.C., B.A. and M.S. writing—original draft preparation: L.C. writing—review and editing B.A. and L.P. supervision and funding acquisition: L.P. All authors have read and agreed to the published version of the manuscript.

Funding: This work was supported by the São Paulo Research Foundation (FAPESP, Brazil) project FAPESP 2014/12616-9, Fundecitrus, PRIMA Foundation (Reference number: 2132) through 1.3. 1-2021 PROMEDLIFE and the Rosty-COST Actions CA18210 for networking. LC was funded by FAPESP grant (2014/23447-3).

Institutional Review Board Statement: Not applicable.

Informed Consent Statement: Not applicable.

Data Availability Statement: The data are contained within the article and supplementary material.

Acknowledgments: We appreciate the help of Citrosuco for allowing us accessing their orchards and using their fruits. The technical support of Tatiane M. Malara is also acknowledged.

Conflicts of Interest: The authors declare that this research was conducted in the absence of any commercial or financial relationships that could be construed as a potential conflict of interest.

References

1. Oz, A.T.; Kafkas, E. Phytochemicals in Fruits and Vegetables. In *Superfood and Functional Food. An Overview of Their Processing and Utilization*; IntechOpen: London, UK, 2017; pp. 175–184.
2. Liu, R.H. Health-promoting components of fruits and vegetables in the diet. *Adv. Nutr.* **2013**, *4*, 384S–392S. [CrossRef] [PubMed]
3. Yahia, E.M.; Celis, M.E.M.; Svendsen, M. The contribution of fruit and vegetable consumption to human health. In *Fruit and Vegetable Phytochemicals: Chemistry and Human Health*, 2nd ed.; John Wiley & Sons, Ltd.: Hoboken, NJ, USA, 2017; Volume 1, pp. 3–52.

4. Kumar, S.; Pandey, A.K. Chemistry and biological activities of flavonoids: An overview. *Sci. World J.* **2013**, *2013*, 162750. [CrossRef] [PubMed]
5. Bueno, J.M.; Sáez-Plaza, P.; Ramos-Escudero, F.; Jiménez, A.M.; Fett, R.; Asuero, A.G. Analysis and antioxidant capacity of anthocyanin pigments. Part II: Chemical structure, color, and intake of anthocyanins. *Crit. Rev. Anal. Chem.* **2012**, *42*, 126–151. [CrossRef]
6. Zou, Z.; Xi, W.; Hu, Y.; Nie, C.; Zhou, Z. Antioxidant activity of Citrus fruits. *Food Chem.* **2016**, *196*, 885–896. [CrossRef]
7. Addi, M.; Elbouzidi, A.; Abid, M.; Tungmunnithum, D.; Elamrani, A.; Hano, C. An overview of bioactive flavonoids from Citrus fruits. *Appl. Sci.* **2021**, *12*, 29. [CrossRef]
8. Miles, E.A.; Calder, P.C. Effects of Citrus fruit juices and their bioactive components on inflammation and immunity: A narrative review. *Front. Immunol.* **2021**, *12*, 712608. [CrossRef]
9. Wang, Y.; Qian, J.; Cao, J.; Wang, D.; Liu, C.; Yang, R.; Li, X.; Sun, C. Antioxidant capacity, anticancer ability and flavonoids composition of 35 citrus (*Citrus reticulata* Blanco) varieties. *Molecules* **2017**, *22*, 1114–1134. [CrossRef]
10. Giuffrè, A.M.; Zappia, C.; Capocasale, M. Physicochemical stability of blood orange juice during frozen storage. *Int. J. Food Prop.* **2017**, *20*, 1930–1943.
11. Giuffrè, A.M. Bergamot (*Citrus bergamia*, Risso): The effects of cultivar and harvest date on functional properties of juice and cloudy juice. *Antioxidants* **2019**, *8*, 221. [CrossRef]
12. Xi, W.; Lu, J.; Qun, J.; Jiao, B. Characterization of phenolic profile and antioxidant capacity of different fruit part from lemon (*Citrus limon* Burm.) cultivars. *J. Food Sci. Technol.* **2017**, *54*, 1108–1118. [CrossRef]
13. Lv, X.; Zhao, S.; Ning, Z.; Zeng, H.; Shu, Y.; Tao, O.; Xiao, C.; Lu, C.; Liu, Y. Citrus fruits as a treasure trove of active natural metabolites that potentially provide benefits for human health. *Food Chem.* **2015**, *9*, 1–14. [CrossRef] [PubMed]
14. Simpson, E.; Mendis, B.; Macdonald, I. Orange juice consumption and its effect on blood lipid profile and indices of the metabolic syndrome; A randomised, controlled trial in an at-risk population. *Food Funct.* **2016**, *7*, 1884–1891. [CrossRef] [PubMed]
15. Manach, C.; Scalbert, A.; Morand, C.; Rémésy, C.; Jiménez, L. Polyphenols: Food sources and bioavailability. *Am. J. Clin. Nutr.* **2004**, *79*, 727–747. [CrossRef] [PubMed]
16. Rapisarda, P.; Tomaino, A.; Lo Cascio, R.; Bonina, F.; De Pasquale, A.; Saija, A. Antioxidant effectiveness as influenced by phenolic content of fresh orange juices. *J. Agric. Food Chem.* **1999**, *47*, 4718–4723. [CrossRef] [PubMed]
17. Legua, P.; Modica, G.; Porras, I.; Conesa, A.; Continella, A. Bioactive compounds, antioxidant activity and fruit quality evaluation of eleven blood orange cultivars. *J. Sci. Food Agric.* **2021**, 1–12. [CrossRef]
18. Grosso, G.; Galvano, F.; Mistretta, A.; Marventano, S.; Nolfo, F.; Calabrese, G.; Buscemi, S.; Drago, F.; Veronesi, U.; Scuderi, A. Red orange: Experimental models and epidemiological evidence of its benefits on human health. *Oxid. Med. Cell. Longev.* **2013**, *2013*, 157240. [CrossRef]
19. Tena, N.; Martín, J.; Asuero, A.G. State of the art of anthocyanins: Antioxidant activity, sources, bioavailability, and therapeutic effect in human health. *Antioxidants* **2020**, *9*, 451. [CrossRef]
20. Blesso, C.N. Dietary Anthocyanins and Human Health. *Nutrients* **2019**, *11*, 2107. [CrossRef]
21. Speer, H.; D'Cunha, N.M.; Alexopoulos, N.I.; McKune, A.J.; Naumovski, N. Anthocyanins and human healt. A focus on oxidative stress, inflammation and disease. *Antioxidants* **2020**, *9*, 366. [CrossRef]
22. Li, L.; Lyall, G.K.; Alberto Martinez-Blazquez, J.; Vallejo, F.; Tomas-Barberan, F.A.; Birch, K.M.; Boesch, C. Blood orange juice consumption increases flow-mediated dilation in adults with overweight and obesity: A randomized controlled trial. *J. Nutr.* **2020**, *150*, 2287–2294. [CrossRef]
23. Butelli, E.; Licciardello, C.; Zhang, Y.; Liu, J.; Mackay, S.; Bailey, P.; Reforgiato-Recupero, G.; Martin, C. Retrotransposons control fruit-specific, cold-dependent accumulation of anthocyanins in blood oranges. *Plant Cell* **2012**, *24*, 1242–1255. [CrossRef] [PubMed]
24. Continella, A.; Pannitteri, C.; La Malfa, S.; Legua, P.; Distefano, G.; Nicolosi, E.; Gentile, A. Influence of different rootstocks on yield precocity and fruit quality of 'Tarocco Scirè' pigmented sweet orange. *Sci. Hortic.* **2018**, *230*, 62–67. [CrossRef]
25. Rapisarda, P.; Bellomo, S.E.; Intelisano, S. Storage temperature effects on blood orange fruit quality. *J. Agric. Food Chem.* **2001**, *49*, 3230–3235. [CrossRef] [PubMed]
26. Lo Piero, A.R.; Puglisi, I.; Rapisarda, P.; Petrone, G. Anthocyanins accumulation and related gene expression in red orange fruit induced by low temperature storage. *J. Agric. Food Chem.* **2005**, *53*, 9083–9088. [CrossRef] [PubMed]
27. Lo Piero, A.R. The state of the art in biosynthesis of anthocyanins and its regulation in pigmented sweet oranges [(*Citrus sinensis*) L. Osbeck]. *J. Agric. Food Chem.* **2015**, *63*, 4031–4041. [CrossRef] [PubMed]
28. Crifò, T.; Puglisi, I.; Petrone, G.; Recupero, G.R.; Lo Piero, A.R. Expression analysis in response to low temperature stress in blood oranges: Implication of the flavonoid biosynthetic pathway. *Gene* **2011**, *476*, 1–9. [CrossRef]
29. Pannitteri, C.; Continella, A.; Lo Cicero, L.; Gentile, A.; La Malfa, S.; Sperlinga, E.; Napoli, E.M.; Strano, T.; Ruberto, G.; Siracusa, L. Influence of postharvest treatments on qualitative and chemical parameters of Tarocco blood orange fruits to be used for fresh chilled juice. *Food Chem.* **2017**, *230*, 441–447. [CrossRef]
30. Carmona, L.; Alquézar, B.; Tárraga, S.; Peña, L. Protein analysis of Moro blood orange pulp during storage at low temperatures. *Food Chem.* **2019**, *277*, 75–83. [CrossRef]
31. Carmona, L.; Alquézar, B.; Marques, V.V.; Peña, L. Anthocyanin biosynthesis and accumulation in blood oranges during postharvest storage at different low temperatures. *Food Chem.* **2017**, *237*, 7–14. [CrossRef]

32. Carmona, L.; Alquézar, B.; Diretto, G.; Sevi, F.; Malara, T.; Lafuente, M.T.; Peña, L. Curing and low-temperature combined post-harvest storage enhances anthocyanin biosynthesis in blood oranges. *Food Chem.* **2020**, *342*, 128334. [CrossRef]
33. Grierson, W.; Ben-Yehoshua, S. Storage of Citrus Fruits. In *Fresh Citrus Fruits*; Springer: New York, NY, USA, 1986; pp. 479–507.
34. El-Otmani, M.; Ait-Oubahou, A.; Zacarías, L. *Citrus* spp.: Orange, Mandarin, Tangerine, Clementine, Grapefruit, Pomelo, Lemon and Lime. In *Postharvest Biology and Technology of Tropical and Subtropical Fruits*; Elsevier: Amsterdam, The Netherlands, 2011; pp. 437e–516e.
35. Licciardello, F.; Arena, E.; Rizzo, V.; Fallico, B. Contribution of blood orange-based beverages to bioactive compounds intake. *Front. Chem.* **2018**, *6*, 374. [CrossRef] [PubMed]
36. Fundecitrus Harvest Data (2021–2022). Available online: https://www.fundecitrus.com.br/pes/estimativa (accessed on 19 June 2021).
37. Rolim, G.S.; Camargo, M.B.P.; Grosselilania, D.; Moraes, J.F.L. Climatic classification of köppen and thornthwaite sistems and their applicability in the determination of agroclimatic zonning for the state of São Paulo, Brazil. *Bragantia* **2007**, *66*, 711–720. [CrossRef]
38. Rossi, M. *Mapa Pedológico do Estado de São Paulo: Revisado e Ampliado*; Instituto Florestal: Sao Paulo, Brazil, 2017.
39. Diretto, G.; Jin, X.; Capell, T.; Zhu, C.; Gomez-Gomez, L. Differential accumulation of pelargonidin glycosides in petals at three different developmental stages of the orange-flowered gentian (*Gentiana lutea* L. var. aurantiaca). *PLoS ONE* **2019**, *14*, e0212062. [CrossRef] [PubMed]
40. Zheng, G.D.; Yang, X.J.; Chen, B.Z.; Chao, Y.X.; Hu, P.J.; Cai, Y.; Wu, B.; Wei, M.Y. Identification and determination of chemical constituents of *Citrus reticulata* semen through ultra high performance liquid chromatography combined with Q Exactive Orbitrap tandem mass spectrometry. *J. Sep. Sci.* **2020**, *43*, 438–451. [CrossRef] [PubMed]
41. Livak, K.J.; Schmittgen, T.D. Analysis of relative gene expression data using real- time quantitative PCR and the 2 -dd CT Method. *Methods* **2001**, *25*, 402–408. [CrossRef] [PubMed]
42. Morrow, N.M.; Burke, A.C.; Samsoondar, J.P.; Seigel, K.E.; Wang, A.; Telford, D.E.; Sutherland, B.G.; O'Dwyer, C.; Steinberg, G.R.; Fullerton, M.D.; et al. The citrus flavonoid nobiletin confers protection from metabolic dysregulation in high-fat-fed mice independent of AMPK. *J. Lipid Res.* **2020**, *61*, 387–402. [CrossRef]
43. Parihar, A.; Grotewold, E.; Doseff, A.I. Flavonoid dietetics: Mechanisms and Emerging Roles of Plant Nutraceuticals. In *Pigments in Fruits and Vegetables*; Springer: New York, NY, USA, 2015; pp. 93–126.
44. Muscatello, M.R.A.; Zoccali, R.A.; Bruno, A. Citrus fruit polyphenols and flavonoids: Applications to psychiatric disorders. In *Polyphenols: Mechanisms of Action in Human Health and Disease*; Academic Press: Cambridge, MA, USA, 2018; pp. 119–131.
45. Panche, A.N.; Diwan, A.D.; Chandra, S.R. Flavonoids: An overview. *J. Nutr. Sci.* **2016**, *5*, e47. [CrossRef]
46. Gironés-Vilaplana, A.; Moreno, D.A.; García-Viguera, C. Phytochemistry and biological activity of Spanish Citrus fruits. *Food Funct.* **2014**, *5*, 764–772. [CrossRef]
47. Zhang, Y.; Butelli, E.; Martin, C. Engineering anthocyanin biosynthesis in plants. *Curr. Opin. Plant Biol.* **2014**, *19*, 81–90. [CrossRef]
48. Fabroni, S.; Ballistreri, G.; Amenta, M.; Rapisarda, P. Anthocyanins in different Citrus species: An UHPLC-PDA-ESI/MS n-assisted qualitative and quantitative investigation. *J. Sci. Food Agric.* **2016**, *96*, 4797–4808. [CrossRef]
49. Wang, S.Y.; Jiao, H. Scavenging capacity of berry crops on superoxide radicals, hydrogen peroxide, hydroxyl radicals, and singlet oxygen. *J. Agric. Food Chem.* **2000**, *48*, 5677–5684. [CrossRef] [PubMed]
50. Habibi, F.; Ramezanian, A.; Guillén, F.; Castillo, S.; Serrano, M.; Valero, D. Changes in bioactive compounds, antioxidant activity, and nutritional quality of blood orange cultivars at different storage temperatures. *Antioxidants* **2020**, *9*, 1016. [CrossRef] [PubMed]
51. Khan, M.K.; Zill-E-Huma; Dangles, O. A comprehensive review on flavanones, the major citrus polyphenols. *J. Food Compos. Anal.* **2014**, *33*, 85–104. [CrossRef]
52. Wang, S.; Yang, C.; Tu, H.; Zhou, J.; Liu, X.; Cheng, Y.; Luo, J.; Deng, X.; Zhang, H.; Xu, J. Characterization and metabolic diversity of favonoids in Citrus species. *Sci. Rep.* **2017**, *7*, 10549. [CrossRef]
53. Yi, L.; Ma, S.; Ren, D. Phytochemistry and bioactivity of Citrus flavonoids: A focus on antioxidant, anti-inflammatory, anticancer and cardiovascular protection activities. *Phytochem. Rev.* **2017**, *16*, 479–511. [CrossRef]
54. Lafuente, M.T.; Ballester, A.R.; Calejero, J.; González-Candelas, L. Effect of high-temperature-conditioning treatments on quality, flavonoid composition and vitamin C of cold stored "Fortune" mandarins. *Food Chem.* **2011**, *128*, 1080–1086. [CrossRef]
55. Rapisarda, P.; Lo Bianco, M.; Pannuzzo, P.; Timpanaro, N. Effect of cold storage on vitamin C, phenolics and antioxidant activity of five orange genotypes [*Citrus sinensis* (L.) Osbeck]. *Postharvest Biol. Technol.* **2008**, *49*, 348–354. [CrossRef]
56. He, D.; Shan, Y.; Wu, Y.; Liu, G.; Chen, B.; Yao, S. Simultaneous determination of flavanones, hydroxycinnamic acids and alkaloids in citrus fruits by HPLC-DAD-ESI/MS. *Food Chem.* **2011**, *127*, 880–885. [CrossRef]
57. Barreca, D.; Gattuso, G.; Bellocco, E.; Calderaro, A.; Trombetta, D.; Smeriglio, A.; Laganà, G.; Daglia, M.; Meneghini, S.; Nabavi, S.M. Flavanones: Citrus phytochemical with health-promoting properties. *BioFactors* **2017**, *43*, 495–506. [CrossRef]
58. Di Majo, D.; Giammanco, M.; La Guardia, M.; Tripoli, E.; Giammanco, S.; Finotti, E. Flavanones in Citrus fruit: Structure–antioxidant activity relationships. *Food Res. Int.* **2005**, *38*, 1161–1166. [CrossRef]
59. Li, Y.; Du, Y.; Yang, J.; Xiu, Z.; Yang, N.; Zhang, J.; Gao, Y.; Li, B.; Shi, H. Narirutin produces antidepressant-like effects in a chronic unpredictable mild stress mouse model. *Neuroreport* **2018**, *29*, 1264–1268. [CrossRef] [PubMed]

60. Chakraborty, S.; Basu, S. Multi-functional activities of citrus flavonoid narirutin in Alzheimer's disease therapeutics: An integrated screening approach and in vitro validation. *Int. J. Biol. Macromol.* **2017**, *103*, 733–743. [CrossRef] [PubMed]
61. Valls, R.M.; Pedret, A.; Calderón-Pérez, L.; Llauradó, E.; Pla-Pagà, L.; Companys, J.; Moragas, A.; Martín-Luján, F.; Ortega, Y.; Giralt, M.; et al. Effects of hesperidin in orange juice on blood and pulse pressures in mildly hypertensive individuals: A randomized controlled trial (Citrus study). *Eur. J. Nutr.* **2021**, *60*, 1277–1288. [CrossRef] [PubMed]
62. Yang, H.L.; Chen, S.C.; Senthil Kumar, K.J.; Yu, K.N.; Lee Chao, P.D.; Tsai, S.Y.; Hou, Y.C.; Hseu, Y.C. Antioxidant and anti-inflammatory potential of hesperetin metabolites obtained from hesperetin-administered rat serum: An *ex vivo* approach. *J. Agric. Food Chem.* **2012**, *60*, 522–532. [CrossRef]
63. Kuwano, T.; Watanabe, M.; Kagawa, D.; Murase, T. Hydrolyzed methylhesperidin induces antioxidantenzyme expression via the Nrf2-ARE pathway in normal human epidermal keratinocytes. *J. Agric. Food Chem.* **2015**, *63*, 7937–7944. [CrossRef]
64. Erdoğan, Ş.; Özbakır Işın, D. A DFT study on OH radical scavenging activities of eriodictyol, Isosakuranetin and pinocembrin. *J. Biomol. Struct. Dyn.* **2021**, 1–10. [CrossRef]
65. Salehi, B.; Fokou, P.V.T.; Sharifi-Rad, M.; Zucca, P.; Pezzani, R.; Martins, N.; Sharifi-Rad, J. The therapeutic potential of naringenin: A review of clinical trials. *Pharmaceuticals* **2019**, *12*, 11. [CrossRef]
66. Duque, A.L.R.F.; Monteiro, M.; Adorno, M.A.T.; Sakamoto, I.K.; Sivieri, K. An exploratory study on the influence of orange juice on gut microbiota using a dynamic colonic model. *Food Res. Int.* **2016**, *84*, 160–169. [CrossRef]
67. Barreca, D.; Mandalari, G.; Calderaro, A.; Smeriglio, A.; Trombetta, D.; Felice, M.R.; Gattuso, G. Citrus flavones: An update on sources, biological functions, and health promoting properties. *Plants* **2020**, *9*, 23. [CrossRef]
68. Gentile, D.; Fornai, M.; Pellegrini, C.; Colucci, R.; Benvenuti, L.; Duranti, E.; Masi, S.; Carpi, S.; Nieri, P.; Nericcio, A.; et al. Luteolin prevents cardiometabolic alterations and vascular dysfunction in mice with HFD-induced obesity. *Front. Pharmacol.* **2018**, *9*, 13. [CrossRef]
69. Luo, Y.; Shang, P.; Li, D. Luteolin: A Flavonoid that has multiple cardio-protective effects and its molecular mechanisms. *Front. Pharmacol.* **2017**, *8*, 692. [CrossRef] [PubMed]
70. Kang, K.A.; Piao, M.J.; Ryu, Y.S.; Hyun, Y.J.; Park, J.E.; Shilnikova, K.; Zhen, A.X.; Kang, H.K.; Koh, Y.S.; Jeong, Y.J.; et al. Luteolin induces apoptotic cell death via antioxidant activity in human colon cancer cells. *Int. J. Oncol.* **2017**, *51*, 1169–1178. [CrossRef] [PubMed]
71. Johann, S.; de Oliveira, V.L.; Pizzolatti, M.G.; Schripsema, J.; Braz-Filho, R.; Branco, A.; Smânia, A., Jr. Antimicrobial activity of wax and hexane extracts from Citrus spp. peels. *Mem. Inst. Oswaldo Cruz* **2007**, *102*, 681–685. [CrossRef] [PubMed]
72. Wang, M.; Firrman, J.; Liu, L.S.; Yam, K. A review on flavonoid apigenin: Dietary intake, ADME, antimicrobial effects, and interactions with human gut microbiota. *BioMed Res. Int.* **2019**, *2019*, 7010467. [CrossRef] [PubMed]
73. Holland, T.M.; Agarwal, P.; Wang, Y.; Leurgans, S.E.; Bennett, D.A.; Booth, S.L.; Morris, M.C. Dietary flavonols and risk of Alzheimer dementia. *Neurology* **2020**, *94*, E1749–E1756. [CrossRef] [PubMed]
74. Luo, X.; Wang, J.; Chen, H.; Zhou, A.; Song, M.; Zhong, Q.; Chen, H.; Cao, Y. Identification of flavoanoids from finger citron and evaluation on their antioxidative and antiaging activities. *Front. Nutr.* **2020**, *7*, 207–219. [CrossRef]
75. Zhang, M.; Swarts, S.G.; Yin, L.; Liu, C.; Tian, Y.; Cao, Y.; Swarts, M.; Yang, S.; Zhang, S.B.; Zhang, K.; et al. Antioxidant Properties of Quercetin. In *Advances in Experimental Medicine and Biology*; Springer: Boston, MA, USA, 2011; Volume 701, pp. 283–289.
76. Ozgen, S.; Kilinc, O.K.; Selamoğlu, Z. Antioxidant activity of quercetin: A mechanistic review. *Turk. J. Agric. Food Sci. Technol.* **2016**, *4*, 1134–1138. [CrossRef]

Article

Preharvest Application of Phenylalanine Induces Red Color in Mango and Apple Fruit's Skin

Michal Fanyuk [1,2], Manish Kumar Patel [1], Rinat Ovadia [3], Dalia Maurer [1], Oleg Feygenberg [1], Michal Oren-Shamir [3] and Noam Alkan [1,*]

1. Department of Postharvest Science of Fresh Produce, Agricultural Research Organization (ARO), Volcani Institute, Rishon LeZion 7505101, Israel; michal.fanyuk@mail.huji.ac.il (M.F.); patelm1402@gmail.com (M.K.P.); daliam@volcani.agri.gov.il (D.M.); fgboleg@volcani.agri.gov.il (O.F.)
2. Robert H. Smith Faculty of Agriculture, Food and Environment, The Hebrew University of Jerusalem, Rehovot 76100, Israel
3. Department of Plant Science, Agricultural Research Organization (ARO), Volcani Institute, Rishon LeZion 7505101, Israel; rinat@volcani.agri.gov.il (R.O.); vhshamir@volcani.agri.gov.il (M.O.-S.)
* Correspondence: noamal@volcani.agri.gov.il; Tel.: +972-3-9683605

Citation: Fanyuk, M.; Kumar Patel, M.; Ovadia, R.; Maurer, D.; Feygenberg, O.; Oren-Shamir, M.; Alkan, N. Preharvest Application of Phenylalanine Induces Red Color in Mango and Apple Fruit's Skin. *Antioxidants* 2022, 11, 491. https://doi.org/10.3390/antiox11030491

Academic Editor: Stanley Omaye

Received: 24 January 2022
Accepted: 26 February 2022
Published: 28 February 2022

Publisher's Note: MDPI stays neutral with regard to jurisdictional claims in published maps and institutional affiliations.

Copyright: © 2022 by the authors. Licensee MDPI, Basel, Switzerland. This article is an open access article distributed under the terms and conditions of the Creative Commons Attribution (CC BY) license (https://creativecommons.org/licenses/by/4.0/).

Abstract: Anthocyanins are secondary metabolites responsible for the red coloration of mango and apple. The red color of the peel is essential for the fruit's marketability. Anthocyanins and flavonols are synthesized via the flavonoid pathway initiated from phenylalanine (Phe). Anthocyanins and flavonols have antioxidant, antifungal, and health-promoting properties. To determine if the external treatment of apple and mango trees with Phe can induce the red color of the fruit peel, the orchards were sprayed 1 to 4 weeks before the harvest of mango (cv. Kent, Shelly, and Tommy Atkins) and apple fruit (cv. Cripps pink, Gala and Starking Delicious). Preharvest Phe treatment increased the red coloring intensity and red surface area of both mango and apple fruit that was exposed to sunlight at the orchard. The best application of Phe was 2–4 weeks preharvest at a concentration of 0.12%, while a higher concentration did not have an additive effect. A combination of Phe and the positive control of prohydrojasmon (PDJ) or several applications of Phe did not have a significant added value on the increase in red color. Phe treatment increased total flavonoid, anthocyanin contents, and antioxidant activity in treated fruit compared to control fruits. High Performance Liquid Chromatography analysis of the peel of Phe treated 'Cripps pink' apples showed an increase in total flavonols and anthocyanins with no effect on the compound composition. HPLC analysis of 'Kent' mango fruit peel showed that Phe treatment had almost no effect on total flavonols content while significantly increasing the level of anthocyanins was observed. Thus preharvest application of Phe combined with sunlight exposure offers an eco–friendly, alternative treatment to improve one of the most essential quality traits—fruit color.

Keywords: anthocyanin; phenylalanine; prohydrojasmon; flavonoids; mango; apples; preharvest; red color; fruit quality

1. Introduction

Fruits and vegetables are lost or get a lower price due to poor appearance, quality problems, and consumer preferences [1]. Fruit is downgraded if it does not meet a very high standard of quality requirements, causing loss of profitability [2] and possibly food loss. Color and appearance, nutritional value, texture, and aroma are the main factors determining fruit and vegetable quality [3].

Fruit quality and maturity stage can often be indicated by the color of its skin [4]. The consumer prefers red-colored fruit. Therefore, the red color in fruit peel is a major contributor to the acceptance of fruit, which allows it to be priced higher and sold more easily [5]. Red color of fruit peel is important in terms of appearance and constitutes an advantage in resistance to pathogens and cold. For example, mango fruit exposed to sunlight developed red color, had higher anthocyanin concentration in the fruit peel, and was more resistant to pathogenic fungi including *Colletotrichum gloeosporioides*, *Alternaria alternata*, and *Lasiodiplodia theobromae* in comparison with green fruit from the same tree [6,7]. Red-colored mango fruit (cv. Shelly) is more resistant to cold than green fruit and, therefore, could be stored for more extended periods [6,8].

Anthocyanins are the main compounds responsible for the red coloring of mango and apple fruit peel [9]. Anthocyanins and flavonoids are natural compounds synthesized via the phenylpropanoid pathway, which have antioxidant and antifungal qualities, as well as health-promoting properties for humans [7,10]. Anthocyanins and flavonoids synthesis is induced in response to biotic or abiotic stresses, such as light, temperature, drought, and pathogen attacks [11,12].

Phenylalanine (Phe), the precursor of the phenylpropanoid pathway, is an aromatic amino acid existing naturally in plants and derived from the shikimate pathway [13]. Flavonoids, including anthocyanins and lignin, are secondary metabolites of the phenylpropanoid pathway, which is one of the main plant defense mechanisms [14]. Application of Phe to petunia, *Arabidopsis*, and cut chrysanthemum flowers, increased their resistance to *B. cinerea* by activating the phenylpropanoid pathway and flavonoids with antifungal activity [15,16]. Similarly, applying phenylalanine pre- or postharvest reduces decay in mango, avocado, strawberry, and citrus fruit, which is caused by various pathogenic fungi [17].

Red color induction methods, including application of plant growth regulators, in fruit have been studied for many years. In particular, previous studies have shown that mango and apple fruit have stronger red pigmentation and higher anthocyanin levels when exposed to direct sunlight, when grown outside of the tree canopy, compared with fruit from the inner parts of the canopy [6,18]. Pre-harvest application of harpin proteins or inactivated yeast treatments also showed an enhanced red color and anthocyanin accumulation [19,20]. Moreover, it was found that exogenous application of Prohydrojasmon (PDJ), methyl jasmonate (MJ), or Abscisic acid (ABA), along with direct sunlight exposure, promotes the phenylpropanoid pathway in mango fruit and contributes to the accumulation of red skin on the fruit [4,21].

Mango (cv. Kent, Shelly, and Tommy Atkins) and apple (cv. Cripps pink, Gala, Starking Delicious, and Anna) are fruits known to have light-red fruit peel and were chosen for this study. Our research goal is to determine if preharvest Phe application to mango and apple orchards will induce the red color in the fruit peel. Preharvest application of Phe could provide a healthy, safe, and relatively cheap means to increase fruit red color, reduce postharvest decay, and improve the profitability of the cultivar.

2. Materials and Methods

2.1. Plant Material, Preharvest Spray, Harvest, and Storage

The orchards used for this study were twenty-year-old mango orchards (cv. Kent and Shelly and Tommy Atkins), in Kibbutz Ravid (32°51′03″ N 35°27′52″ E; elevation +165 m), 15-years old apple orchards (cv. Cripps pink, Starking Delicious, and Gala) in Merom Golan (33°07′59″ N 35°46′33″ E; elevation +977 m), and a 15-years old apple orchard cv. Anna was grown in Arugot (31°44′5″ N 34°46′15″ E). Average-size mango trees (cv. Shelly and Kent) in the season of August 2020 (three trees per repetition, three repetitions per treatment) and apple trees (cv. Gala and Cripps Pink) in August 2020 and November 2020 (three trees per repetitions, four repetitions per treatment) were untreated (control) or sprayed with 0.12% phenylalanine (Phe) or with 0.2% prohydrojasmon (PDJ) on different weeks (1, 2, or 3) before harvest. The experiments were repeated on mango (cv. Shelly and Tommy Atkins) and apple (Cripps pink, Anna and Starking Delicious) fruit from June to November 2021. With the exception of 'Starking Delicious' apple cultivar, the mango and apple fruits were harvested from the outside of the tree canopy, which was exposed to direct sunlight. After harvest, control, and treated mango and apple fruits were transported (up to 2 h) to the ARO Volcani Center, Israel, and immediately stored at 12 °C and 2 °C for 21–28 days, respectively, followed by 7 days of shelf-life storage at 22 °C. Gala apple fruit was stored at 2 °C for 2 months, followed by 13 days of shelf-life storage at 22 °C. The fruits were evaluated at harvest, after cold storage, and after shelf life.

2.2. Measurements of Fruit Skin Color

2.2.1. Red Color Evaluation

We assessed the percentage of the red-colored surface area of each fruit from individual mango and apple fruits in each treatment at various time points (at harvest and after shelf life). Apple fruits were also evaluated after cold storage. Furthermore, the intensity of the red color was graded according to an index (0–5; 0—no red color, 1—faint red color, and 5—very intense red color). Data were collected and analyzed from 14 and 28 mango and apple fruits, respectively, per treatment.

2.2.2. Chroma Measurement

The skin color (Hue) was measured at the reddest point of 14 and 26 mango and apple fruits, respectively, for each treatment using a CR-400/410 Chromometer (Konica Minolta, Osaka, Japan). The hue angle (h°) measures color according to the wheel of colors, where 120° angle represents green color; 40°–60° angle represents orange-yellow color and 0°–40° represents red color. The transition between red color to green color is represented by a* value, where value of +60° correlates to full red color and −60° correlates to full green color.

2.2.3. Estimation of Chlorophyll, Anthocyanins, and Flavonoids in Fruit Skin

Fruit peel fluorescence was evaluated to measure chlorophyll (SFR_R), flavonoid (FLAV), and anthocyanin (ANTH_RG) signals using a Multiplex III fluorescence detector (Force A, Orsay, France), based on fluorescence signal ratios between excitations and emissions correlated to these compounds. Data were collected and evaluated from 14 and 26 fruits of each treatment in mango and apple fruit, respectively.

2.3. Decay Evaluation

Stem end rot (SER), side decay, and total rotten fruits of mango and total decay of apples were evaluated after shelf-life storage, and apples (cv. Gala) were also evaluated after cold storage. We evaluated decay incidence (percent of fruit) and severity (index 0–10; 0—no decay, 1—mild decay; 5—moderate decay; 10—severe decay) for each treatment per box.

2.4. Extract Preparation

Fruit peel samples from each treatment were harvested and stored at −80 °C. Samples were ground (IKA A11 basic, Germany) in liquid N_2, and 0.5 g were transferred into methanol solution (70%) for extraction and kept on the shaker overnight (250–300 rpm). The extraction mixture was centrifuged at 4100 rpm for 20 min at 20 °C (NF 8000R, Nuve, Turkey), and the supernatant was collected. The extraction was repeated. The collected supernatant liquid was concentrated in a CentriVap Concentrator (Labconco, Kansas, MO, USA) at room temperature until the extraction solution reached a volume of 0.5 mL. The concentrated samples were centrifuged at 12,000 rpm for 10 min at 20 °C, diluted 1:4 with distilled water, and stored at 4 °C until further use.

2.4.1. DPPH Assay

The free radical scavenging activity was measured by 2,2′- diphenyl-1-picrylhydrazyl (DPPH) assay as previously described [22,23] with slight modifications. The stock solution was prepared (0.24% in absolute methanol) and diluted with methanol until the absorbance was 0.98 ± 0.02 at 517 nm. For each treatment, the diluted extracts (1:4) of mango and apple peels were mixed with 1 mL of DPPH solution and incubated for 10 min at room temperature. The absorbance was measured at 517 nm, and scavenging activities were calculated according to the following equation:

$$scavenging\ activity\ (\%) = [\frac{OD_{517\ control} - OD_{517\ treatment}}{OD_{517\ control}}] \times 100$$

2.4.2. Total Phenolic Content

Total phenolic content of mango and apple fruit peels was evaluated with 0.2 N Folin–Ciocalteu (FC) reagent, and gallic acid was used as a reference standard [24,25]. For each treatment, the diluted mango and apple peel extracts were mixed with 750 µL of FC reagent and incubated for 5 min. 500 µL of sodium carbonate (Na_2CO_3, 75 gL^{-1}) were added to the reaction mixture followed by 15 min incubation. The absorbance was measured at 760 nm and total phenolic content was calculated as mg gallic acid equivalent/g of fresh weight (gFW).

2.4.3. Total Flavonoid Content

To determine the total flavonoid content, the diluted mango and apple peel extracts from each treatment were mixed with 200 µL of sodium nitrite ($NaNO_2$, 5%) and incubated for 5 min at room temperature. Then, 200 µL of aluminum chloride ($AlCl_3$, 10%) and 1.5 mL of sodium hydroxide (NaOH, 1 M) were added to the reaction mixture and absorbance was measured at 510 nm. The total flavonoid content was calculated as mg quercetin equivalent/gFW [26].

2.5. Flavonoid and Anthocyanin Characterization (HPLC)

Analysis was conducted on peels of mango (cv. Kent) and apple (cv. Cripps Pink) fruit after cold storage, to identify and quantify the flavonoids and anthocyanins in the fruit peel. 0.3 g of peel samples were used for extraction (2 mL, methanol:water:acetic acid, 11:5:1, v/v) [27]. Anthocyanins and flavonols were quantified as described by [28].

The experiment was conducted according to [29]. High-performance liquid chromatography (HPLC) (Shimatzu, Kyoto, Japan) equipped with an LC-10AT pump, an SCL-10A controller, and an SPD-M10AVP photodiode array detector. Extracts were loaded onto an RP-18 column (Vydac 201TP54) and separated at 27 °C with the following solutions: (A) H2O, pH 2.3, and (B) H2O:MeCN:HOAc (107:50:40), pH 2.3. Solutions were applied as a linear gradient from a ratio of 4:1 (A:B) to 3:7 over 45 min and held at a ratio of 3:7 for an additional 10 min at a flow rate of 0.5 mL/min. Anthocyanidins and flavonoids were identified by comparing both the retention time and the absorption spectrum at 250–650 nm with those of standard purified anthocyanins and flavonols (ChemFaces, Wuhan), China and Extrasynthase (Genay, France). Identification and quantification of anthocyanins and flavonols were done using reference standards, and concentrations were expressed as peak area/gram of FW.

2.6. Statistical Analysis

Data are represented as mean value ± standard error (SE). Multifactorial analysis of variance (One-way ANOVA, Tukey–Kramer HSD test) and Wilcoxon non-parametric comparison were performed using JMP (JMP Pro 15 software, SAS Institute, Cary, NC, USA). Different letters represent a statistically significant difference ($p \leq 0.05$) among different treatments at the same time point.

3. Results

The effect of preharvest Phe spraying on the redness of mango (cv. Kent and Shelly) and apple (cv. Cripps Pink and Gala) fruit peel was evaluated in comparison to untreated control fruit and the positive control of PDJ treated fruit. Fruit treated with either Phe or PDJ at harvest and after shelf-life storage had a significantly higher red color intensity of the peel in mango and apple fruit compared to the control (Figures 1A, 2A, S1A and S2A).

The percentage of surface coverage of the red color in the mango (cv. Kent and Shelly) fruit peel at all evaluated time points was significantly higher in most treatments (Figures 1B and S1B). Phe and PDJ significantly increased the red color surface coverage in almost all treated apples (cv. Gala) and increased the red color surface in treated 'Cripps pink' apples (Figures 2B and S2B). Hue values at the reddest point on the fruit peel were evaluated at several time points (harvest, after CS, and after SL) for mango (cv. Kent and Shelly) and apple (cv. Cripps pink and Gala) fruit. In general, Phe and PDJ treatments decreased hue values (wheel of colors, where 120 is for green color; 0–40 represents red color) in almost all treated fruit (Figures 1C, 2C, S1C, S2C and S3A,C). This trend was especially apparent in mango (cv. Kent) and apple (cv. Gala) fruit, where hue values of treated fruit significantly decreased compared to control fruit both at harvest and after SL (Figures 1C, S2A and S3A). In mango (cv. Shelly), fruit peel color varied similarly at harvest from 70.3 in control to 43.9–21.2 in treated fruit at this time point (Figure S1C). In apple (cv. Cripps pink), even though treated apples had significantly lower hue values at their reddest point compared to the control, all the treated and untreated fruit were within the orange color range (Figures 2C and S3C). The Red-Green color range, which is represented by a* value, is directly proportional to the red-green color intensity. Higher and positive a* values correlate with more intense red color, whereas lower positive and negative values correlate with yellow and green color intensities, respectively. Phe and PDJ significantly increased a* values in most mango (cv. Kent and Shelly) and apple (cv. Gala) treated fruit. a* value of apple (cv. Cripps pink) also increased after Phe treatment compared to the control (Figures 1D, Figure 2D, S1D, S2D, S3B,D).

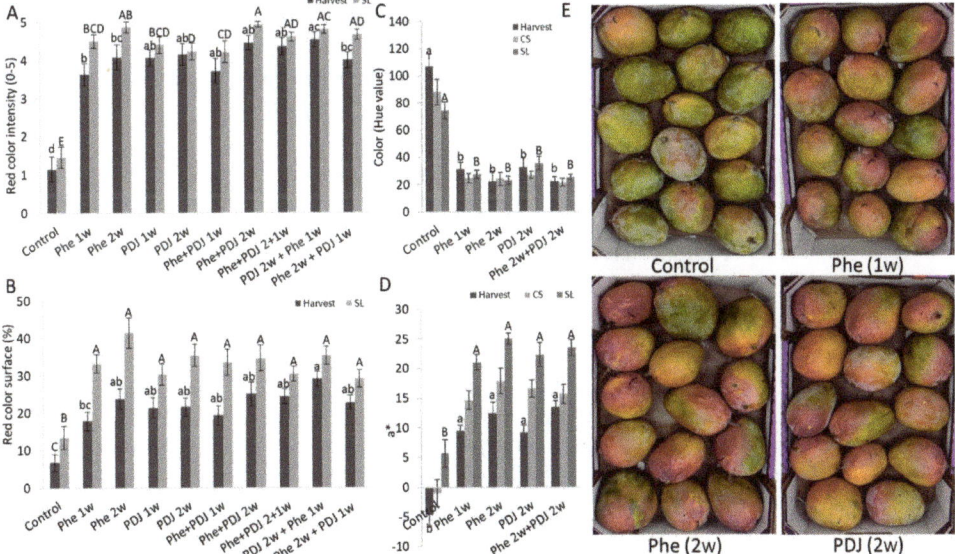

Figure 1. Red color evaluation of 'Kent' mango fruit peel. 'Kent' mango orchards were sprayed with 0.12% phenylalanine (Phe) or 0.2% prohydrojasmon (PDJ), one or two weeks preharvest. The fruit was evaluated at harvest (T0), after cold storage (CS, 3 weeks at 12 °C), and after shelf life (SL, 7 days at 22 °C). (**A**) Red color intensity (index 0–5). (**B**) Red surface area (% of fruit coverage). (**C**) Color (Hue value of the reddest point). (**D**) Green-red color range (a* value). (**E**) Representative pictures of 'Kent' mango boxes after shelf-life storage. Mean values and standard errors are presented. Statistical analysis was conducted for each time point separately (small or capital letters). Different letters represent a significant difference ($p \leq 0.05$).

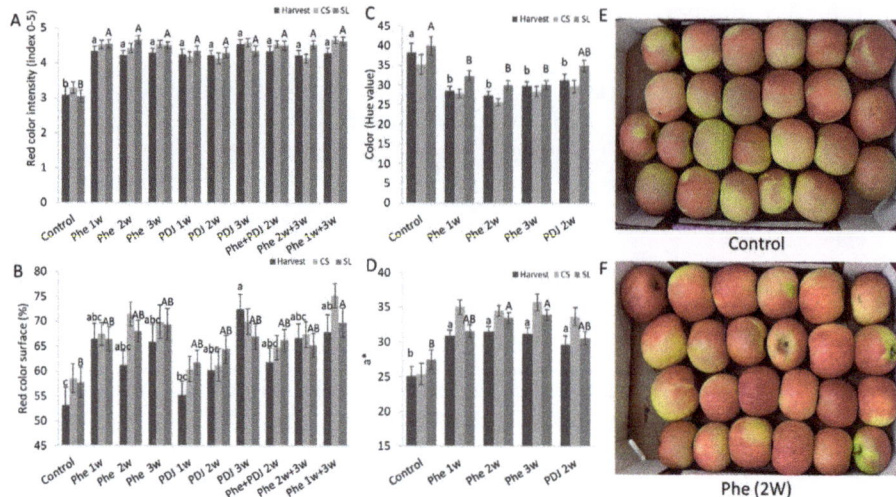

Figure 2. Red color evaluation of apple (cv. Cripps pink) peel. Apple orchard was sprayed preharvest with 0.12% phenylalanine (Phe) and/or 0.2% prohydrojasmon (PDJ). The fruit was evaluated at 3-time points: at harvest, after cold storage (CS, 3 weeks at 2 °C), and after shelf life (SL, 7 days at 20 °C). (**A**) Red color intensity (index 0–5). (**B**) Red surface area (% of fruit coverage). (**C**) Color (hue value at the reddest point). (**D**) Green-red color range (a* value at the reddest point). (**E,F**) Representative pictures of Cripps pink apple box after shelf-life storage. Mean values and standard errors are presented. Statistical analysis was conducted for each time point separately (small or capital letters). Different letters represent a significant difference ($p \leq 0.05$).

The results described above were experiments conducted on apple cultivars (cv. 'Anna,' 'Starking Delicious, and 'Cripps Pink') and mango cultivars (cv. 'Shelly' and 'Tommy Atkins') in the following year, which showed a similar trend (Tables S1–S3). The fruit treated preharvest with Phe showed a significant increase in red color intensity and red color surface of almost all treatments compared to control. Moreover, hue values generally decreased and a* value was usually increased (Tables S1 and S2).

Preharvest application of Phe in various concentrations showed that an increase in Phe concentration better induced the red color until optimum results at 0.12% Phe (Tables S1, S2 and S6). Phe at 0.12% increased the red color area and intensity in the fruit peel significantly better compared to lower concentrations as Phe at 0.01%, which was applied to apple (cv. Starking Delicious and Cripps pink) and mango (cv. Shelly) orchards (Table S1). Higher concentration (0.24%) did not contribute to better results of red color accumulation (Tables S1–S3). In the examination of the best time to apply Phe preharvest, it was seen that in most experiments 2 weeks preharvest led to the best induction of red peel color (Tables S1–S3). While in Starking apples the application of 4 weeks preharvest was optimal for inducing red color (Table S4). A combination of several applications usually did not further induce the red color (Tables S2–S4). This induction of red color by preharvest application of Phe was correlated to a small decrease in decay incidence and severity in mango fruit (cv. Kent, Shelly, and Tommy) and inconclusive results in apple fruit (Tables S6 and S7).

Preharvest application of Phe increased the antioxidant activity in mango (cv. Kent and Shelly) and apple (cv. Cripps pink) fruit peels (Figures 3A,D and S4A). Phe also increased total phenolics and flavonoid content in most treatments in mango (cv. Kent and Shelly) and apple (cv. Cripps pink) fruit peels (Figures 3B,C,E,F and S4B,C). Trends of total phenolic and flavonoid contents correlate to the level of antioxidant activity in mango fruit (cv. Kent and Shelly) both after CS and after SL, and in apple (cv. Cripps pink) after SL (Figures 3 and S4). In both mango cultivars (cv. Kent and Shelly), combined treatment of

PDJ 2w + Phe 2w had the highest antioxidant activity (Figures 3A and S4A). However, in apples (cv. Cripps pink), Phe 2w and Phe 3w treatments had the highest antioxidant activity at harvest and after CS and SL, respectively (Figure 3D). The highest quantity of flavonoid content in mango (cv. Kent) was measured in the PDJ 2w treatment and PDJ 2w + Phe 2w treatment in mango (cv. Shelly) (Figures 3C and S4C). Quantification of total phenolic content showed a similar trend in 'Kent' mango and 'Shelly' mango after CS (Figures 3B and S4B). In apple (cv. Cripps pink), total phenolics and flavonoid content were highest in Phe 2w or 3w treatments, similar to its antioxidant evaluation (Figure 3).

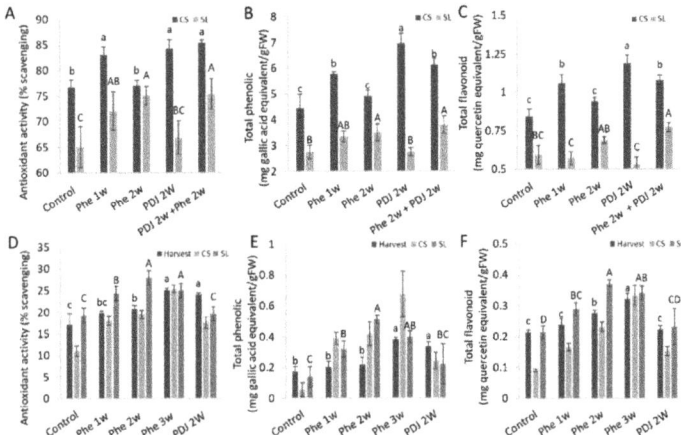

Figure 3. Effect of preharvest treatment with 0.12% Phe or 0.2% PDJ on antioxidant, phenolic, and flavonoid content in mango (cv. Kent) and apple fruit (cv. Cripps pink) at harvest, after cold storage (CS), and shelf-life (SL). Antioxidant activity, total phenolic, and total flavonoid content were evaluated from mango peels (cv. Kent) (**A–C**), apple (cv. Cripps pink) (**D–F**) peels. Mean values and standard errors are presented. Statistical analysis was conducted for each time point separately (small or capital letters). Different letters represent significant differences ($p \leq 0.05$).

The fluorescence of chlorophyll, flavonoids, and anthocyanins in the fruit peel was evaluated at the reddest point of all mango (cv. Kent and Shelly) and apple (cv. Cripps pink and Gala) fruit. Chlorophyll fluorescence decreased in almost all the Phe treated mango (cv. Kent and Shelly) and apple (cv. Cripps pink and Gala) fruits, and almost all treatments presented statistical significance (Figures 4A,D, S5A,D and S6A,D).

'Kent' mango had a significant increase in flavonoid fluorescence in all Phe treatments compared to the control both after harvest and after SL (Figures 4B and S6B). Similarly, 'Shelly' mango also showed an increase in flavonoids in all Phe treatments at harvest and after CS, however, after SL storage, no difference between the treatments was observed (Figure S6B). Apple (cv. Cripps pink and Gala) showed increased flavonoid fluorescence in treated fruit at harvest, while flavonoid fluorescence did not increase in treated fruit after SL storage (Figures 4F, S5E and S6E). Phe and PDJ increased anthocyanin fluorescence in all treated fruit. The highest level of anthocyanins was detected mainly in Phe 2w and Phe 3w and their combination with other treatments in both mango (cv. Kent and Shelly) and apple (cv. Cripps pink) after SL. Apple (cv. Gala) presented the highest level of anthocyanins in Phe 1w treatment and its combinations (Figures 4C,F, S5C,F and S6C,F).

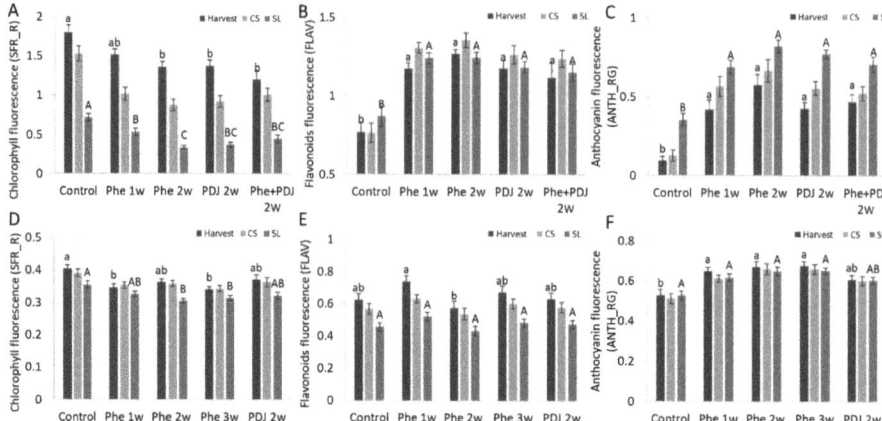

Figure 4. Effect of preharvest treatment of Phe and PDJ on chlorophyll, anthocyanin, and flavonoids fluorescence. Mango (cv. Kent, **A–C**) and apple (cv. Cripps pink, **D–F**) orchards were treated 1–3 weeks (1W, 2W, 3W) preharvest with 0.12% Phe or 0.2% PDJ and chlorophyll, anthocyanin, and flavonoids fluorescence at the reddest point of the fruit peel was analyzed at 3-time points: after harvest, after cold storage (CS), and after shelf-life (SL). (**A,D**) Chlorophyll fluorescence (SFR_R). (**B,E**) Flavonoids fluorescence (FLAV). (**C,F**) Anthocyanin fluorescence (ANTH_RG). Mean values and standard errors are presented. Statistical analysis was conducted for each time point separately (small and capital letters). Different letters represent significant differences ($p \leq 0.05$).

Quantification and identification of metabolites in mango (cv. Kent) and apple (cv. Cripps pink) fruits treated with Phe or PDJ after cold storage was performed using HPLC analysis (Figure 5). In apple (cv. Cripps pink) peel, three types of compounds were detected related to the phenylpropanoid pathway: anthocyanins, flavonols, and dihydrochalcones, while in mango peel only anthocyanins and flavonols were detected (Figure 5). Both apple and mango peels from fruit treated with Phe had significantly higher levels of anthocyanins in the sample, compared to the control (Figure 5A,D). In mango, the anthocyanins content increased by 11.7 fold in the Phe 2w treatment. While the highest levels of anthocyanins and flavonols in apple (cv. Cripps pink) peel were observed in Phe 1w treatment, with an increase in anthocyanins and flavonols by 9.3 and 3.3 folds, respectively, compared to control (Figure 5). These levels of increase suggest a major shift in biosynthetic activity. Similarly, the total amount of dihydrochalcones detected in Phe treated apple samples was higher compared to the control (Figure 5C). Unlike mango, apple peel samples contained a significantly higher amount of flavonols in Phe treated fruit compared to the control (Figure 5B,E).

Eight flavonols were detected in mango (cv. Kent) peel: quercetin-3-O-galactoside (26.1 min, Qu-gal), quercetin-3-O-glucoside (27.2 min, Qu-glc), quercetin-3-O-xyloside (28.4 min, Qu-xyl), quercetin-3-O-arabinopyranoside (29.6 min, Qu-arap), quercetin-3-O-arabinofuranoside (30.4 min, Qu-araf), quercetin-3-O-rhamnoside (31.8 min, Qu-rha), kaempferol-3-O-glucoside (32.8 min, Ka-glc) and unknown compound (47.1 min) (Figure 5E). The ratios between the different flavonols within each treatment were similar, indicating that the treatment did not change the synthesis of individual compounds (Figures 5E and S7). In apple peels, seven flavonols were revealed (Figures 5B and S8). The chromatogram As opposed to mango (cv. Kent) peel analysis, several minor changes in composition levels were detected between treated fruit to control in apple (cv. Cripps pink) peel. For example, quercetin-3-O-galactoside (Qu-gal, 26.4 min) and quercetin-3-glucoside (Qu-glc, 27.6 min) showed a slightly higher level in treatments compared to the control, while quercetin-3-rhamnoside (Qu-rha, 32.3 min) and quercetin 3-xyloside

(Qu-xyl, 28.8 min) showed lower levels in treated fruit compared to control. Most of the other compounds had a more or less similar ratio to the control (Figure 5B).

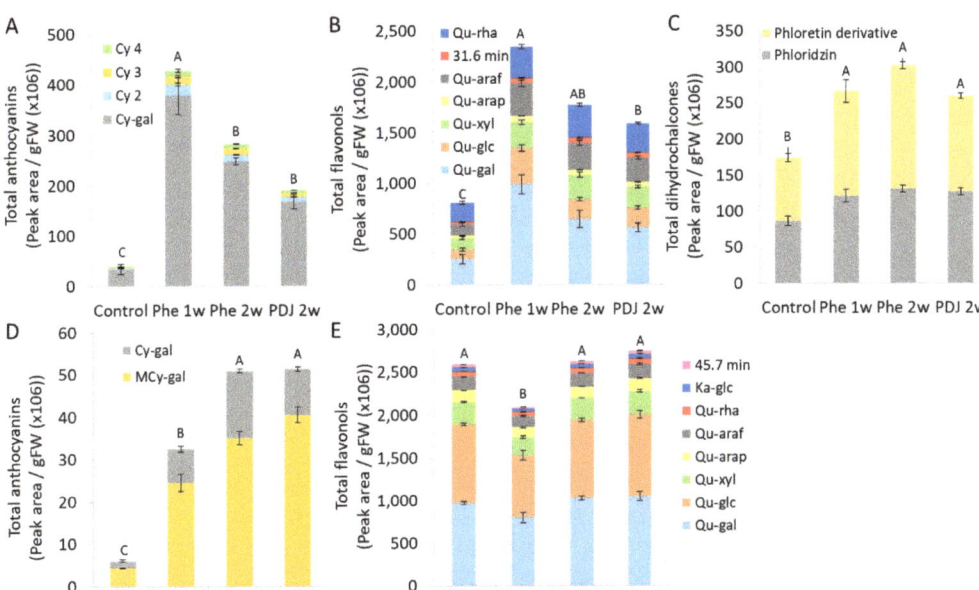

Figure 5. Quantification of metabolites using HPLC in mango and apple fruit peel. Mango (cv. Kent) and apple (cv. Cripps pink) were sprayed with 0.12% phenylalanine (Phe) or 0.2% prohydrojasmon (PDJ) one or two weeks preharvest. Apple (**A–C**) and mango (**D,E**) peels were analyzed after 3 weeks of cold storage at 2 °C and 12 °C, respectively. (**A,D**) Anthocyanin content in apple and mango, respectively. (**B,E**) Flavonol content in apple and mango, respectively. (**C**) Dihydrochalcones in apple. All the values expressed as Peak area/gFW ($\times 10^6$). Mean values and standard errors are presented. Different letters represent significant differences for total content ($p < 0.05$).

Mango contained only two anthocyanins: cyaniding-3-*O*-β-D-galactoside (11.3 min, Cy-gal) and 7-*O*-methylcyanidin 3-*O*-β-D-galactopyranoside (20.9 min, MCy-gal) (Figure 5D). 7-*O*-methylcyanidin 3-*O*-β-D-galactopyranoside was the main anthocyanin compound in mango fruit (Figure 5D and Figure S7). Anthocyanins apple (cv. Cripps pink) samples contained four cyanidin derivatives: cyanidin 3-galactoside (11.4 min, Cy-gal), and three unknown derivatives of cyanidin (18.0 min, 22.7 min, and 24.1 min) were detected. According to the literature, the most abundant peak in apple peels is cyanidin galactoside which has been identified before in 'Cripps Pink' [30,31]. The other cyanidin derivative peaks (Cy2, Cy3 and Cy4) in our study on apple peels could probably be designated as cyanidin 3-arabinoside, cyanidin 3-glucoside, cyanidin 3-xyloside [32,33]. All treatments presented similar ratios between the anthocyanin compounds, with cyanidin 3-galactoside being the main derivative detected (Figure 5). Dihydrochalcones, which were detected in apple peel as phloretin derivatives, also had a similar ratio in the control and treated fruit, including both phloridzin (35.1 min) and an unknown derivative of phloretin (31.7 min) (Figure 5C). The phloretin derivative could probably be designated as phloretin 2′-xyloglucoside [34]. In summary, it seems that preharvest Phe treatment increases the biosynthesis of flavonoids, anthocyanins, and dihydrochalcones, while keeping a similar ratio of the compounds as in the control (Figure 5).

4. Discussion

Red color is important for the marketability of fruit. Redder fruit can help decrease food loss and also contribute to customer health [10]. Red color of mango, apple, and other fruit peels is associated with the secondary metabolites named anthocyanins. In this work, we evaluated the effect of preharvest Phe application as a relatively cheap and eco-friendly method [17] on the induction of the red color of mango (cv. Kent, Tommy Atkins, and Shelly) and apple (cv. Starking Delicious, Gala and Cripps Pink) fruit peel exposed to direct sunlight. The main finding of this research is that preharvest application of phenylalanine in combination with sunlight radiation, results in the induction of anthocyanin biosynthesis, enhanced red color of the mango and apple fruit peel, and an increase in antioxidant activity.

Phenylalanine, a naturally occurring aromatic amino acid derived from the shikimate pathway [13], is the precursor for the phenylpropanoid biosynthetic pathway, where anthocyanins are among the secondary metabolites of the downstream pathway, which contribute to the accumulation of red color in the fruit peel [9]. Phe treatment increased flavonols and fragrance related to upstream of the phenylpropanoid pathway in various flowers but did not lead to the accumulation of anthocyanins and red color in flowers as well as Arabidopsis, tomato, chrysanthemum and petunia leaves [15,16,35]. Similarly, postharvest application to various fruits increased flavonols but did not increase the red color of fruit [17].

Flavonoid and phenylpropanoid pathways are defense mechanisms in the plant which are tightly regulated and induced in response to biotic or abiotic stress, that is, pathogens or sunlight [11,12]. Thus, Phe application induces flavonols production but does not increase the production of anthocyanins unless the downstream of the pathway is induced. Therefore, in this work, we show that preharvest treatment with phenylalanine on mango or apple fruit accompanied by sunlight radiation led to an induction of red color coverage and intensity of the peel. The increase in red coloration occurred in fruit on the outer side of the canopy, due to sunlight exposure. Indeed, sunlight radiation is known to induce the phenylpropanoid pathway in various plants, including mango and apple fruit [6,18,21,36].

The current study also applied prohydrojasmon (PDJ), an analog of Jasmonic acid phytohormone as a positive control, after having been described as effective in red color induction of the fruit peel [29]. Although Phe showed similar results as did PDJ, the mode of action of Phe seems to be as a precursor of phenylpropanoid pathway and not in a hormone-like manner. Phenylalanine also has an economic advantage that makes it more affordable compared to PDJ.

Previous studies reported that preharvest treatment of mango fruit (cv. Kent, Shelly, and Maya) with PDJ and ABA exposed to sunlight, increased the red color intensity and coverage of the fruit peel [21,29]. Similarly, both in apple and mango fruit, the red color surface coverage and intensity of the peel have increased due to the preharvest application of phenylalanine in combination with sunlight (Figures 1 and 2). Shafiq and Singh have shown that phenylalanine at a concentration of 0.01% which was applied about 4 weeks preharvest increased the red color and anthocyanins of 'Cripps Pink' apples [37]. Here, various concentrations of Phe on both mango and apple fruit were tested on a larger scale in both mango and apples and at different years, finding that a higher concentration of 0.12% was much more efficient in inducing red color than the application of 0.01% of Phe (Tables S1, S3 and S4). It seems that increase concertation up to 0.12% led to an increase in red color in mango and apple peel (Tables S1 and S2).

The intensity of the red color, a^*, and hue value measurements showed that Phe treated fruit at 0.12% had a redder and more intense red peel color, whereas the fruit in the control group had a lighter intensity red color varying from orange to yellow colors (Figures 1 and 2). Similarly, preharvest applications of Methyl Jasmonate (MJ) and prohydrojasmon (PDJ) were reported to increase a^* values and reduce hue values in mango fruit (cv. Mahachanok) [4,21].

The phenylpropanoid pathway is responsible for the synthesis of flavonoids and anthocyanins [38]. Preharvest treatment with Phe significantly increased anthocyanin levels both in mango (cv. Kent) and in apple (cv. Cripps pink) fruit, while flavonoid levels were either increased or unchanged (Figures S5 and S6). Further phenylpropanoid metabolic analysis was done by HPLC, which showed a significant increase in total anthocyanin and flavonols levels compared to control in apple (cv. Cripps pink) and in anthocyanin levels in mango (cv. Kent) with almost no effect on the compound composition (Figure 5). As a comparison, mango fruit exposed to sunlight showed an increase of the same flavonoids in comparison to fruit from the inside of the tree canopy [8]. Due to the increase in flavonols and anthocyanins, which have antioxidative properties, it has been observed that preharvest application with phenylalanine also improved antioxidant activity both in mango and apple fruit (Figures 3–5). Indeed, a positive correlation between total phenolic and flavonoid content and the level of antioxidant activity was found in different apple varieties [39].

5. Conclusions

Preharvest treatment of Phenylalanine in combination with sunlight radiation increased the phenylpropanoid biosynthesis pathway, leading to an increase in the coverage and intensity of red-colored peel fruit in various mango and apple cultivars. Phe application increased the phenolic and flavonoid contents due to the activation of the phenylpropanoid pathway, which in combination with sunlight radiation led to the biosynthesis of anthocyanins that directly contribute to the red color of the peel and have health benefits. This study provides an effective new method of Phe spray at the orchard to improve the appearance of the fruit and its health benefits.

6. Patents

Provisional Patent Application No. 63/134,403 and 63/164,051, titled "Methods for improving fruit quality."

Supplementary Materials: The following supporting information can be downloaded at: https://www.mdpi.com/article/10.3390/antiox11030491/s1.

Author Contributions: Conceptualization, N.A.; methodology, M.F. and M.K.P.; validation, N.A.; formal analysis, M.F., M.K.P., R.O., D.M. and O.F.; resources, N.A. and M.O.-S.; data curation, M.F., R.O., D.M. and O.F.; writing—original draft preparation, M.F.; writing—review and editing, M.F., M.K.P., R.O. and N.A.; supervision; project administration; funding acquisition. All authors have read and agreed to the published version of the manuscript.

Funding: This research was funded by Copia and PostBoost, grant number 430-0875-Y-88.

Institutional Review Board Statement: Not applicable.

Informed Consent Statement: Not applicable.

Data Availability Statement: Data is contained within the article or supplementary material.

Acknowledgments: We thank Yigal Elad for his advices. We thank Copia Company, Galit Sharabani, Eyal Cohen and Ohad Zuckerman for their consult and support. A special thanks for Amos Ovadia for conducting all the field experiment.

Conflicts of Interest: The authors declare no conflict of interest.

References

1. Porter, S.D.; Reay, D.S.; Bomberg, E.; Higgins, P. Avoidable food losses and associated production-phase greenhouse gas emissions arising from application of cosmetic standards to fresh fruit and vegetables in Europe and the UK. *J. Clean. Prod* **2018**, *201*, 869–878. [CrossRef]
2. Dar, J.A.; Wani, A.A.; Ahmed, M.; Nazir, R.; Zargar, S.M.; Javaid, K. Peel colour in apple (*Malus domestica* Borkh.): An economic quality parameter in fruit market. *Sci. Hortic.* **2019**, *244*, 50–60. [CrossRef]
3. Barrett, D.M.; Beaulieu, J.C.; Shewfelt, R. Color, flavor, texture, and nutritional quality of fresh-cut fruits and vegetables: Desirable levels, instrumental and sensory measurement, and the effects of processing. *Crit. Rev. Food Sci. Nutr.* **2010**, *50*, 369–389. [CrossRef]

4. Muengkaew, R.; Chaiprasart, P.; Warrington, I. Changing of physiochemical properties and color development of mango fruit sprayed methyl jasmonate. *Sci. Hortic.* **2016**, *198*, 70–77. [CrossRef]
5. Saure, M.C. External control of anthocyanin formation in apple. *Sci. Hortic.* **1990**, *42*, 181–218. [CrossRef]
6. Sivankalyani, V.; Feygenberg, O.; Diskin, S.; Wright, B.; Alkan, N. Increased anthocyanin and flavonoids in mango fruit peel are associated with cold and pathogen resistance. *Postharvest Biol. Technol.* **2016**, *111*, 132–139. [CrossRef]
7. Sudheeran, P.K.; Ovadia, R.; Galsarker, O.; Maoz, I.; Sela, N.; Maurer, D.; Alkan, N. Glycosylated flavonoids: Fruit's concealed antifungal arsenal. *New Phytol.* **2020**, *225*, 1788–1798. [CrossRef]
8. Sudheeran, P.K.; Feygenberg, O.; Maurer, D.; Alkan, N. Improved cold tolerance of mango fruit with enhanced anthocyanin and flavonoid contents. *Molecules* **2018**, *23*, 1832. [CrossRef]
9. Fernández-López, J.A.; Fernández-Lledó, V.; Angosto, J.M. New insights into red plant pigments: More than just natural colorants. *RSC Adv.* **2020**, *10*, 24669–24682. [CrossRef]
10. Khoo, H.E.; Azlan, A.; Tang, S.T.; Lim, S.M. Anthocyanidins and anthocyanins: Colored pigments as food, pharmaceutical ingredients, and the potential health benefits. *Food Nutr. Res.* **2017**, *61*, 1361779. [CrossRef]
11. Treutter, D. Significance of flavonoids in plant resistance: A review. *Environ. Chem. Lett.* **2006**, *4*, 147–157. [CrossRef]
12. Naing, A.H.; Kim, C.K. Abiotic stress-induced anthocyanins in plants: Their role in tolerance to abiotic stresses. *Physiol. Plant.* **2021**, *172*, 1711–1723. [CrossRef] [PubMed]
13. Fraser, C.M.; Chapple, C. *The Arabidopsis Book*; American Society of Plant Biologists: Rockville, MD, USA, 2011; p. 9.
14. Dixon, R.A.; Achnine, L.; Kota, P.; Liu, C.J.; Reddy, M.S.; Wang, L. The phenylpropanoid pathway and plant defense—A genomics perspective. *Mol. Plant Pathol.* **2002**, *3*, 371–390. [CrossRef] [PubMed]
15. Oliva, M.; Hatan, E.; Kumar, V.; Galsurker, O.; Nisim-Levi, A.; Ovadia, R.; Oren-Shamir, M. Increased phenylalanine levels in plant leaves reduces susceptibility to *Botrytis cinerea*. *Plant Sci.* **2020**, *290*, 110289. [CrossRef] [PubMed]
16. Kumar, V.; Hatan, E.; Bar, E.; Davidovich-Rikanati, R.; Doron-Faigenboim, A.; Elad, Y.; Alkan, N.; Lewinsohn, E.; Spitzer-Rimon, B.; Oren-Shamir, M. Phenylalanine increases Chrysanthemum flower immunity against Botrytis cinerea attack. *Plant J.* **2020**, *104*, 226–240. [CrossRef]
17. Patel, M.K.; Maurer, D.; Feygenberg, O.; Ovadia, A.; Elad, Y.; Oren-Shamir, M.; Alkan, N. Phenylalanine: A promising inducer of fruit resistance to postharvest pathogens. *Foods* **2020**, *9*, 646. [CrossRef] [PubMed]
18. Hamadziripi, E.T.; Theron, K.I.; Muller, M.; Steyn, W.J. Apple compositional and peel color differences resulting from canopy microclimate affect consumer preference for eating quality and appearance. *Hortic. Sci.* **2014**, *49*, 384–392. [CrossRef]
19. Crupi, P.; Alba, V.; Masi, G.; Caputo, A.R.; Tarricone, L. Effect of two exogenous plant growth regulators on the color and quality parameters of seedless table grape berries. *Food Res. Int.* **2019**, *126*, 108667. [CrossRef]
20. Crupi, P.; Palattella, D.; Corbo, F.; Clodoveo, M.L.; Masi, G.; Caputo, A.R.; Battista, F.; Tarricone, L. Effect of pre-harvest inactivated yeast treatment on the anthocyanin content and quality of table grapes. *Food Chem.* **2021**, *337*, 128006. [CrossRef]
21. Sudheeran, P.K.; Maurer, D.; Feygenberg, O.; Love, C.; Alkan, N. Improving the red color and fruit quality of 'Kent' mango fruit by pruning and preharvest spraying of prohydrojasmon or abscisic acid. *Agronomy* **2020**, *10*, 944.
22. Saeed, N.; Khan, M.R.; Shabbir, M. Antioxidant activity, total phenolic and total flavonoid contents of whole plant extracts *Torilis leptophylla* L. *BMC Complement Altern. Med.* **2012**, *12*, 221. [CrossRef]
23. Patel, M.K.; Mishra, A.; Jaiswar, S.; Jha, B. Metabolic profiling and scavenging activities of developing circumscissile fruit of psyllium (*Plantago ovata* Forssk.) reveal variation in primary and secondary metabolites. *BMC Plant Biol.* **2020**, *20*, 116. [CrossRef] [PubMed]
24. Hazra, B.; Biswas, S.; Mandal, N. Antioxidant and free radical scavenging activity of Spondias pinnata. *BMC Complement. Altern. Med.* **2008**, *8*, 63. [CrossRef] [PubMed]
25. Patel, M.K.; Mishra, A.; Jha, B. Non-targeted metabolite profiling and scavenging activity unveil the nutraceutical potential of psyllium (*Plantago ovata* Forsk). *Front. Plant Sci.* **2016**, *7*, 431. [CrossRef]
26. Zhishen, J.; Mengcheng, T.; Jianming, W. The determination of flavonoid in mulberry and their scavenging effects on superoxide radicals. *Food Chem.* **1999**, *64*, 555–559. [CrossRef]
27. Markham, K.R.; Ofman, D.J. Lisianthus flavonoid pigments and factors influencing their expression in flower colour. *Phytochemistry* **1993**, *34*, 679–685. [CrossRef]
28. Sapir, M.; Oren-Shamir, M.; Ovadia, R.; Reuveni, M.; Evenor, D.; Tadmor, Y.; Levin, I. Molecular aspects of Anthocyanin fruit tomato in relation to high pigment-1. *J. Hered* **2008**, *99*, 292–303. [CrossRef] [PubMed]
29. Sudheeran, P.K.; Love, C.; Feygenberg, O.; Maurer, D.; Ovadia, R.; Oren-Shamir, M.; Alkan, N. Induction of red skin and improvement of fruit quality in 'Kent', 'Maya' mangoes by preharvest spraying of prohydrojasmon at the orchard. *Postharvest Biol. Technol.* **2019**, *149*, 18–26. [CrossRef]
30. Wan Sembok, W.Z.B. Regulation of Fruit Colour Development, Quality, and Storage Life of 'Cripps Pink' Apples with Deficit Irrigation and Plant Bioregulators. Ph.D. Thesis, Curtin University of Technology, Perth, Australia, 2009; pp. 1–242.
31. Hoang, N.T.; Golding, J.B.; Wilkes, M.A. The effect of postharvest 1-MCP treatment and storage atmosphere on 'Cripps Pink' apple phenolics and antioxidant activity. *Food Chem.* **2011**, *127*, 1249–1256. [CrossRef]
32. Honda, C.; Kotoda, N.; Wada, M.; Kondo, S.; Kobayashi, S.; Soejima, J.; Zhang, Z.; Tsuda, T.; Moriguchi, T. Anthocyanin biosynthetic genes are coordinately expressed during red coloration in apple skin. *Plant Physiol. Biochem.* **2002**, *40*, 955–962. [CrossRef]

33. Vrhovsek, U.; Rigo, A.; Tonon, D.; Mattivi, F. Quantitation of polyphenols in different apple varieties. *J. Agric. Food Chem.* **2004**, *52*, 6532–6538. [CrossRef] [PubMed]
34. Wojdyło, A.; Oszmiański, J.; Laskowski, P. Polyphenolic compounds and antioxidant activity of new and old apple varieties. *J. Agric. Food Chem.* **2008**, *56*, 6520–6530. [CrossRef] [PubMed]
35. Kumar, V.; Elazari, Y.; Ovadia, R.; Bar, E.; Nissim-Levi, A.; Carmi, N.; Lewinsohn, E.; Oren-Shamir, M. Phenylalanine treatment generates scent in flowers by increased production of phenylpropanoid-benzenoid volatiles. *Postharvest Biol. Technol.* **2021**, *181*, 111657. [CrossRef]
36. Matus, J.T.; Loyola, R.; Vega, A.; Peña-Neira, A.; Bordeu, E.; Arce-Johnson, P.; Alcalde, J.A. Post-veraison sunlight exposure induces MYB-mediated transcriptional regulation of anthocyanin and flavonol synthesis in berry skins of *Vitis vinifera*. *J. Exp. Bot.* **2009**, *60*, 853–867. [CrossRef] [PubMed]
37. Shafiq, M.; Singh, Z. Preharvest spray application of phenylpropanoids influences accumulation of anthocyanin and flavonoids in 'Cripps Pink' apple skin. *Sci. Hortic.* **2018**, *233*, 141–148. [CrossRef]
38. Davies, K.M.; Jibran, R.; Zhou, Y.; Albert, N.W.; Brummell, D.A.; Jordan, B.R.; Bowman, J.L.; Schwinn, K.E. The evolution of flavonoid biosynthesis: A bryophyte perspective. *Front. Plant Sci.* **2020**, *11*, 7. [CrossRef] [PubMed]
39. Vieira, F.G.K.; Borges, G.D.S.C.; Copetti, C.; Di Pietro, P.F.; da Costa Nunes, E.; Fett, R. Phenolic compounds and antioxidant activity of the apple flesh and peel of eleven cultivars grown in Brazil. *Sci. Hortic.* **2011**, *128*, 261–266. [CrossRef]

Article

Phytogenotypic Anthocyanin Profiles and Antioxidant Activity Variation in Fruit Samples of the American Cranberry (*Vaccinium macrocarpon* Aiton)

Rima Urbstaite [1,*], Lina Raudone [1,2] and Valdimaras Janulis [1]

1. Department of Pharmacognosy, Faculty of Pharmacy, Lithuanian University of Health Sciences, 50166 Kaunas, Lithuania; lina.raudone@lsmuni.lt (L.R.); valdimaras.janulis@lsmuni.lt (V.J.)
2. Laboratory of Biopharmaceutical Research, Institute of Pharmaceutical Technologies, Lithuanian University of Health Sciences, 50166 Kaunas, Lithuania
* Correspondence: rima.urbstaite@lsmu.lt; Tel.: +370-696-779-17

Abstract: In this study, we conducted an analysis of the qualitative and quantitative composition of anthocyanins and anthocyanidins in different cultivars and genetic clones of American cranberries grown in Lithuanian climatic conditions. Four anthocyanin compounds predominated in fruit samples of American cranberry cultivars: cyanidin-3-galactoside, cyanidin-3-arabinoside, peonidin-3-galactoside, and peonidin-3-arabinoside. They accounted for 91.66 ± 2.79% of the total amount of the identified anthocyanins. The total anthocyanin content detected via the pH differential method was found to be by about 1.6 times lower than that detected via the UPLC method. Hierarchical cluster analysis and principal component analysis showed that the 'Woolman' cultivar distinguished from other cranberry cultivars in that its samples contained two times the average total amount of anthocyanins (8.13 ± 0.09 mg/g). The group of American cranberry cultivars 'Howes', 'Le Munyon', and 'BL-8' was found to have higher than average levels of anthocyanidin galactosides (means 3.536 ± 0.05 mg/g), anthocyanidins (means 0.319 ± 0.01 mg/g), and total anthocyanins (means 6.549 ± 0.09 mg/g). The evaluation of the antioxidant effect of cranberry fruit sample extracts showed that the greatest radical scavenging activity of the cranberry fruit extracts was determined in the fruit samples of 'Woolman' (849.75 ± 10.88 µmol TE/g) and the greatest reducing activity was determined in 'Le Munyon' (528.05 ± 12.16 µmol TE/g). The study showed a correlation between the total anthocyanin content and the antiradical and reductive activity of the extracts in vitro (respectively, R = 0.635 and R = 0.507, $p < 0.05$).

Keywords: cranberry; anthocyanidin; antioxidant activity; UPLC; *Vaccinium macrocarpon*

---

1. Introduction

The American cranberry (*Vaccinium macrocarpon* Aiton) is a perennial evergreen plant of the *Ericaceae* A.L. de Jussie family growing in natural habitats in North America [1–3]. The selection of cranberry cultivars began in the U.S. in the early 1800s, in studies with cranberry plants growing in natural cenopopulations [4]. More than 200 cultivars of American cranberries are cultivated worldwide [5]. According to the data for 2019, the United States, Canada, and Chile provide 98% of world cranberry production [6].

In the climatic conditions of Lithuania, small cranberries (*Vaccinium oxycoccos* L.) grow in the natural cenopopulations of raised bogs and intermediate-type wetlands [7]. During the land reclamation works in Lithuania, those raised bogs and intermediate-type wetlands were drained, which decreased the areas of small cranberry habitats [7]. In about 1967, the selection and introduction of the first cultivars of American cranberries as a perennial berry culture began in Lithuania [8]. Recently, the cultivation of the introduced cranberry cultivars in Lithuania has gained popularity [7].

Citation: Urbstaite, R.; Raudone, L.; Janulis, V. Phytogenotypic Anthocyanin Profiles and Antioxidant Activity Variation in Fruit Samples of the American Cranberry (*Vaccinium macrocarpon* Aiton). *Antioxidants* **2022**, *11*, 250. https://doi.org/10.3390/antiox11020250

Academic Editors: Agustín G. Asuero and Noelia Tena

Received: 27 December 2021
Accepted: 25 January 2022
Published: 27 January 2022

Publisher's Note: MDPI stays neutral with regard to jurisdictional claims in published maps and institutional affiliations.

Copyright: © 2022 by the authors. Licensee MDPI, Basel, Switzerland. This article is an open access article distributed under the terms and conditions of the Creative Commons Attribution (CC BY) license (https://creativecommons.org/licenses/by/4.0/).

The most important groups of biologically active compounds found in cranberry fruits are flavonols (derivatives of quercetin and myricetin) [9], flavan-3-ols, anthocyanins [10], phenolic acids [11], and triterpenoids [12]. The cranberry-specific proantocyanidin and flavonol complex possess antiadhesive activity to the uropatogenic strains of *Escherichia coli* [13]. The effects of these biologically active compounds in cranberry fruit determine the use of the fruit in the prevention and treatment of urinary tract infections [14]. Studies on the effects of cranberry fruit extracts showed that quercetin in those extracts inhibited cell proliferation and reduced the growth of bladder [15] and ovarian [16] cancer cells [17].

Anthocyanins and anthocyanidins are one of the most important groups of biologically active compounds in cranberry fruit. Pappas and Schaich have found that in samples of freshly harvested cranberries, their levels could vary from 13.6 to 171 mg/100 g [18]. About 90% of the total amount of anthocyanins in cranberry samples are cyanidin and peonidin glycosides (cyanidin-3-arabinoside, cyanidin-3-galactoside, peonidine-3-arabinoside, and peonidine-3-galactoside) [19]. The other components of the anthocyanin complex in cranberry fruit make up a small percentage, amounting to about 10%. Of these, glycosides of delphinidin, cyanidin, petunidin, and malvidin are worth mentioning, as their molecules contain various monosaccharides [19].

Yan et al. found that cyanidin-3-galactoside isolated from cranberry fruit extract was capable of scavenging free radicals and inhibiting the oxidation of low-density lipoproteins [20]. Ho et al. performed studies in mice and found that peonidin-3-glucoside reduced the number of the metastases of lung carcinoma cells [21]. Smeriglio et al. found a glycemia-lowering effect of cyanidin-3-glucoside [22].

Seeram et al. found that the anthocyanin fraction in cranberry fruit extracts inhibited inflammatory processes [23], and cranberry fruit extract reduced inflammatory processes in the liver [24]. Santana et al. conducted studies in mice and found that cranberry fruit extract was effective in the treatment of acute pancreatitis [25]. Seeram et al. conducted studies with cancer cell lines. The studies showed that anthocyanins, proanthocyanidins, and flavonol glycosides isolated from cranberry fruit extracts had antiproliferative effects on oral, colon, and prostate cancer cells [26].

Cranberry fruits, both fresh and processed, are used in the production of food supplements, juice, and in the confectionery industry [7,8]. Cranberry fruits and their extracts could be used as natural preservatives, since the biologically active compounds (anthocyanins and phenolic compounds) found in them have antioxidant and antimicrobial effects. Phenolic compounds have antimicrobial activity against bacterial and fungal strains that can cause food spoilage and poisoning [27]. Anthocyanins obtained from cranberry fruit are used as food colorants in confectionery [28]. The ability of anthocyanins to change color from red to blue and other shades depending on the pH of the environment can be used to determine the quality and shelf life of perishable foods and to create smart food packaging [29,30]. The biologically active compounds of cranberry fruit—anthocyanins—can have antimicrobial, antioxidant, and other effects on disease prevention and health in the development of healthy and ecological food products.

The studies of the qualitative and quantitative composition of anthocyanins in cranberry samples are important for the evaluation of the quality of food products, food supplements, and pharmaceuticals. The preparation of high commercial value raw material of cranberry fruit should be based on the evaluation of phytochemical profiles, especially anthocyanins and their distribution in cultivated or wild cranberries.

Determination of anthocyanins content in cranberry fruit are important for the evaluation of fruit samples of small cranberries (*Vaccinium oxycoccos*) and American cranberries (*Vaccinium macrocarpon*) and their cultivars, determining the regularities of the accumulation of biologically active compounds, the optimal time of berry harvesting, and the selection of the most promising cranberry cultivars for introduction.

This study aimed to determine the anthocyanin profiles and antioxidant activity of the fruits of *Vaccinium macrocarpon* cultivars. The performed research will provide new knowledge about the variation in the qualitative and quantitative composition of anthocyanins in cranberry cultivars and is important in determining the most promising cranberry cultivars for cultivation in Lithuanian climatic conditions.

2. Materials and Methods

2.1. Reagents

Acetonitrile (manufacturer: Sigma-Aldrich, Steinheim, Germany), methanol (manufacturer: Sigma-Aldrich, Steinheim, Germany), formic acid (manufacturer: Merck, Darmstadt, Germany), reference standards for delphinidin-3-galactoside, cyaniding-3-galactoside, cyaniding-3-glucoside, cyaniding-3-arabinoside, peonidin-3-galactoside, peonidin-3-arabinoside, peonidin-3-glucoside, malvidin-3-galactoside, malvidin-3-arabinoside, cyanidin chloride, peonidin chloride, and malvidin chloride were purchased from Extrasynthese (Genay, France). ABTS 2,2′-Azino-bis(3-ethylbenzothiazoline-6-sulfonic acid), Trolox (6-hydroxy-2,5,7,8-tetramethylchroman-2-carboxylic acid), potassium peroxydisulfate, sodium acetate (manufacturer: Scharlau Sentmenat, Barcelona, Spain), ferric (III) chloride hexahydrate (manufacturer: Vaseline-Fabrik Rhenania, Bonn, Germany), TPTZ (2,4,6-Tris(2-pyridyl)-s-triazine) (manufacturer: Carl Roth, (Karlsruhe, Germany), acetic acid (manufacturer: Lach Ner, Neratovice, Czech Republic), hydrochloric acid (manufacturer Sigma-Aldrich, Steinheim, Germany), potassium chloride, and ethanol 96% (v/v) (manufacturer: AB Stumbras, Kaunas, Lithuania) were also acquired.

2.2. Raw Material

The fruits examined in the present study were mature and ripe fruit of different cultivars of American cranberry (*Vaccinium macrocarpon* Aiton) (Table 1) grown in Lithuanian climatic conditions, in the collection of the Institute of Botany of the Nature Research Center, Mažieji Gulbinai, Vilnius (54°41′36.6″ N 25°21′56.0″ E). Dynamics of meteorological factors (precipitation (mm), sunshine duration (h), and temperature (°C)) in region of Vilnius in 2020 are presented in Figure 1 [31]. The collection time was September 2020. Cranberry fruits were ground and frozen at −60 °C in an ultra-low-temperature freezer (CVF330/86, ClimasLab SL, Barcelona, Spain). Cranberry fruits were freeze-dried according to the methodology described by Gudžinskaitė et al. [32]. The fruits were powdered in a Retsch GM 200 electric mill (Retsh GmbH, Hahn, Germany). Loss on drying was determined using the method described in the European Pharmacopoeia Ph.Eur.01/2008: 20232 [33].

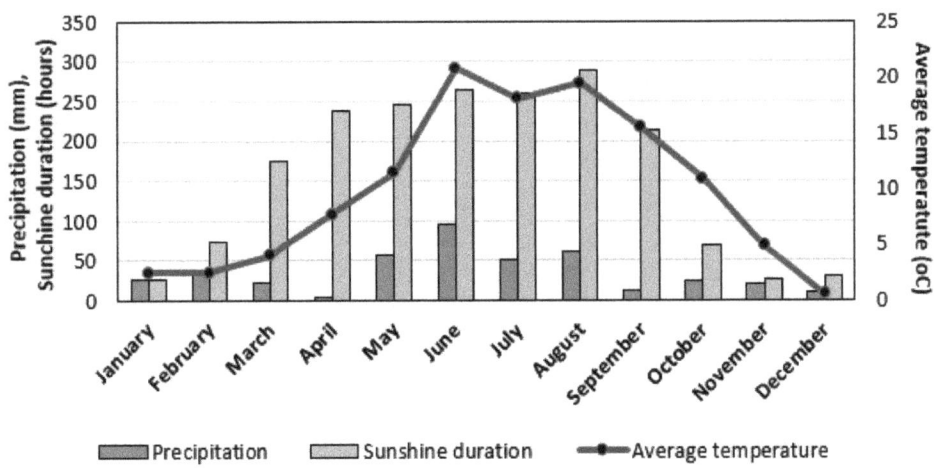

Figure 1. General climate condition (temperature, precipitation, and sunshine duration) in region of Vilnius in 2020.

Table 1. Characteristics typical of American cranberry cultivars; ND (no data).

No.	Cranberry Cultivar	Country of Origin and Year	Cultivar Characteristics	Reference
1	'Baifay'	ND	ND	
2	'Bergman'	1961	'Early Black' × 'Searles' hybrid, abundantly growing, medium-early, fertile. The berries are suitable for storage.	[4,8]
3	'Crowley'	1961	'McFarlin' × 'Prolific' hybrid. The berries are of medium size, dark red, suitable for processing.	[4,8]
4	'Early Black'	1835, MA, USA	Native Selection. The berries are of medium size, sometimes small, blackish red. It grows abundantly and is moderately fertile.	[4,8,34]
5	'Early Richard'	1870, NJ, USA	A medium early cultivar. Medium-sized, dark red berries. Grows abundantly. The berries are not very suitable for storage.	[8,35]
6	'Habelman'	ND	ND	
7	'Holliston'	1885, MA, USA	Native Selection, a medium-early cultivar.	[36]
8	'Howes'	1843, MA, USA	Native Selection. A late cultivar, disease resistant, fertile. The berries are small, medium-sized, store a lot of pectin, are suitable for storage, and are resistant to frost.	[4,8,34]
9	'Le Munyon'	1960, NJ, USA	Native Selection, a very early fertile cultivar with medium-sized large dark red berries.	[4,8,34]
10	'McFarlin'	1874, MA, USA	Native Selection.	[4,34]
11	'Prolific'	1900, MI, USA	Native Selection.	[4,34]
12	'Searles'	1893, WI, USA	Native Selection. A medium-early cultivar. The berries are large, red, not suitable for storage. The cultivar is fertile, with high matching stems.	[4,8,34]
13	'Woolman'	1897, NJ, USA	An early cultivar.	[8,35]
14	'Ar-2'	ND	A late cultivar.	[8]
15	'Bain-6'	WI, USA	Native Selection.	[36]
16	'Bain-MC'	ND	A late cultivar.	[8]
17	'BL-8'	ND	Is characterized by early berry ripening, lush growth, and medium-sized, large, dark red, oval berries.	[8]
18	'BL-12'	ND	ND	
19	'BL-15'	ND	Is characterized by early berry ripening, lush growth, and medium-sized, large, dark red, oval berries.	[8]
20	'BL-22'	ND	ND	

2.3. Preparation of Cranberry Extracts

About 1 g (precise weight) of the lyophilized cranberry powder was extracted with 20 mL of 70% (v/v) ethanol containing 1% hydrochloric acid in an ultrasonic bath for 15 min at 80 kHz and 565 W at room temperature. The extract was collected and filtered through a membrane filter with a pore size of 0.20 µm. The prepared extracts were stored in dark glass vials at $-20\ °C$.

2.4. Spectrophotometric Studies

2.4.1. Determination of Antioxidant Activity

An ABTS·+ radical cation decolorization assay was applied according to the methodology described by Re et al., 1999 [37] and modified by Raudone et al. [38]. A volume of 3 mL of ABTS$^+$ solution (absorbance 0.800 ± 0.02) was mixed with 5 µL of the 5-fold diluted cranberry extract. A decrease in absorbance was measured at a wavelength of 734 nm after keeping the samples for 30 min in the dark. A standard curve ($y = 0.000008x -$

0.0909; $R^2 = 0.997$) was produced by using standard Trolox solutions of 4000–48,000 µmol/L concentration.

The ferric reducing antioxidant power (FRAP) assay was carried out as described by Raudone et al. [39]. The working FRAP solution included TPTZ (0.01 M dissolved in 0.04 M HCl), $FeCl_3 \cdot 6H_2O$ (0.02 M in water), and an acetate buffer (0.3 M, pH 3.6) at the ratio of 1:1:10. A volume of 3 mL of a freshly prepared FRAP reagent was mixed with 10 µL of the cranberry extract. After 30 min, the absorbance was read at 593 nm using a UV-vis spectrophotometer. A standard curve ($y = 0.0000166x + 0.000950$; $R^2 = 0.993$) was produced by using standard Trolox solutions of 400–24,000 µmol/L concentration.

2.4.2. Total Anthocyanins Content Determination

The total anthocyanin content was determined according to the pH differential method as described by Lee et al. [40]. During the evaluation, 0.625 mL of cranberry extracts were diluted in 25 mL of two different buffers: 0.025 M potassium chloride (pH = 1.0) and 0.4 M sodium acetate (pH = 4.5). The samples were kept in the dark for 20 to 30 min and then the absorption (A) was measured at $\lambda = 520$ nm and $\lambda = 700$ nm. The diluted test portions were read versus a blank cell filled with distilled water.

The anthocyanin pigment concentration, expressed as cyanidin-3-glucoside equivalents, was calculated as follows:

$$\text{Anthocyanin content (mg/L)} = A \times MW \times DF \times 1000/(\varepsilon \times l)$$

where: $A = (A520 - A700)_{pH1.0} - (A520 - A700)_{pH4.5}$; MW (molecular weight) = 449.2 g mol^{-1} for cyanidin-3-glucoside; DF = dilution factor; l = cuvette path length in cm (1 cm); ε = 26,900 mol extinction coefficient, in L × mol^{-1} × cm^{-1} for cyanidin-3-glucoside.

2.5. The UPLC-PDA Method

The analysis of the qualitative and quantitative composition of anthocyanins in cranberry fruit was performed using the UPLC methodology validated by Vilkickyte et al. [19]. The identification of the peaks was performed by comparing the UV absorption spectrum of the reference standard with the UV absorption spectrum of the matrix peaks of American cranberry, using the same retention time.

2.6. Statistical Analysis

Data analysis was performed using Microsoft Excel 2016 (Microsoft, Santa Rosa, CA, USA) and SPSS Statistics 27 (IBM, Armonk, NY, USA). All experiments were carried out in triplicate, and presented as the mean value ± SD. Significant differences between samples were determined using ANOVA with Tukey's test for multiple comparisons. The variability of the results were evaluated using coefficients of variation (CV). Principal component analysis (PCA) and hierarchical cluster analysis applying the between-groups clustering method with Euclidean distances were performed to elucidate the groupings of cranberries. Correlational analysis was performed using Pearson coefficient. Level of significance $\alpha = 0.05$.

3. Results and Discussion

3.1. Determination of the Qualitative and Quantitative Composition of Anthocyanins in Cranberry Fruit Samples via the UPLC Method

Increasing consumer awareness creates the demand for the products with health-promoting effects and capabilities of prevention of various pathological processes. To assure the quality of such products, it is important to study the chemical composition of cranberry fruits and to conduct qualitative and quantitative evaluation of their biologically active compounds [41]. The chromatogram of anthocyanins and anthocyanidins in cranberry fruit samples identified via the UPLC-DAD method is shown in Figure 2.

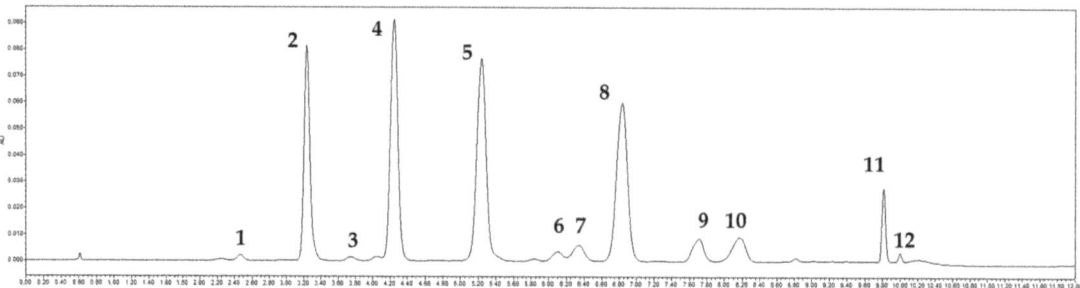

Figure 2. Chromatogram profile of anthocyanins in fruit samples of American cranberries at 520 nm: 1—Delphinidin-3-galactoside, 2—Cyanidin-3-galactoside, 3—Cyanidin-3-glucoside, 4—Cyanidin-3-arabinoside, 5—Peonidin-3-galactoside, 6—Peonidin-3-glucoside, 7—Malvidin-3-galactoside, 8—Peonidin-3-arabinoside, 9—Cyanidin, 10—Malvidin-3-arabinoside, 11—Peonidin, 12—Malvidin.

Four anthocyanin glycosides, namely cyanidin-3-galactoside, cyanidin-3-arabinoside, peonidin-3-galactoside, and peonidin-3-arabinoside, predominated in the studied samples of cranberry cultivars grown in Lithuanian (Figure 2) in a range of 15.12–25.41%, 15.29–25.30%, 15.05–37.98%, and 13.25–25.92%, respectively (Figure 3). They accounted for $91.66 \pm 2.79\%$ of the total amount of anthocyanins identified. The qualitative profiles determined in our study are consistent with the results of Česonienė et al. and Viskelis et al. [7,42] and these compounds could be regarded as anthocyanin marker compounds. However, their quantitative profiles were variable. Česonienė et al. have determined that cyanidin-3-galactoside accounted for 24.11%, cyanidin-3-arabinoside for 18.73%, peonidin-3-galactoside for 33.29%, and peonidin-3-arabinoside for 16.7% of total determined anthocyanins [42]. Vikelis et al. have determined the following marker compound composition in fruit samples of 'Stevens', 'Pilgrim', 'Ben Lear', and 'Black Veil': cyanidin-3-galactoside—20.5%, cyanidin-3-arabinoside—19%, peonidin-3-galactoside—32.7%, and peonidin-3-arabinoside—6.7% of total anthocyanin content [7]. Furthermore, Zhang et al. have studied the composition of fruit wine of cranberry cultivars 'Stevens', 'Pilgrim', and 'Bergman'. The anthocyanin content of cranberry wine has been found to be about 50% cyanidin-3-arabinoside and about 27% peonidin-3-arabinoside. The anthocyanin content in the wine of the fruits of other cranberry cultivars have been determined with different compositions, namely of 4% of cyanidin-3-galactoside, 9% of peonidin-3-galactoside, 1.5% of cyanidin-3-glucoside, and 7% of peonidin-3-glucoside [43].

In our study, the highest amounts of cyanidin-3-galactoside (1.92 ± 0.02 mg/g) were determined in cranberry samples of the 'Woolman' cultivar ($p < 0.05$). Statistically significantly lower levels of cyanidin-3-galactoside (1.36 ± 0.03 mg/g, 1.28 ± 0.04 mg/g, and 1.21 ± 0.02 mg/g) were found in fruit samples of cranberry cultivars 'Bergman', 'Howes', and 'Le Munyon', respectively. The lowest cyanidin-3-galctoside levels (0.29 ± 0.01 mg/g and 0.38 ± 0.01 mg/g) were determined in the samples of the 'Early Black' cultivar and the 'BL-22' genetic clone, respectively ($p < 0.05$).

The highest content of cyanidin-3-arabinoside (1.51 ± 0.01 mg/g) was determined in cranberry samples of the 'Woolman' cultivar ($p < 0.05$). Lower levels of cyanidin-3-arabinoside (1.24 ± 0.02 mg/g, 1.25 ± 0.02 mg/g, 1.22 ± 0.01 mg/g, and 1.21 ± 0.02 mg/g) were found in fruit samples of cranberry cultivars 'Crowley', 'Howes', 'Le Munyon', and 'Mc Farlin', respectively. The lowest cyanidin-3-arabinoside content (0.40 ± 0.02 mg/g) was found in fruit samples of the 'BL-22' genetic clone ($p < 0.05$).

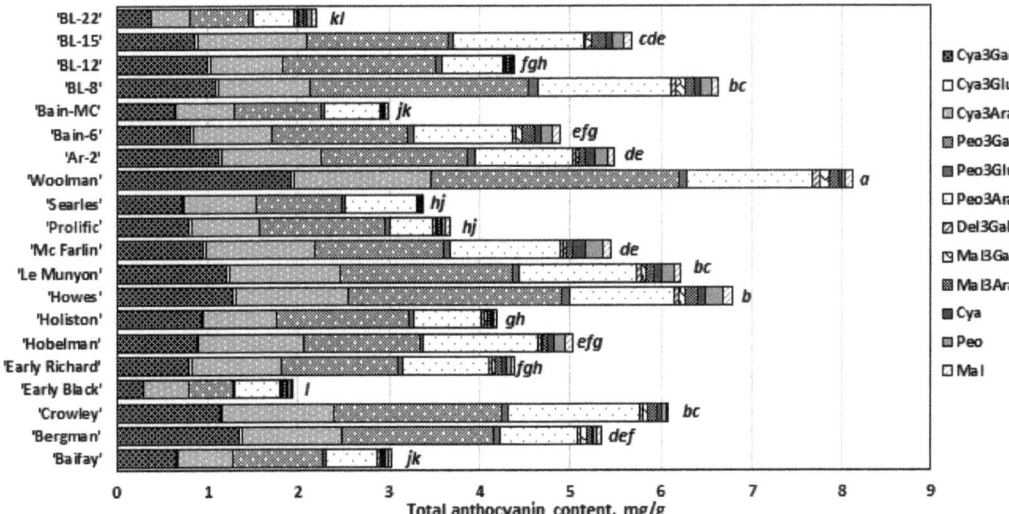

Figure 3. Variation of anthocyanin content in the cranberries of American cultivars. Statistically significant differences between the total anthocyanin content in fruit samples of cranberry cultivars are marked by different letters ($p < 0.05$): Cya3Gal—Cyanidin-3-galactoside, Cya3Glu—Cyanidin-3-glucoside, Cya3Ara—Cyanidin-3-arabinoside, Peo3Gal—Peonidin-3-galactoside, Peo3Glu—Peonidin-3-glucoside, Peo3Ara—Peonidin-3-arabinoside, Del3Gal—Delphinidin-3-galactoside, Mal3Gal—Malvidin-3-galactoside, Mal3Ara—Malvidin-3-arabinoside, Cya—Cyanidin, Peo—Peonidin, Mal—Malvidin.

The highest content of peonidin-3-galactoside (2.74 ± 0.03 mg/g) was determined in fruit samples of the 'Woolman' cultivar ($p < 0.05$). Statistically significantly lower levels of peonidin-3-galactoside (2.35 ± 0.02 mg/g and 2.41 ± 0.03 mg/g) were determined in cranberry samples of the 'Howes' cultivar and the 'BL-8' genetic clone, respectively. The lowest content of peonidin-3-galactoside (0.49 ± 0.02 mg/g) was determined in samples of the 'Early Black' cultivar ($p < 0.05$).

The highest levels of peonidine-3-arabinoside (1.48 ± 0.02 mg/g, 1.38 ± 0.02 mg/g, and 1.46 ± 0.02 mg/g) were determined in fruit samples of the genetic clone 'BL-8' and 'Woolman' and 'Crowley' cultivars, respectively ($p < 0.05$). The lowest peonidine-3-arabinoside levels (0.50 ± 0.01 mg/g and 0.47 ± 0.02 mg/g) were determined in cranberries of the 'Early Black' cultivar and the 'BL-22' genetic clone, respectively ($p < 0.05$).

The analysis of the quantitative composition of the anthocyanin complex in cranberry fruit samples showed that aglycones formed a small part of the anthocyanin complex. The highest content of one of the anthocyanidins, cyanidin (0.133 ± 0.01 mg/g), was determined in fruit samples of the 'McFarlin' cultivar ($p < 0.05$), while no cyanidin was detected in fruit samples of the 'Searles' cultivar. Peonidin content ranged from 0.17% to 3.53% in cranberry fruit samples. The highest levels of peonidin (0.184 ± 0.01 mg/g and 0.192 ± 0.01 mg/g) were found in fruit samples of 'Howes' and 'McFarlin' cultivars, respectively ($p < 0.05$). The lowest aglycone peonidin content (0.006 ± 0.00 mg/g) was determined in cranberry samples of the 'Searles' cultivar ($p < 0.05$). Malvidin content ranged from 0.21% to 2.60% in cranberry samples. The highest content of the anthocyanidin malvidin (0.107 ± 0.01 mg/g) was found in fruit samples of the 'Howes' cultivar, and the lowest (0.012 ± 0.00 mg/g) in fruit samples of the 'Crowley' cultivar ($p < 0.05$). These results suggest that genotypes have characteristic variation in the amounts of individual anthocyanins and anthocyanidins.

The highest total anthocyanin content (8.13 ± 0.09 mg/g) was determined in cranberry samples of the 'Woolman' cultivar ($p < 0.05$). Lower total anthocyanin levels (6.90 ± 0.12 mg/g, 6.22 ± 0.15 mg/g, 6.07 ± 0.08 mg/g, and 6.63 ± 0.11 mg/g) were found in cranberry samples of 'Howes', 'Le Munyon', and 'Crowley' cultivars and in the 'BL-8' genetic clone, respectively. The lowest total anthocyanin levels (1.95 ± 0.11 mg/g and 2.21 ± 0.15 mg/g) were found in fruit samples of the 'Early Black' cultivar and the 'BL-22' genetic clone, respectively ($p < 0.05$).

Gardana et al. have studied the anthocyanin content in fruit samples of 'Ben Lear', 'Howes', 'Stevens', and 'Bergman' cultivars and found that the total anthocyanin content ranged from 2.4 mg/g to 5.6 mg/g [44]. The total anthocyanin levels (5.07 mg/g in cranberry samples of the 'Howes' cultivar and 3.36 mg/g in cranberry samples of the 'Bergman' cultivar) that were found by the authors were lower than those found in our study (6.80 mg/g and 5.34 mg/g in fruit samples of the 'Howes' and 'Bergman' cultivars, respectively) [44]. Brown et al. have examined the anthocyanin composition in fruit samples of cultivars 'Ben Lear', 'Bergman', 'GH 1', 'Pilgrims', and 'Stevens' and found that the total anthocyanin content ranged from 2.81 mg/g to 7.98 mg/g [45]. The total anthocyanin content that were found by the authors in fruit samples of the 'Bergman' cultivar (7.02 mg/g) were higher than that found in our study (5.34 mg/g) [45]. The total anthocyanin content in samples of fresh cranberry fruit of the 'Bergman' and 'Early Richard' cultivars (0.48 mg/g and 0.52 mg/g, respectively) were found in a study conducted by Narwojsz et al. to be about 10 times lower than the total anthocyanin levels in the lyophilized raw material of 'Bergman' and 'Early Richard' cultivars found in our study (4.38 mg/g and 5.34 mg/g, respectively) [46].

The quantified mean amounts of anthocyanins and anthocyanidins in fruit samples of cranberry cultivars can be presented in the following decreasing order: peonidin-3-galactoside > peonidin-3-arabinoside > cyanidin-3-arabinoside > cyanidin-3-galactoside > peonidin > malvidin-3-arabinoside > peonidin-3-glucoside > malvidin > cyanidin > malvidin-3-galactoside > delphinidin-3-galactoside > cyanidin-3-glucoside. The chromatographic profile of cranberry anthocyanins is characteristic, and thus its identification can be applied for establishing the authenticity of the cranberry plant raw material and identifying possible falsifications with other anthocyanin-accumulating botanical raw materials [47].

3.2. Quantification of the Total Anthocyanin Content in Cranberry Samples

Spectrophotometric analysis is used to evaluate the quality of herbal raw materials and their preparations. This method is used to assess the quantitative composition of the groups of biologically active compounds. The spectrophotometric pH differential method is simple, fast, and economical. This method is often used in practice to determine the total anthocyanin content in a sample [40,48,49]. The total anthocyanin content determined via the spectrophotometric pH differential method should be evaluated individually for each plant raw material, as the test results are influenced by the methodology used [50]. The total amount of anthocyanins is usually expressed in CGE (cyanidin-3-glucoside equivalent), as cyanidin-3-glucoside is the predominant anthocyanin in many fruits and vegetables, yet it does not necessarily reflect the anthocyanin composition of the raw material under study [51,52].

The quantitative composition of anthocyanins determined by pH differential method were analyzed in comparison with the total identified anthocyanin origin compounds determined by the ultra-high performance liquid chromatography method (Table 2).

The total anthocyanin content determined in the studied fruit samples of American cranberries via the pH differential method ranged from 1.13 ± 0.02 mg CGE/g to 5.09 ± 0.24 mg CGE/g. The lowest total anthocyanin content (1.13 ± 0.02 mg CGE/g) was found in fruit samples of the 'Early Black' cultivar, and the maximum total content of anthocyanins (5.09 ± 0.24 mg) was found in fruit samples of the 'Woolman' cultivar. The total anthocyanin content in the fruit samples of American cranberry cultivars 'Bergman', 'Crowley', 'Howes', 'Le Munyon', and 'McFarlin' grown in New Zealand's climatic conditions ranged from 1.34 mg CGE/g to 1.90 mg CGE/g and were about 1.5 times lower than the total anthocyanin content found in our study in samples of the abovementioned cultivars [53].

Table 2. Total anthocyanin content in fruit samples of American cranberry cultivars determined via UPLC and the spectrophotometric pH differential methods.

Cultivar	UPLC mg/g	pH Differential Method
'Baifay'	3.02 ± 0.052	1.54 ± 0.018
'Bergman'	5.34 ± 0.076	3.36 ± 0.137
'Crowley'	6.07 ± 0.057	3.72 ± 0.050
'Early Black'	1.95 ± 0.110	1.13 ± 0.021
'Early Richard'	4.38 ± 0.053	2.67 ± 0.145
Hobelman'	5.03 ± 0.002	3.10 ± 0.123
'Holiston'	4.19 ± 0.030	2.30 ± 0.066
'Howes'	6.79 ± 0.106	4.37 ± 0.101
'Le Munyon'	6.22 ± 0.077	3.60 ± 0.149
'Mc Farlin'	5.44 ± 0.026	3.41 ± 0.116
'Prolific'	3.66 ± 0.002	1.90 ± 0.089
'Searles'	3.37 ± 0.001	1.86 ± 0.103
'Woolman'	8.13 ± 0.093	5.09 ± 0.244
'Ar-2'	5.49 ± 0.010	3.18 ± 0.010
'Bain-6'	4.89 ± 0.079	2.88 ± 0.101
'Bain-MC'	2.99 ± 0.016	1.53 ± 0.073
'BL-8'	6.63 ± 0.070	3.88 ± 0.030
'BL-12'	4.39 ± 0.001	2.53 ± 0.078
'BL-15'	5.67 ± 0.073	3.57 ± 0.244
'BL-22'	2.21 ± 0.088	1.41 ± 0.002

The total anthocyanin content in cranberry fruit samples determined via the pH differential method was about 1.6 times lower than the total anthocyanin content determined via the UPLC method. The coefficient of variation of the total anthocyanin content in cranberry fruit samples determined via the UPLC method was 33.57%, while that determined via the pH differential method was 37.45%.

Grace et al. in their study have found that the total anthocyanin content (expressed as CGE) in American cranberry fruit samples detected via the application of the pH differential method was 1.4 times lower than that detected via the HPLC method [54]. Lee et al. have found no statistically significant difference between the anthocyanin content in cranberry samples determined via the pH differential method (1.95 ± 0.14 mg CGE/g) and the UPLC method (2.06 ± 0.26 mg CGE/g) [55]. The authors did not find any differences between the compared methods because the total amount of anthocyanins was calculated based on the four predominant anthocyanins, namely cyanidin-3-galactoside, cyanidin-3-arabinoside, peonidin-3-galactoside, and peonidin-3-arabinoside, whereas in our study, the amount of other anthocyanins detected in cranberry fruit samples ranged from 3.71% to 13.23% [55].

The results obtained using pH differential methods and UPLC were highly corresponding and correlated (R = 0.975, $p < 0.05$). A very strong significant correlation between the total anthocyanin content in cranberry fruit samples assessed via the pH differential method and the UPLC method was also found in other studies, the correlation coefficients being r = 0.98, r = 0.925, and r ≥ 0.99 ($p < 0.05$) [50,54,55]. The very strong correlation between the pH differential method and the UPLC method indicates that the anthocyanin content determined by both methods is similar, but the results obtained should be evaluated on a case-by-case basis, depending on the raw material and the nature of the study.

3.3. Hierarchical Cluster Analysis and Principal Component Analysis of the Distribution of Anthocyanin Content in Fruit Samples of American Cranberry

The analysis of the similarity of the composition of anthocyanins in American cranberry cultivars introduced and grown in Lithuanian climatic conditions was based on the quantitative composition of anthocyanins determined in the samples of different cultivars and was carried out by performing a hierarchical cluster and principal component analysis. The cluster analysis of the samples of different cultivars of American cranberry was performed on the basis of the quantitative composition of anthocyanins delphinidin-3-galactoside, cyanidin-3-galactoside, cyanidin-3-glucoside, cyanidin-3-arabinoside, peonidine-3-galactoside, peonidin-3-glucoside, peonidin-3-galactoside, peonidine-3-arabinoside, cyanidin, malvidin-3-arabinoside, peonidin, and malvidin. Fruit samples of cranberry cultivars were divided into three clusters (Figure 4).

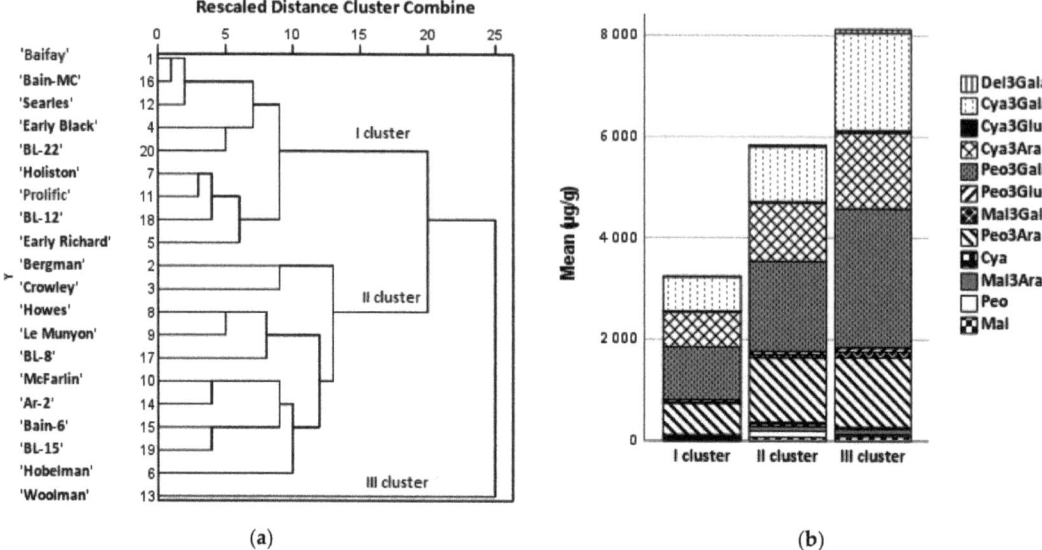

Figure 4. A dendrogram of the distribution of American cranberry cultivars into similar clusters according to the anthocyanin content in fruit samples (**a**); a diagram of the distribution of mean anthocyanin content in clusters (**b**). Cya3Gala—Cyanidin-3-galactoside, Cya3Glu—Cyanidin-3-glucoside, Cya3Ara—Cyanidin-3-arabinoside, Peo3Gala—Peonidin-3-galactoside, Peo3Glu—Peonidin-3-glucoside, Peo3Ara—Peonidin-3-arabinoside, Del3Gala—Delphinidin-3-galactoside, Mal3Gala—Malvidin-3-galactoside, Mal3Ara—Malvidin-3-arabinoside, Cya—Cyanidin, Peo—Peonidin, Mal—Malvidin.

Fruit samples of cranberry cultivars 'Baifay', 'Early Black', 'Early Richard', 'Holiston', 'Prolific', 'Searles', 'Bain-6', 'BL-12', and 'BL-22' were assigned to cluster I because their total anthocyanin content was by 1.7 times lower than the average total anthocyanin content. The total amount of anthocyanins found in the fruit samples of the cluster II cultivars ('Bergman' 'Crowley', 'Hobelman', 'Howes', 'Le Munyon', 'Mc Farlin', 'Ar-2', 'Bain-MC', 'BL-8', and 'BL-15') was close to the mean anthocyanin content. Cluster II is considered to consist of cranberry cultivar samples which have higher than average levels of anthocyanidins (cyanidin, peonidin, and malvidin). Cluster III consisted of one American cranberry cultivar 'Woolman'. The total anthocyanin content in cranberry fruit samples of the 'Woolman' cultivar was twice the mean total anthocyanin content in the studied cultivars.

A principal component analysis (PCA) was performed to evaluate the quantitative variation of the identified anthocyanin group compounds in different American cranberry cultivars (Figure 5). Two main components were used for the analysis, explaining 85.82% of the total data variance. The first component (PC I), which explained 52.18% of the total data variance, had a significant correlation with the anthocyanin galactosides cyanidin-3-galactoside (0.958), peonidin-3-galactoside (0.942), delphinidin-3-galactoside (0.926), and malvidin-3-galactoside (0.881), and a strong positive correlation with cyanidin-3-arabinoside (0.786), cyanidin-3-glucoside (0.781), peonidin-3-glucoside (0.759), and peonidin-3-arabinoside (0.658). The second component (PC II), which explained 33.64% of the total data variance, had a significant correlation with cyanidin (0.973) and peonidin (0.973), and a positive correlation with malvidin (0.738) and malvidin-3-arabinoside (0.613).

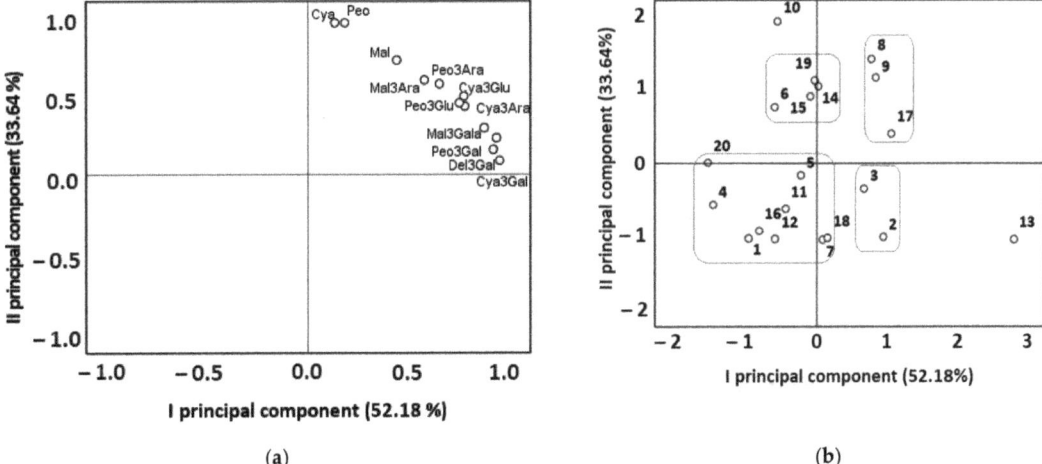

Figure 5. Principal component analysis loading (**a**) and score (**b**) plots of different cranberry fruit samples: (**a**) Cya3Gala—Cyanidin-3-galactoside, Cya3Glu—Cyanidin-3-glucoside, Cya3Ara—Cyanidin-3-arabinoside, Peo3Gal—Peonidin-3-galactoside, Peo3Glu—Peonidin-3-glucoside, Peo3Ara—Peonidin-3-arabinoside, Del3Gal—Delphinidin-3-galactoside, Mal3Gala—Malvidin-3-galactoside, Mal3Ara—Malvidin-3-arabinoside, Cya—Cyanidin, Peo—Peonidin, Mal—Malvidin; (**b**) 1—'Baifay', 2—'Bergman', 3—'Crowley', 4—'Early Black', 5—'Early Richard', 6—'Hobelman', 7—'Holiston', 8—'Howes', 9—'Le Munyon', 10—'Mc Farlin', 11—'Prolific', 12—'Searles', 13—'Woolman', 14—'Ar-2', 15—'Bain-MC', 16—'Bain-6', 17—'BL-8', 18—'BL-12', 19—'BL-15', 20—'BL-22'.

American cranberry cultivars located in the negative squares of PC I and PC II ('Baifay', 'Early Black', 'Early Richard', 'Holiston', 'Prolific', 'Searles', 'Bain-6', 'BL-12', and 'BL-22') coincided with cluster I in the cluster analysis, which had a lower-than-average total anthocyanin content. The cranberry cultivars 'Bergman' and 'Crowley' formed the second group of similar cultivars with above-average total anthocyanin content and below-average total anthocyanidin (cyanidin, peonidin, and malvidin) content. The total anthocyanin content in the samples of the hybrid cultivars 'Bergman' (a hybrid of 'Early Black' and 'Searles') and 'Crowley' (a hybrid of 'Mc Farlin' and 'Prolific') was higher than that in samples of the precursor cultivars 'Early Black' and 'Searles', and 'Mc Farlin' and 'Prolific', respectively. Diaz-Garcia et al. found that fruit samples of the second- and third-generation cycle cultivars contained higher levels of anthocyanins than the samples of the early selection attempts did [56].

The American cranberry cultivar 'Woolman' was far removed from other cranberry cultivars in the PCA chart because the total anthocyanin content found in it was almost two times higher than the mean total anthocyanin content. Exceptionally high levels of cyanidin-3-galactoside and peonidin-3-galactoside were determined in samples of the cranberry cultivar 'Woolman', but the total content of anthocyanidins (cyanidin, peonidin, and malvidin) was lower than their mean total content in cultivar samples. In our previously published study, samples of the cranberry cultivar 'Woolman' differed from samples of cultivars 'Baiwfay', 'Drever', 'Bain', 'Bergman', 'Searles', 'Holliston', and 'Piligrim' in that they contained high levels of phenolic acids and dihydrochalcones and low levels of flavonols and flavan-3-ols [32].

The total content of anthocyanidin galactosides in fruit samples of the group consisting of the cranberry cultivar 'Hobelman' and genetic clones 'Ar-2', 'Bain-MC', and 'BL-15' was close to the mean total amount of galactosides in the studied cultivars, and the total content of anthocyanidins (cyanidin, peonidin, and malvidin) was higher than the mean total amount of anthocyanidins. The cranberry cultivar 'Mc Farlin' was far from other cultivars in the PCA chart because its fruit samples contained two times the average amount of aglycones (cyanidin, peonidin, and malvidin). The group of American cranberry cultivars 'Howes', 'Le Munyon', and 'BL-8' located in the positive quadrant of PC I and PC II was characterized by higher anthocyanidin galactoside and total anthocyanidin contents compared to the mean total anthocyanidin and galactoside contents.

3.4. Determination of Antioxidant Activity

Recently, many studies have been conducted on phenolic compounds in plant matrices as natural antioxidants with protective antioxidant properties as well as on the application of these phenolic compounds for prophylactic purposes [57]. The antioxidant effects of anthocyanins occur through a number of complex mechanisms by which anthocyanins can directly scavenge free radicals, prevent the formation of reactive oxygen species (by forming chelating compounds with metals, they inhibit redox reactions and inhibit xanthine oxidase and NADPH oxidase), or promote the release of antioxidant enzymes [58,59]. The antioxidant properties of phenolic compounds are associated with biological effects such as anti-inflammatory, anticancer, and antimicrobial effects. In vitro assays for antiradical and reductive activity are expedient to determine the potency of the potential antioxidant activity of the extracts of fruit samples of the tested cultivars.

The strongest antiradical activity detected by using the ABTS method (849.75 ± 10.88 μmol TE/g) was found in samples of the cranberry cultivar 'Woolman' (Figure 6). The strongest antioxidant activity determined by using the FRAP method (528.05 ± 12.16 μmol TE/g) was determined in samples of the cranberry cultivar 'Le Munyon'. The weakest antiradical activity detected by applying the ABTS method (203.20 ± 9.19 μmol TE/g) was determined in samples of the cranberry cultivar 'Baifay'. The weakest reducing activity (215.23 ± 3.24 μmol TE/g) was found in fruit samples of the 'Prolific' cultivar.

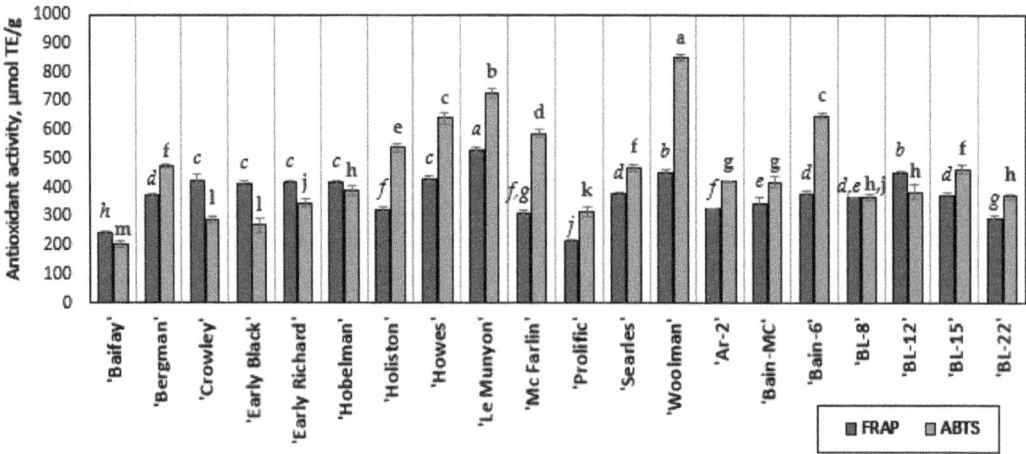

Figure 6. Determination of the antioxidant activity of cranberry extracts. The letters of different fonts indicate statistically significant differences in the antioxidant activity of the fruit extracts of cranberry cultivars determined by the FRAP and ABTS methods ($p < 0.05$).

Oszmiański et al. have investigated the antioxidant activity of fruit extracts of 'Howes' by applying the ABTS and FRAP methods [5]. However, in our study, the reducing and radical scavenging activities of 'Howes' fruit extracts were 4-fold and 3-fold greater, respectively.

The correlation between the total anthocyanin content and antioxidant capacity was determined for the ABTS method (r = 0.635; $p < 0.001$), and for the FRAP method (r = 0.507; $p < 0.001$). Oszmiański et al. determined similar correlation coefficients for the ABTS (r = 0.675), FRAP (r = 0.614), and DPPH (r = 0.602) methods. Chaves et al. have found that the total anthocyanin content determined in the samples of blackberries, blueberries, and strawberries correlated with the antiradical capacity of the extracts determined by the DPPH and ABTS methods (respectively, r = 0.86, r = 0.82) [60]. The positive correlation of anthocyanin content with antioxidant activity suggests that these compounds may contribute to the antioxidant effects of fruit extracts. The American cranberries, besides the great fraction of anthocyanins, also contain other phenolic origin compounds such as flavonoids, phenolic acids, and proanthocyanidins [47]. The anthocyanins together with other phenolic compounds present in the extracts express the total antioxidant activity of the extract.

Scientific studies confirm that antioxidant activity mechanisms are interrelated with the biological activities expressed in the body and substantiates that the bioavailability of phenolic compounds is highly important for their bioactivities in vivo [61]. The data in the literature show that, after ingestion, anthocyanins are quickly absorbed through the stomach and the small intestine by various mechanisms [62]. Nevertheless, small amounts of anthocyanins are absorbed, and plasma levels of cranberry anthocyanins appear to be too low to compete effectively with antioxidants such as ascorbate and glutathione [63]. However, the unabsorbed part of anthocyanins reach the colon, where they can modulate the microbiota [62,63]. Consumption of anthocyanins can modify the colonization of the gut microbiota by the stimulation of the growth of beneficial bacteria, such as *Bifidobaterium* spp. and *Lactobacillus* spp., and the inhibition the growth of pathogenic bacteria, such as *Staphylococcus aureus* and *Salmonella typhimurium* [62,64]. Gutpaper et al. showed that obese mice with a diet of 200 mg/kg cranberry extract per 8 weeks increased the *Akkermansia* population and induced features of the prevention of metabolic phenotypes linked to obesity. Furthermore, treatment with cranberry extract was found to lower content of intestinal triglyceride and to reduce inflammation and oxidative stress of the intestinal tract [65]. The beneficial effects of anthocyanins could be related to the effects on the intestinal tract, which are important for various biochemical processes in the body.

4. Conclusions

Cranberry's phytochemical profile is genotype-dependent and this study provides new knowledge on the variation in the composition of anthocyanins and anthocyanidins in different cranberry cultivars. Moreover, this study allows for the assessment of the quality of cranberry fruit and the use of high-quality cranberry raw material for food and health promotion. Four anthocyanin compounds predominated in fruit samples of American cranberry cultivars: cyanidin-3-galactoside, cyanidin-3-arabinoside, peonidin-3-galactoside, and peonidin-3-arabinoside, and they can be regarded as qualitative profile markers of American cranberry products. The chromatographic profiles are species-characteristic and can be applied for cranberry authenticity studies.

The fruits of American cranberry cultivars 'Woolman' and 'Le Munyon' distinguished with the highest content of anthocyanins, and their extracts had the strongest antioxidant activity. The cultivation of cranberry cultivars 'Woolman', 'Howes', and 'Le Munyon' in Lithuanian climatic conditions could be carried out in gardens and collections. The results of this study allow for evaluating the qualitative and quantitative composition of anthocyanins and anthocyanidins in fruit samples of the grown cultivars and for ensuring the preparation of high-quality cranberry raw material. High-quality phytochemically characterized American cranberry extracts rich in natural antioxidants can be used in the production of high-added-value products, food supplements, and other functional ingredients for nutrition and health.

Author Contributions: Conceptualization V.J. and L.R.; methodology L.R. and R.U.; software R.U. and L.R.; formal analysis R.U., investigation R.U., L.R. and V.J.; resources R.U., L.R. and V.J.; data curation V.J. and L.R.; writing—original draft preparation R.U.; writing—review and editing V.J. and L.R., visualization R.U.; supervision V.J. and L.R.; project administration V.J. All authors have read and agreed to the published version of the manuscript.

Funding: This research received no external funding.

Institutional Review Board Statement: Not applicable.

Informed Consent Statement: Not applicable.

Data Availability Statement: All datasets generated for this study are included in the article.

Conflicts of Interest: The authors declare no conflict of interest.

References

1. McMahan, E.; Guédot, C. Development of *Sparganothis sulfureana* (Lepidoptera: Tortricidae) on cranberry cultivars. *Insects* **2018**, *9*, 4. [CrossRef] [PubMed]
2. Jurikova, T.; Skrovankova, S.; Mlcek, J.; Balla, S.; Snopek, L. Bioactive compounds, antioxidant activity, and biological effects of european cranberry (*Vaccinium oxycoccos*). *Molecules* **2018**, *24*, 24. [CrossRef] [PubMed]
3. Hoekstra, B.R.; Neill, C.; Kennedy, C.D. Trends in the Massachusetts cranberry industry create opportunities for the restoration of cultivated riparian wetlands. *Restor. Ecol.* **2020**, *28*, 185–195. [CrossRef]
4. Vorsa, N.; Zalapa, J. Domestication, genetics, and genomics of the American cranberry. *Plant Breed. Rev.* **2019**, *43*, 279–315.
5. Oszmiański, J.; Kolniak-Ostek, J.; Lachowicz, S.; Gorzelany, J.; Matłok, N. Phytochemical compounds and antioxidant activity in different cultivars of cranberry (*Vaccinium macrocarpon* L.). *J. Food Sci.* **2017**, *82*, 2569–2575. [CrossRef]
6. Cranberry Production in 2019. UN Food and Agriculture Organization, Corporate Statistical Database. 2020. Available online: https://www.fao.org/faostat/en/#compare (accessed on 20 December 2021).
7. Viskelis, P.; Rubinskienė, M.; Jasutienė, I.; Šarkinas, A.; Daubaras, R.; Česonienė, L. Anthocyanins, antioxidative, and antimicrobial properties of American cranberry (*Vaccinium macrocarpon* Ait.) and their press cakes. *J. Food Sci.* **2009**, *74*, 157–161. [CrossRef]
8. Daubaras, R.; Česonienė, L. *Stambiauogių Spanguolių Pramoninių Plantacijų Eksploatavimas*; Versus Aureus: Vilnius, Lithuania, 2015; pp. 1–40.
9. Gupta, P.; Song, B.; Neto, C.; Camesano, T.A. Atomic force microscopy-guided fractionation reveals the influence of cranberry phytochemicals on adhesion of *Escherichia coli*. *Food Funct.* **2016**, *7*, 2655–2666. [CrossRef]
10. Xue, H.; Tan, J.; Li, Q.; Cai, X.; Tang, J. Optimization ultrasound-assisted extraction of anthocyanins from cranberry using response surface methodology coupled with genetic algorithm and identification anthocyanins with HPLC-MS. *J. Food Process. Preserv.* **2021**, *45*, 15378. [CrossRef]
11. Abeywickrama, G.; Debnath, S.C.; Ambigaipalan, P.; Shahidi, F. Phenolics of selected cranberry genotypes (*Vaccinium macrocarpon* Ait.) and their antioxidant efficacy. *J. Agric. Food Chem.* **2016**, *64*, 9342–9351. [CrossRef]
12. Murphy, B.T.; MacKinnon, S.L.; Yan, X.; Hammond, G.B.; Vaisberg, A.J.; Neto, C.C. Identification of triterpene hydroxycinnamates with in vitro antitumor activity from whole cranberry fruit (*Vaccinium macrocarpon*). *J. Agric. Food Chem.* **2003**, *51*, 3541–3545. [CrossRef]
13. Liu, H.; Howell, A.B.; Zhang, D.J.; Khoo, C. A randomized, double-blind, placebo-controlled pilot study to assess bacterial anti-adhesive activity in human urine following consumption of a cranberry supplement. *Food Funct.* **2019**, *10*, 7645–7652. [CrossRef] [PubMed]
14. Shaheen, G.; Ahmad, I.; Mehmood, A.; Akhter, N.; Usmanghani, K.; Shamim, T.; Shah, S.M.A.; Sumreen, L.; Akram, M. Monograph of *Vaccinium macrocarpon*. *J. Med. Plants Res.* **2011**, *5*, 5340–5346.
15. Prasain, J.K.; Grubbs, C.; Barnes, S. Cranberry anti-cancer compounds and their uptake and metabolism: An updated review. *J. Berry Res.* **2020**, *10*, 1–10. [CrossRef]
16. Wang, Y.; Han, A.; Chen, E.; Singh, R.K.; Chichester, C.O.; Moore, R.G.; Singh, A.P.; Vorsa, N. The cranberry flavonoids PAC DP-9 and quercetin aglycone induce cytotoxicity and cell cycle arrest and increase cisplatin sensitivity in ovarian cancer cells. *Int. J. Oncol.* **2015**, *46*, 1924–1934. [CrossRef] [PubMed]
17. Rauf, A.; Imran, M.; Khan, I.A.; Ur-Rehman, M.; Gilani, S.A.; Mehmood, Z.; Mubarak, M.S. Anticancer potential of quercetin: A comprehensive review: Quercetin as an anticancer agent. *Phytother. Res.* **2018**, *32*, 2109–2130. [CrossRef] [PubMed]
18. Pappas, E.; Schaich, K.M. Phytochemicals of cranberries and cranberry products: Characterization, potential health effects, and processing stability. *Crit. Rev. Food Sci. Nutr.* **2009**, *49*, 741–781. [CrossRef] [PubMed]
19. Vilkickyte, G.; Motiekaityte, V.; Vainoriene, R.; Liaudanskas, M.; Raudone, L. Development, validation, and application of UPLC-PDA method for anthocyanins profiling in *Vaccinium* L. berries. *J. Berry Res.* **2021**, *11*, 583–599. [CrossRef]
20. Yan, X.; Murphy, B.T.; Hammond, G.B.; Vinson, J.A.; Neto, C.C. Antioxidant activities and antitumor screening of extracts from cranberry fruit (*Vaccinium macrocarpon*). *J. Agric. Food Chem.* **2002**, *50*, 5844–5849. [CrossRef]
21. Ho, M.-L.; Chen, P.-N.; Chu, S.-C.; Kuo, D.-Y.; Kuo, W.-H.; Chen, J.-Y.; Hsieh, Y.-S. Peonidin 3-glucoside inhibits lung cancer metastasis by downregulation of proteinases activities and mapk pathway. *Nutr. Cancer* **2010**, *62*, 505–516. [CrossRef]
22. Smeriglio, A.; Barreca, D.; Bellocco, E.; Trombetta, D. Chemistry, pharmacology and health benefits of anthocyanins: Anthocyanins and human health. *Phytother. Res.* **2016**, *30*, 1265–1286. [CrossRef]
23. Seeram, N. Cyclooxygenase inhibitory and antioxidant cyanidin glycosides in cherries and berries. *Phytomedicine* **2001**, *8*, 362–369. [CrossRef] [PubMed]

24. Glisan, S.L.; Ryan, C.; Neilson, A.P.; Lambert, J.D. Cranberry extract attenuates hepatic inflammation in high-fat-fed obese mice. *J. Nutr. Biochem.* **2016**, *37*, 60–66. [CrossRef] [PubMed]
25. Santana, D.G.; Oliveira, A.S.; de Santana Souza, M.T.; do Carmo Santos, J.T.; Hassimotto, N.M.A.; de Oliveira e Silva, A.M.; Grespan, R.; Camargo, E.A. *Vaccinium macrocarpon* Aiton extract ameliorates inflammation and hyperalgesia through oxidative stress inhibition in experimental acute pancreatitis. *Evid.-Based Complement. Altern. Med.* **2018**, *2018*, 1–13. [CrossRef] [PubMed]
26. Seeram, N.P.; Adams, L.S.; Hardy, M.L.; Heber, D. Total cranberry extract versus its phytochemical constituents: Antiproliferative and synergistic effects against human tumor cell lines. *J. Agric. Food Chem.* **2004**, *52*, 2512–2517. [CrossRef] [PubMed]
27. Stobnicka, A.; Gniewosz, M. Antimicrobial protection of minced pork meat with the use of swamp cranberry (*Vaccinium oxycoccos* L.) fruit and pomace extracts. *J. Food Sci. Technol.* **2018**, *55*, 62–71. [CrossRef]
28. Cortez, R.; Luna-Vital, D.A.; Margulis, D.; Gonzalez de Mejia, E. Natural Pigments: Stabilization Methods of Anthocyanins for Food Applications. *Compr. Rev. Food Sci. Food Saf.* **2017**, *16*, 180–198. [CrossRef]
29. Bechtold, T.; Mahmud-Ali, A.; Mussak, R. Anthocyanin dyes extracted from grape pomace for the purpose of textile dyeing. *J. Sci. Food Agric.* **2007**, *87*, 2589–2595. [CrossRef]
30. Singh, S.; Gaikwad, K.K.; Lee, Y.S. Anthocyanin—A natural dye for smart food packaging systems. *Korean J. Packag. Sci. Technol.* **2018**, *24*, 167–180. [CrossRef]
31. The Meteorological Data in 2020, Lithuanian Hydrometeorological Service, Monthly Overview of Hydrometeorological Conditions. Available online: http://www.meteo.lt/lt/2021-gruodis (accessed on 22 January 2021).
32. Gudžinskaitė, I.; Stackevičienė, E.; Liaudanskas, M.; Zymonė, K.; Žvikas, V.; Viškelis, J.; Urbštaitė, R.; Janulis, V. Variability in the qualitative and quantitative composition and content of phenolic compounds in the fruit of introduced American cranberry (*Vaccinium macrocarpon* Aiton). *Plants* **2020**, *9*, 1379. [CrossRef]
33. Council of Europe. *European Pharmacopoeia*, 10th ed.; Council of Europe: Strasbourg, France, 2019; p. 51.
34. Fajardo, D.; Morales, J.; Zhu, H.; Steffan, S.; Harbut, R.; Bassil, N.; Hummer, K.; Polashock, J.; Vorsa, N.; Zalapa, J. Discrimination of American cranberry cultivars and assessment of clonal heterogeneity using microsatellite markers. *Plant Mol. Biol. Rep.* **2013**, *31*, 264–271. [CrossRef]
35. Novy, R.G.; Vorsa, N.; Kobak, C.; Goffreda, J. RAPDs identify varietal misclassification and regional divergence in cranberry [*Vaccinium macrocarpon* (Ait.) Pursh]. *Theor. Appl. Genet.* **1994**, *88*, 1004–1010. [CrossRef] [PubMed]
36. Schlautman, B.; Covarrubias-Pazaran, G.; Rodriguez-Bonilla, L.; Hummer, K.; Bassil, N.; Smith, T.; Zalapa, J. Genetic diversity and cultivar variants in the NCGR cranberry (*Vaccinium macrocarpon* Aiton) collection. *J. Genet.* **2018**, *97*, 1339–1351. [CrossRef]
37. Re, R.; Pellegrini, N.; Proteggente, A.; Pannala, A.; Yang, M.; Rice-Evans, C. Antioxidant activity applying an improved ABTS radical cation decolorization assay. *Free Radic. Biol. Med.* **1999**, *26*, 1231–1237. [CrossRef]
38. Raudone, L.; Vilkickyte, G.; Pitkauskaite, L.; Raudonis, R.; Vainoriene, R.; Motiekaityte, V. Antioxidant activities of *Vaccinium vitisidaea* L. leaves within cultivars and their phenolic compounds. *Molecules* **2019**, *24*, 844. [CrossRef] [PubMed]
39. Raudone, L.; Raudonis, R.; Liaudanskas, M.; Janulis, V.; Viskelis, P. Phenolic antioxidant profiles in the whole fruit, flesh and peel of apple cultivars grown in Lithuania. *Sci. Hortic.* **2017**, *216*, 186–192. [CrossRef]
40. Lee, J.; Durst, R.W.; Wrolstad, R.E. Determination of total monomeric anthocyanin pigment content of fruit juices, beverages, natural colorants, and wines by the ph differential method: Collaborative study. *J. AOAC Int.* **2005**, *88*, 1269–1278. [CrossRef]
41. Pourmasoumi, M.; Hadi, A.; Najafgholizadeh, A.; Joukar, F.; Mansour-Ghanaei, F. The effects of cranberry on cardiovascular metabolic risk factors: A systematic review and meta-analysis. *Clin. Nutr.* **2020**, *39*, 774–788. [CrossRef]
42. Česonienė, L.; Daubaras, R.; Jasutienė, I., Venclovienė, J.; Miliauskienė, I. Evaluation of the biochemical components and chromatic properties of the juice of *Vaccinium macrocarpon* Aiton and *Vaccinium oxycoccos* L. *Plant. Foods Hum. Nutr.* **2011**, *66*, 238–244. [CrossRef]
43. Zhang, J.; Chen, D.; Chen, X.; Kilmartin, P.; Quek, S.Y. The influence of vinification methods and cultivars on the volatile and phenolic profiles of fermented alcoholic beverages from cranberry. *Antioxidants* **2019**, *8*, 144. [CrossRef]
44. Gardana, C.; Scialpi, A.; Fachechi, C.; Simonetti, P. Identification of markers for the authentication of cranberry extract and cranberry-based food supplements. *Heliyon* **2020**, *6*, e03863. [CrossRef]
45. Brown, P.N.; Murch, S.J.; Shipley, P. Phytochemical diversity of cranberry (*Vaccinium macrocarpon* Aiton) cultivars by anthocyanin determination and metabolomic profiling with chemometric analysis. *J. Agric. Food Chem.* **2012**, *60*, 261–271. [CrossRef] [PubMed]
46. Narwojsz, A.; Tańska, M.; Mazur, B.; Borowska, E.J. Fruit physical features, phenolic compounds profile and inhibition activities of cranberry cultivars (*Vaccinium macrocarpon*) compared to wild-grown cranberry (*Vaccinium oxycoccus*). *Plant Foods Hum. Nutr.* **2019**, *74*, 300–306. [CrossRef] [PubMed]
47. Mannino, G.; Di Stefano, V.; Lauria, A.; Pitonzo, R.; Gentile, C. *Vaccinium macrocarpon* (cranberry)-based dietary supplements: Variation in mass uniformity, proanthocyanidin dosage and anthocyanin profile demonstrates quality control standard needed. *Nutrients* **2020**, *12*, 992. [CrossRef] [PubMed]
48. Wang, S.Y.; Stretch, A.W. Antioxidant capacity in cranberry is influenced by cultivar and storage temperature. *J. Agric. Food Chem.* **2001**, *49*, 969–974. [CrossRef]
49. Caminiti, I.M.; Noci, F.; Muñoz, A.; Whyte, P.; Morgan, D.J.; Cronin, D.A.; Lyng, J.G. Impact of selected combinations of non-thermal processing technologies on the quality of an apple and cranberry juice blend. *Food Chem.* **2011**, *124*, 1387–1392. [CrossRef]

50. Lee, J.; Rennaker, C.; Wrolstad, R.E. Correlation of two anthocyanin quantification methods: HPLC and spectrophotometric methods. *Food Chem.* **2008**, *110*, 782–786. [CrossRef]
51. Tonutare, T.; Moor, U.; Szajdak, L. Strawberry anthocyanin determination by ph differential spectroscopic method—How to get true results? *Acta Sci. Pol. Hortorum Cultus* **2014**, *13*, 35–47.
52. Li, H.; Deng, Z.; Zhu, H.; Hu, C.; Liu, R.; Young, J.C.; Tsao, R. Highly pigmented vegetables: Anthocyanin compositions and their role in antioxidant activities. *Food Res. Int.* **2012**, *46*, 250–259. [CrossRef]
53. Lu, Y.; Pekerti, B.N.; Toh, Z.S.; Broom, F.; Savage, G.; Liu, S.Q.; Huang, D. Physico-chemical parameters and proanthocyanidin profiles of cranberries cultivated in New Zealand. *J. Food Compos. Anal.* **2017**, *63*, 1–7. [CrossRef]
54. Grace, M.H.; Massey, A.R.; Mbeunkui, F.; Yousef, G.G.; Lila, M.A. Comparison of health-relevant flavonoids in commonly consumed cranberry products. *J. Food Sci.* **2012**, *77*, 176–183. [CrossRef]
55. Lee, S.G.; Vance, T.M.; Nam, T.-G.; Kim, D.-O.; Koo, S.I.; Chun, O.K. Evaluation of pH differential and HPLC methods expressed as cyanidin-3-glucoside equivalent for measuring the total anthocyanin contents of berries. *J. Food Meas. Charact.* **2016**, *10*, 562–568. [CrossRef]
56. Diaz-Garcia, L.; Covarrubias-Pazaran, G.; Johnson-Cicalese, J.; Vorsa, N.; Zalapa, J. Genotyping-by-sequencing identifies historical breeding stages of the recently domesticated American cranberry. *Front. Plant Sci.* **2020**, *11*, 607770. [CrossRef] [PubMed]
57. Martins, N.; Barros, L.; Ferreira, I.C.F.R. In vivo antioxidant activity of phenolic compounds: Facts and gaps. *Trends Food Sci. Technol.* **2016**, *48*, 1–12. [CrossRef]
58. Martín, J.; Kuskoski, E.M.; Navas, M.J.; Asuero, A.G. Antioxidant Capacity of Anthocyanin Pigments. In *Flavonoids—From Biosynthesis to Human Health*; Justino, C.G., Ed.; Science, Technology and Medicine Open Access: Rijeka, Croatia, 2017; pp. 205–255.
59. Reis, J.F.; Monteiro, V.V.S.; de Souza Gomes, R.; do Carmo, M.M.; da Costa, G.V.; Ribera, P.C.; Monteiro, M.C. Action mechanism and cardiovascular effect of anthocyanins: A systematic review of animal and human studies. *J. Transl. Med.* **2016**, *14*, 315. [CrossRef] [PubMed]
60. Chaves, V.C.; Boff, L.; Vizzotto, M.; Calvete, E.; Reginatto, F.H.; Simões, C.M. Berries grown in Brazil: Anthocyanin profiles and biological properties: Chemical composition of berries grown in Brazil. *J. Sci. Food Agric.* **2018**, *98*, 4331–4338. [CrossRef]
61. Zeng, Y.; Song, J.; Zhang, M.; Wang, H.; Zhang, Y.; Suo, H. Comparison of in vitro and in vivo antioxidant activities of six flavonoids with similar structures. *Antioxidants* **2020**, *9*, 732. [CrossRef]
62. Tian, L.; Tan, Y.; Chen, G.; Wang, G.; Sun, J.; Ou, S.; Chen, W.; Bai, W. Metabolism of anthocyanins and consequent effects on the gut microbiota. *Crit. Rev. Food Sci. Nutr.* **2019**, *59*, 982–991. [CrossRef]
63. Milbury, P.E.; Vita, J.A.; Blumberg, J.B. Anthocyanins are bioavailable in humans following an acute dose of cranberry juice. *J. Nutr.* **2010**, *140*, 1099–1104. [CrossRef]
64. Lavefve, L.; Howard, L.R.; Carbonero, F. Berry polyphenols metabolism and impact on human gut microbiota and health. *Food Funct.* **2020**, *11*, 45–65. [CrossRef]
65. Anhê, F.F.; Roy, D.; Pilon, G.; Dudonné, S.; Matamoros, S.; Varin, T.V.; Garofalo, C.; Moine, Q.; Desjardins, Y.; Levy, E.; et al. A polyphenol-rich cranberry extract protects from diet-induced obesity, insulin resistance and intestinal inflammation in association with increased *Akkermansia* spp. population in the gut microbiota of mice. *Gut* **2015**, *64*, 872–883. [CrossRef]

Article

Development of Optimized Ultrasound-Assisted Extraction Methods for the Recovery of Total Phenolic Compounds and Anthocyanins from Onion Bulbs

Ana V. González-de-Peredo, Mercedes Vázquez-Espinosa, Estrella Espada-Bellido, Marta Ferreiro-González, Ceferino Carrera, Gerardo F. Barbero * and Miguel Palma

Department of Analytical Chemistry, Faculty of Sciences, Agrifood Campus of International Excellence (ceiA3), IVAGRO, University of Cadiz, Puerto Real, 11510 Cadiz, Spain; ana.velascogope@uca.es (A.V.G.-d.-P.); mercedes.vazquez@uca.es (M.V.-E.); estrella.espada@uca.es (E.E.-B.); marta.ferreiro@uca.es (M.F.-G.); ceferino.carrera@uca.es (C.C.); miguel.palma@uca.es (M.P.)
* Correspondence: gerardo.fernandez@uca.es; Tel.: +34-956-01-6355

Abstract: *Allium cepa* L. is one of the most abundant vegetable crops worldwide. In addition to its versatile culinary uses, onion also exhibits quite interesting medicinal uses. Bulbs have a high content of bioactive compounds that are beneficial for human health. This study intends to develop and optimize two appropriate ultrasound-assisted methods for the extraction of the phenolic compounds and anthocyanins present in red onion. A response surface methodology was employed and, specifically, a Box–Behnken design, for the optimization of the methods. The optimal conditions for the extraction of the phenolic compounds were the follows: 53% MeOH as solvent, pH 2.6, 60 °C temperature, 30.1% amplitude, 0.43 s cycle, and 0.2:11 g sample/mL solvent ratio. On the other hand, the optimal conditions for the anthocyanins were as follows: 57% MeOH as solvent, pH 2, 60 °C temperature, 90% amplitude, 0.64 s cycle, and 0.2:15 g sample/mL solvent ratio. Both methods presented high repeatability and intermediate precision, as well as short extraction times with good recovery yields. These results illustrate that the use of ultrasound-assisted extraction, when properly optimized, is suitable for the extraction and quantification of the compounds of interest to determine and improve the quality of the raw material and its subproducts for consumers.

Keywords: *Allium cepa* L.; anthocyanins; Box–Behnken; onion; phenolic compounds; UHPLC; ultrasound-assisted extraction

1. Introduction

The genus *Allium* spans to more than 750 species that can be found all over the Northern Hemisphere [1]. Among these species, *Allium cepa* L. (common onion) is one of the most ancient crops cultivated worldwide and among the most popular ones [2]. Its great popularity is largely thanks to its versatile culinary uses as a raw food or in different cooked forms: baked, boiled, braised, grilled, fried, and so on [3]. In addition to its extended use as a flavored vegetable or spicy ingredient, onions are also well known for their employment in different forms of traditional medicine [4,5]. Numerous epidemiological studies have confirmed that the regular consumption of onions decreases the occurrence of various forms of cancer or cardiovascular and neurodegenerative diseases [2,6–8]. Onions are rich in antioxidant compounds, such that their consumption represents an interesting provision of antioxidants that contribute to prevent certain diseases associated with oxidative stress. From the article previously published by our research group [9], onion bulbs present a high content of flavonols, which are phenolic compounds with a high antioxidant capacity, as well as other health-promoting properties. However, flavonols are not the only compounds responsible for the antioxidant capacity of this bulb. Thus, in addition to flavonols, onions, particularly its red varieties, are a rich source of anthocyanins. Anthocyanins are natural pigments from the phenolic compounds family. Their interest comes from their capacity to directly scavenge reactive oxygen species (ROS) [10]. In fact, anthocyanin-rich vegetables have been demonstrated to have a health-promoting effect against several disorders or processes such as cancer, neurological diseases, inflammation, diabetes, or bacterial infections [11,12]. Furthermore, although vegetables have generally a lower anthocyanin content than most fruits, root and tuber types of vegetables (such as potatoes, carrots, and onions) present some advantages, such as their lower cost and longer storage periods, which favors a greater consumption [11]. Concerning onions, they have been frequently reported to contain cyanidin derivatives, together with some minor amounts of peonidin derivatives [13–15]. In fact, red onions varieties are the only ones reported to have a significant anthocyanin content, which, together with a greater total phenolic compounds content, implies that these varieties have higher antioxidant capacity [13]. The considerable intake of phenolic compounds and particularly of anthocyanins would make red onions particularly healthy food [16].

For these reasons, a rapid and efficient extraction and analysis method needs to be developed to obtain quality extracts of these bioactive compounds from onions. However, analyzing anthocyanins is a difficult task because of their liability to alkaline pH, light, or temperature [17]. Therefore, to prevent their degradation, the extraction procedure should be as short as possible. Traditionally, anthocyanins have been extracted from onions by means of organic solvents, such as methanol, using long processes at low temperature [18]. Table 1 describes some of the extraction methods that have been used by other authors to extract anthocyanins from onions. Most of the methods involve long periods or shorter periods of stirring, but repeated several cycles to obtain relevant yields. This implies a large consumption of solvents and, therefore, an increment in costs. In this study, ultrasound-assisted extraction (UAE) is presented to improve the methods for the extraction of both anthocyanins and phenolic compounds from red onion. UAE, supported by the phenomenon of cavitation, achieves a greater dispersion of the solid phase into the liquid and enhances contact interface [19]. This means that greater yields can be obtained in a shorter time, thus reducing solvent consumption and costs, which makes UAE a more environmentally friendly technique [20]. Nevertheless, although the use of UAE represents on its own an improvement when compared with traditional techniques, the optimization of several key parameters associated with performance is also a crucial aspect regarding the efficiency. Currently, temperature, time, solvent type, and concentration are some of the variables that greatly affect UAE efficiency [21–23].

Response surface methodology (RSM) is one of the experimental designs most often used for the optimization of the variables involved in extraction processes. One of the advantages of RSM lies in the fact that it allows the evaluation of the actual effect that multiple factors as well as their interactions have on one or more response variables. For this study, a Box–Behnken (BBD) design was chosen because it allows generating higher-order response surfaces using fewer runs than a regular factorial technique [24]. In addition to studying individually the effect on the UAE process from each one of the response variables, a multi-response optimization (MRO) approach with desirability functions was also applied. The possibility of performing a simultaneous analysis of both phenolic compounds and anthocyanins in red onions is also of great interest because of the cost and time saving that this represents.

Having efficient extraction methods would be quite useful, as it should facilitate the extraction and quantification of the compounds of interest for the different industries and would allow better quality raw material and its byproducts to be supplied to consumers. Thus, the concrete aim of the present study is the development of optimized individual and combined UAE methods for the extraction of phenolic compounds and anthocyanins from red onions. The general objective is to highlight how the use of UAE, an advanced extraction technique, duly optimized by RSM and MRO, leads to valuable improvements in relation to yield and extraction times.

Table 1. Extraction methods used by other authors to extract anthocyanins from onions.

Publication Year	Extraction Method	Anthocyanins Analyzed [1]	Onion Variety	Total Anthocyanins Measured (mg g^{-1})	Reference
2020	Homogenization (1 min), sonication (30 min), and centrifugation. The supernatant residue was re-extracted twice	1, 2, 3, 4, 5, 6, 7	Honeysuckle red onions and sweet Italian red onions	Honeysuckle 0.103 ± 2.206 Sweet Italian 0.086 ± 1.843	[25]
2019	Centrifugation at 3214 g. The supernatant residue was re-extracted until the samples turned colourless	1, 2, 4, 5	Red onion	0.056	[26]
2018	Sonication at 60 °C for 1 h	Total anthocyanins measured by colourimetric methods	Red onions from eight different cultivars	0.02 ± 0.01–0.12 ± 0.01	[27]
2018	Three different methods: maceration (24 h), percolation (8 h), reflux and Soxhlet method (2 h). The extractions were repeated three times	1, 4, 5, 8	Bima Brebes and Maja Cipanas	Maceration: 1.463 ± 0.013 and 1.181 ± 0.008 Percolation: 0.328 ± 0.010 and 0.597 ± 0.015 Reflux: 1.415 ± 0.08 and 1.449 ± 0.013 Soxhlet: 0.218 ± 0.021 and 0.342 ± 0.022	[28]
2017	Sonication in an ultrasonic bath at 4 °C for 24 h	1, 4, 11, 12, 13	Dark-red onion cultivar 'Xiu Qiu' and white onion cultivar 'Ring Master'	Xiu Qui: 0.3587 ± 0.0054 Ring Master: 0.0142 ± 0.0087	[29]
2013	Extraction at 4 °C, overnight or for 2 h	Total anthocyanins measured by colourimetric methods	Red onion	0.9966	[30]
2012	The extraction was carried out on a rotary shaker overnight (15 h; 400 rpm) at room temperature Centrifugation at	1, 4, 9, 14	Red onion Pier-C and Red onion Pearl	Red onion Pier-C: 0.0777 ± 0.0038 Red onion Pearl: 0.1895 ± 0.0363	[11]
2011	1200 rpm (3 min) and agitation (15 min). Each homogenate was extracted three	1, 2, 4, 5, 6	Red onion "Vermelha da Povoa"	0.059	[31]
2011	Shaking (15 min) and centrifugation. Two additional extractions were performed for each sample	1, 2, 4, 5, 6	Red onion "Vermelha da Povoa"	0.003 ± 0.016	[32]
2010	Incubating (1 h) at room temperature with alternative shaking and subsequently centrifuged at 4000 rpm for 15 min at 28 °C. Two additional extractions were performed	1, 2, 4, 5, 6, 9, 10, 15	Red Onion Vermelha da Povoa, improved Vermelha da Povoa and Red Creole	Vermelha da Povoa: 0.057 ± 0.018 Improved Vermelha: 0.128 ± 0.046 Red Creole: 0.286 ± 0.08	[14]

1. Compounds analyzed: 1. cyanidin 3-O-glucoside, 2. cyanidin 3-O-laminaribioside, 3. delphinidin 3,5-O-diglucoside, 4. cyanidin 3-O-(6″-malonoylglucoside), 5. cyanidin 3-O-(6″-malonoyl-laminaribioside), 6. Peonidin 3-O-malonoylglucoside, 7. cyanidin 3-O-(malonoyl)-(acetyl)-glucoside, 8. cyanidin 3-O-arabinoside, 9. cyanidin 3-O-(3‴-malonilglucósido), 10. pedonidina 3-O-glucósido, 11. delphinidin 3-O-diglucoside, 12. delphinidin 3-O-glucoside, 13. delphinidin aglycon, 14. cyanidin 3-O-(malonyl)diglucoside, 15. cyanidin 3-O-dimalonylaminaribioside.

2. Materials and Methods

2.1. Biological Material

A stock of red onions was purchased in 2019 from a local market in the province of Cadiz (Spain) to be used as the biological material for this study. More specifically, the bulbs of those red onions were to be used for the analysis. For that purpose, after peeling the hard outer skin from each onion bulb, their pulp was chopped into small pieces using a knife. The chopped onions were lyophilized by means of an LYOALFA freeze dryer (Azbil Telstar Technologies, Terrasa, Barcelona, Spain) and crushed using a knife mill GRINDOMIX GM 200 (Retsch GmbH, Haan, Germany) to finally obtain <300 µm particles. This homogeneous material was stored in a freezer at $-20\ ^\circ C$ before analysis.

2.2. Chemical Reagents

The extraction solvents employed in this study were mixtures of methanol (Fischer Chemical, Loughborough, UK) of HPLC purity and Milli-Q water, obtained from a Milli-Q water purification system (Millipore, Bedford, MA, USA), with different pH values. A sodium hydroxide solution (NaOH, 1 M) and a hydrochloric solution (HCl, 1 M), both from Panreac, Barcelona (Spain), were used to adjust pH.

A variety of chemicals were used to determine bioactive compounds' content. According to the Folin–Ciocalteau spectrophotometric method, anhydrous sodium carbonate (Panreac Química, Castellar del Valles, Barcelona, Spain) and Folin–Ciocalteu (Merck KGaA, EMD Millipore Corporation, Darmstadt, Germany) were employed to measure total phenolic compounds' content. Regarding UHPLC analyses, methanol (Fischer Scientific, Loughborough, UK), Milli-Q water, and formic acid (Scharlau, Barcelona, Spain) were used to determine anthocyanins' content. According to the DPPH assay, DPPH (2,2-diphenyl-1-picrylhydrazyl) radical scavenging (Sigma-Aldrich, San Luis, MO, USA) was used. The standard used for the phenolic compounds was gallic acid, the standard for anthocyanins was cyanidin chloride, and the standard for antioxidant activity was 6-hydroxy-2,5,7,8-tetramethylchroman-2-carboxylic acid (Trolox), all supplied by Sigma-Aldrich Chemical Co. (St. Louis, MO, USA). In all the analyses the extracts were previously filtered through a nylon filter (Membrane Solutions, Dallas, TX, USA) of 0.45 µm for Folin–Ciocalteau and DPPH assay, and of 0.2 µm for UHPLC analysis.

2.3. Extraction of Bioactive Compounds

As mentioned above, the extraction of bioactive compounds (anthocyanins and total phenolic compounds) from red onions was carried out in this study by ultrasound-assisted extraction. Specifically, a Sonopuls HD 2070.2 processor, 20 Hz (BANDELIN electronic GmbH & Co KG, Heinrichstrabe, Berlin, Germany), which allows controlling the cycle, the amplitude, and the working time, was employed. An adjustable double vessel thermostatic bath with temperature control was also used (Frigiterm-10, Selecta, Barcelona, Spain). With respect to the UAE probe, a versus 70 T (BANDELIN electronic GmbH & Co KG, Heinrichstrabe, Berlin, Germany) with the following characteristics: approximately 130 mm in length, 13 mm in diameter, 13 µm in amplitude, and 20–900 mL in volume, was used.

Regarding the experimental protocol applied, about 0.2 g of the lyophilized and homogenized sample was weighed in a Falcon tube and the corresponding solvent volume (at the specific pH and methanol/water ratio) was added. The Falcon tube was placed into the double vessel to maintain the sample at the desired temperature, and the ultrasound probe was submerged into it. The solvent type and volume, the temperature, the cycle and the amplitude were set according to each experiment requirement. The range of UAE conditions for the extractions was as follows: % methanol in water 50–100%, temperature 10–60 $^\circ C$, amplitude 30–90% of the equipment maximum power (70 W), cycle 0.4–1 s, pH 2–7, and sample/solvent ratio of 0.2:10–0.2:20 g sample/mL solvent. The initial extraction time was the only parameter that was set to a constant value of 10 min for all of the experiments, which was followed by a sample cooling time. The extracts obtained were then centrifuged at 1702× g for 5 min. The supernatant was collected, and the precipitate was re-centrifuged under the same conditions after adding 5 mL of the same extraction solvent. The two supernatants that were obtained were mixed together and transferred to a volumetric flask (25 mL), which was filled up with the same solvent. The extracts were kept at $-20\ °C$ for subsequent analysis.

2.4. Analysis of Bioactive Compounds

2.4.1. Analysis of Total Phenolic Compounds

The total phenolic compounds (TPCs) content in red onion was determined by means of a modified Folin–Ciocalteau (FC) method [33]. Folin–Ciocalteu is a colorimetric method based on the fact that phenolic compounds react with Folin–Ciocalteu reagent (a mixture of sodium tungstate and sodium molybdate) with basic pH, which gives rise to a blue color susceptible to be determined spectrophotometrically at 765 nm. Specifically, the FC assay was performed by transferring 0.25 mL of UAE onion extract, 1.25 mL of water, and 1.25 mL of the Folin–Ciocalteu reagent into a 25 mL volumetric flask. Then, 5 mL of aqueous sodium carbonate solution (20% p/v) was also added, and the solution was made up to the mark with water. After 30 min, the solution absorbance was measured at 765 nm. The absorbance was measured on a Cary 4000 UV/Vis (Agilent, Santa Clara, CA, USA). Gallic acid was used as the standard. Therefore, the results are expressed as milligrams of gallic acid equivalent per g of dry weight (mg GAE g^{-1} DW). The linear regression of the standard was constructed using six points (50–0.5 mg L^{-1}) in triplicate. The regression equation (y = 0.0014x + 0.0022) and the determination coefficient (R^2 = 0.9995) were calculated by means of Microsoft Office Excel 2013.

2.4.2. Identification of the Anthocyanins

The anthocyanin content in red onion was determined by liquid chromatography. Firstly, the anthocyanins in the UAE extracts were identified by ultra-high-performance liquid chromatography (UHPLC) coupled to a quadrupole time-of-flight mass spectrometer (Q-ToF-MS) (Xevo G2 QToF, Waters Corp., Milford, MA, USA). The chromatographic separation was performed on a reverse-phase C18 analytical column (1.7 µm, 2.1 mm × 100 mm, made by Acquity UPLC BEH C18, Waters Corp., Milford, MA, USA). The gradient of the UHPLC-Q-ToF-MS method was as follows (time, % solvent B): 0.00 min, 15%; 3.30 min, 20%; 3.86 min, 30%; 5.05 min, 40%; 5.35 min, 55%; 5.64 min, 60%; 5.94 min, 95%; and 7.50 min, 95%. The flow rate was 0.4 mL min^{-1}, the injection volume was 3.0 µL, and the mobile phase was a binary solvent system (2% formic as phase A acid and methanol as phase B). An electrospray operating in positive ionization mode was used to perform the analyses under the following conditions: desolvation gas flow = 700 L h^{-1}, desolvation temperature = 500 °C, cone gas flow = 10 L h^{-1}, source temperature = 150 °C, capillary voltage = 700 V, cone voltage = 30 V, and collision energy = 20 eV. The full-scan mode was used (m/z 100–1200). The following nine anthocyanins were individually identified based on their retention time and molecular weight: cyanidin 3-O-glucoside (3.517 min, m/z 449.1087), cyanidin 3-O-laminaribioside (4.132, m/z 611.1641), cyanidin 3-O-(3″-malonylglucoside) (4.875 min, m/z 535.1069), peonidin 3-O-glucoside (5.384 min, m/z 463.1251), delphinidin 3,5-O-diglucoside (5.721 min, m/z 649.1392), cyanidin 3-O-(6″-malonylglucoside) (5.850, m/z 535.1104), cyanidin 3-O-(6″-malonyl-laminaribioside) (6.052 min, m/z 697.1613), peonidin 3-O-(6″-malonylglucoside) (6.323 min, m/z 549.1255), and delphinidin 3-O-glucoside (6.536, m/z 487.0863). The data regarding the anthocyanins identified in the samples and their mass spectrum are included in the Supplementary Material.

2.4.3. Analysis of the Anthocyanins

Once the anthocyanins were identified, an Elite UHPLC LaChrom Ultra System (Hitachi, Tokyo, Japan) was used to separate and quantify them. The UHPLC system is equipped with an L-2420U UV/Vis detector, an L-2200U autosampler, an L-2300 column oven, and two L-2160 U pumps. The chromatographic separation was performed on a reverse-phase C18 analytical column (2.6 µm, 2.1 mm × 100 mm, made by Phenomenex, Torrance, CA, USA). The gradient and characteristics of the UHPLC method employed in this work were previously published by our research group [34]. Cyanidin chloride was used as the standard to quantify the seven anthocyanins identified in onions. The linear regression for the standard was constructed using six points in triplicate (0.06–35 mg L^{-1}). The regression equation (y = 260,596.88x − 4292.66) and the determination coefficient (R^2 = 0.9999) were also calculated using Microsoft Office Excel 2013. Using the same software, the limit of detection (LOD) (0.113 mg L^{-1}) and the limit of quantification (LOQ) (0.402 mg L^{-1}) were also calculated. Repeatability was also studied using nine replicates on the same day. Specifically, it was evaluated in terms of retention time and area of each of the anthocyanin peaks. The results, expressed as the coefficient of variance (CV), were all less than 10%, which is the acceptable CV limit according to the AOAC manual for peer-verified methods. Specifically, the repeatability takes values within the range of 0.05–0.14% for retention time and 0.59–7.26% for areas.

Once the regression equation of cyanidin chloride was obtained, a calibration curve was plotted for each anthocyanin that was identified in the onion samples. For this purpose, it was assumed that the nine anthocyanins have similar absorbance, and the molecular weight of each anthocyanin was taken into account. The results were expressed as milligrams of each anthocyanin per g of dry weight (mg g^{-1} DW). The final UHPLC chromatogram can be seen in Figure 1.

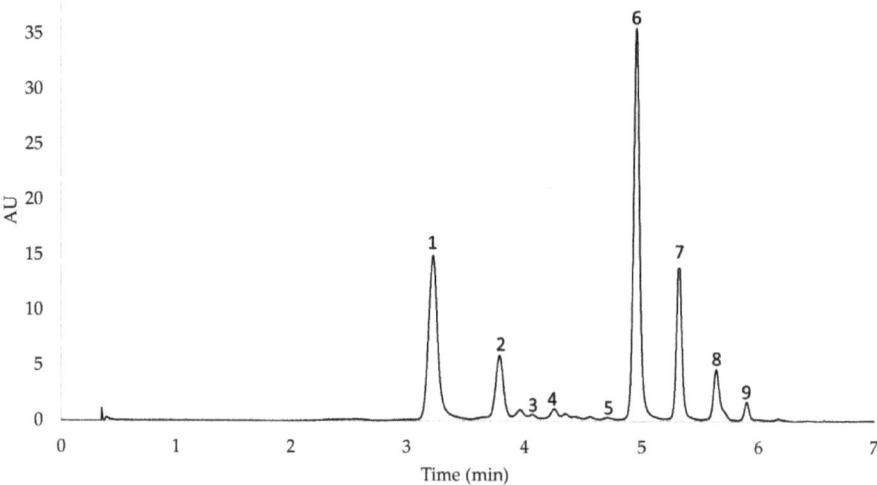

Figure 1. Anthocyanins identified in red onion. Chromatograms peaks corresponding to the nine anthocyanins identified in the UAE extracts from red onion samples. Peak 1. cyanidin 3-O-glucoside, peak 2. cyanidin 3-O-laminaribioside, peak 3. cyanidin 3-O-(3″-malonylglucoside), peak 4. peonidin 3-O-glucoside, peak 5. delphinidin 3,5-O-diglucoside, peak 6. cyanidin 3-O-(6″-malonylglucoside), peak 7. cyanidin 3-O-(6″-malonyl-laminaribioside), peak 8. peonidin 3-O-(6″-malonylglucoside), peak 9. delphinidin 3-O-glucoside.

2.4.4. Determining Antioxidant Activity

A number of different techniques to assess the antioxidant activity of food and plants have been described in the literature. However, preferential attention has been given to the technique that uses 2,2-diphenyl-1-picrylhydrazyl free radical, better known by its acronym DPPH. For this assay, the procedure designed by Brand-Williams et al. [35] and modified by Miliauskas et al. [36] was employed. First, a 6 × 10^{-5} M DPPH solution was prepared in methanol. Then, for each 100 µL of onion extract, 2 mL of the DPPH solution was added to the mixture. The mixture was incubated for 40 min in the absence of light and at room temperature. Then, the absorbance was measured at 515 nm. The results were expressed as mg of Trolox equivalents (TE) per g of dry weight sample. For this purpose, a Trolox calibration curve (y = 88.94x + 0.75; R^2 = 0.9959) was plotted using six points (0–1.4 mM) in triplicate.

2.5. Applying Box–Behnken Design to Optimize the UAE Methods

For the development and optimization of the UAE methods, the spherical response surface Box–Behnken design (BBD) was applied. The Box–Behnken design is characterized by the factor levels being placed at the midpoints of the edges and at the centre of the space [37]. This implies that fewer data points are required when compared with other designs. For example, unlike the central composite design, BBD does not have corner points. This also means that the factors are in no case either all high or all low at the same time, i.e., no extreme combinations take place [38]. Thus, in a BBD, every factor has three levels: a lower level (−1), an intermediate level (0), and an upper level (1) [29].

In this work, six independent factors were considered within the following ranges: composition of the solvent (% methanol in water) (X_1: 50, 75, 100 °C), pH of the solvent (X_2: 2, 4.5, 7), extraction temperature (X_3: 10, 35, 60 °C), ultrasound amplitude (X_4: 30, 60, 90%), ultrasound cycle (X_5; 0.4, 0.7, 1 s), and sample mass/solvent volume ratio (X_6: 0.2:10, 0.2:15, and 0.2:20 g sample/mL solvent). All of these ranges were selected for the study based on the group's previous experience [9,21,23,39]. Thus, the values within these ranges allow performance optimization while no degradation takes place. Temperature was the only factor that was studied separately because phenolic compounds, and especially anthocyanins, may present instability when subjected to high temperature levels [40,41].

In order to determine the temperature for the extractions, several runs were conducted at different temperatures according to the protocol previously explained in Section 2.2. First of all, a control extract was obtained by carrying out an extraction under intermediate conditions (50:50 MeOH/H_2O extraction solvent, 60% ultrasound amplitude, 0.5 s cycle, 0.2:15 g sample/mL solvent, and 20 min extraction time) where no external heat source was applied. Then, 15 mL of the control extract was acquired and subjected to different temperature levels, while the rest of the variables remained constant at the aforementioned intermediate conditions. This procedure was repeated at each one of the temperature levels considered for the study, i.e., 10, 20, 30, 40, 50, 60, and 70 °C. The anthocyanins and phenolic compounds yields obtained under each temperature can be seen in Figure 2a,b.

It can be observed from Figure 2a how the total phenolic compounds' content is lesser at a temperature of 70 °C, while at lower temperature levels, no significant differences in the concentration of bioactive compounds can be noticed. Regarding anthocyanins (Figure 2b), both individual anthocyanins and total anthocyanins were represented, in order to know if there are opposite tendencies depending on each anthocyanin studied. As for phenolic compounds, the total anthocyanins content was lesser at 70 °C and remained constant at the lowest temperature levels. This decrease in the content of total anthocyanins is mainly due to the following anthocyanins that showed a decrease in their content as the temperature increased: cyanidin 3-O-(6″-malonylglucoside), cyanidin 3-O-(6″-malonyl-laminaribioside), and cyanidin 3-O-glucoside. Other anthocyanins such as cyanidin 3-O-(3″-malonylglucoside) or peonidin 3-O-(6″-malonylglucoside) kept their content constant in all the temperatures evaluated. Based on these results, a temperature range from 10 to 60 °C would be used, because, only at temperatures over 60 °C, a decrease in the concentration of both compounds, probably owing to the aforementioned degradation, could be registered.

With regard to the response variables, two responses were considered for this study as follows: total phenolic compounds (TPC) in red onion bulbs, determined by the Folin–Ciocalteau method (Y_{TPC}, mg g^{-1}); and total anthocyanins (TA) in red onion bulbs (Y_{TA}, mg g^{-1}), calculated as the sum of concentrations corresponding to each one of the nine individual anthocyanins quantified by UHPLC. Finally, and according to the specific BBD equation, a design comprising 54 extraction runs, including six repetitions at their centre point to determine the error, was obtained. All the experiments were carried out at random. These experiments and the resulting data can be seen in Table 2.

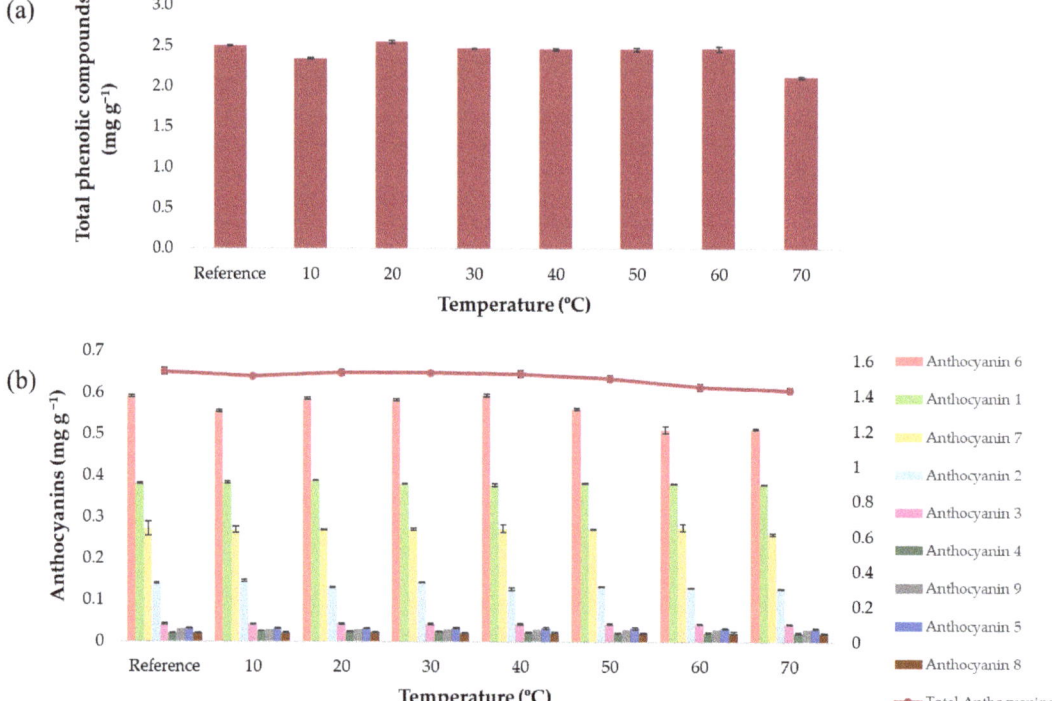

Figure 2. Extract content stability at different temperatures ($n = 3$): (**a**) total phenolic compounds content and (**b**) individual and total anthocyanins, in red onion extracts.

Table 2. Total phenolic compounds and total anthocyanins contents determined by the experiments and predicted values based on the Box–Behnken design.

Run	Factors						Responses			
	X_1	X_2	X_3	X_4	X_5	X_6	Y_{TPC} (mg g^{-1})		Y_{TA} (mg g^{-1})	
							Experimental	Predicted	Experimental	Predicted
1	0	0	−1	0	−1	−1	4.1767	3.5242	2.0668	2.0057
2	0	0	1	0	−1	−1	2.9561	3.2168	1.9910	2.0681
3	0	0	−1	0	1	−1	3.2035	3.0330	1.7513	1.5964
4	0	0	1	0	1	−1	2.9146	2.7132	1.3774	1.3945
5	0	0	−1	0	−1	1	2.2442	2.7397	1.8422	1.8561
6	0	0	1	0	−1	1	2.6180	2.4945	1.9246	2.0485
7	0	0	−1	0	1	1	3.0296	3.0630	1.6633	1.6171
8	0	0	1	0	1	1	2.4469	2.8053	1.5150	1.5452
9	0	−1	0	−1	−1	0	2.7012	2.9196	1.9020	2.0085
10	0	1	0	−1	−1	0	2.5148	2.6226	2.1472	1.9832
11	0	−1	0	1	−1	0	2.4628	2.7319	1.7641	1.9643
12	0	1	0	1	−1	0	2.7902	2.3765	2.1085	1.9085
13	0	−1	0	−1	1	0	2.3645	2.5574	1.3898	1.4864
14	0	1	0	−1	1	0	2.5099	2.4617	1.5738	1.4770
15	0	−1	0	1	1	0	3.0411	2.7125	1.4975	1.5580
16	0	1	0	1	1	0	2.5560	2.5584	1.5213	1.5182
17	−1	0	−1	−1	0	0	4.7011	4.7646	1.3071	1.4164

Table 2. Cont.

Run	Factors						Responses			
	X_1	X_2	X_3	X_4	X_5	X_6	Y_{TPC} (mg g^{-1})		Y_{TA} (mg g^{-1})	
							Experimental	Predicted	Experimental	Predicted
18	1	0	−1	−1	0	0	2.8642	3.0163	0.3299	0.4692
19	−1	0	1	−1	0	0	2.6618	2.4707	1.6671	1.4487
20	1	0	1	−1	0	0	3.3118	3.3064	0.2508	0.4251
21	−1	0	−1	1	0	0	2.7490	3.3144	1.5527	1.3750
22	1	0	−1	1	0	0	3.3056	2.9367	0.2837	0.5053
23	−1	0	1	1	0	0	2.0515	2.4593	1.5521	1.4096
24	1	0	1	1	0	0	5.2890	4.6655	0.5696	0.4635
25	0	−1	−1	0	0	−1	3.6635	3.7722	1.5621	1.5809
26	0	1	−1	0	0	−1	3.9095	4.0100	1.6094	1.6217
27	0	−1	1	0	0	−1	3.2906	3.1606	1.4842	1.5197
28	0	1	1	0	0	−1	4.1254	3.9944	1.5655	1.5434
29	0	−1	−1	0	0	1	4.3195	4.1564	1.5901	1.5813
30	0	1	−1	0	0	1	3.0355	2.8714	1.5588	1.4924
31	0	−1	1	0	0	1	3.4133	3.6069	1.6314	1.6501
32	0	1	1	0	0	1	2.7325	2.9179	1.5319	1.5441
33	−1	−1	0	0	−1	0	2.7751	2.5609	1.9114	1.9364
34	1	−1	0	0	−1	0	3.3414	3.0735	1.4900	1.0493
35	−1	1	0	0	−1	0	2.7997	3.0478	2.1634	2.3055
36	1	1	0	0	−1	0	1.8618	1.9341	0.4220	0.5990
37	−1	−1	0	0	1	0	1.6920	1.8405	1.1957	1.1220
38	1	−1	0	0	1	0	3.4394	3.4122	0.9739	0.9352
39	−1	1	0	0	1	0	2.4817	2.5288	1.1699	1.5071
40	1	1	0	0	1	0	2.4807	2.4741	0.6292	0.5008
41	−1	0	0	−1	0	−1	2.8317	2.8012	1.3718	1.3445
42	1	0	0	−1	0	−1	1.9561	2.1494	0.3941	0.3867
43	−1	0	0	1	0	−1	2.1085	2.0093	1.2244	1.2897
44	1	0	0	1	0	−1	1.9760	2.7280	0.3625	0.4094
45	−1	0	0	−1	0	1	2.3904	2.1984	1.4082	1.3581
46	1	0	0	−1	0	1	2.3983	1.9376	0.4072	0.3451
47	−1	0	0	1	0	1	2.2821	1.5288	1.3218	1.3324
48	1	0	0	1	0	1	2.0481	2.6386	0.3728	0.3969
49	0	0	0	0	0	0	2.3484	2.5702	1.7592	1.7190
50	0	0	0	0	0	0	2.7464	2.5702	1.7587	1.7190
51	0	0	0	0	0	0	2.4622	2.5702	1.7139	1.7190
52	0	0	0	0	0	0	2.6377	2.5702	1.7742	1.7190
53	0	0	0	0	0	0	2.6370	2.5702	1.6840	1.7190
54	0	0	0	0	0	0	2.5892	2.5702	1.6238	1.7190

The advantage of RSM is that it can reduce the prediction error and improve the estimate by means of a polynomial equation [42]. The results from this second-order polynomial equation (Equation (1)) match as closely as possible the actual experimental responses according to the corresponding conditions.

$$Y = \beta_0 + \sum_{i=1}^{k} \beta_i X_i + \beta_{ii} X_i^2 + \sum_{i} \sum_{i=1}^{k} \beta_{ij} X_i X_j + r \quad (1)$$

In this equation, Y represents the responses (Y_{TPC} and Y_{TA}); β_0 is the model constant; X represents each one of the factors considered; β_i is the coefficient of each main effect; β_{ii} is the coefficient of the quadratic factors that represent the curvature of the surface; β_{ij} is the coefficient corresponding to the interactions between factor i and factor j; and r is the residual value (random error). The statistical significance of the polynomial model and the regression terms were evaluated by applying an analysis of variance (ANOVA) following a similar protocol to the one used by Jadhav S.B. et al. [43]. Specifically, the F-test and the 'lack of fit' test were evaluated using the software applications Statgraphics Centurion version XVI (Warrenton, VA, USA) and Design Expert (Version 13, Stat-Ease Inc., Minneapolis, MN, USA).

2.6. Multi-Response Optimization by Desirability Functions

This research work was focused on optimizing two response variables, the extraction of total phenolic compounds and the extraction of anthocyanins. As an alternative to the individual optimization of each response variable, multi-response optimization was proposed. The desirability function is one of the methods that is most frequently used to perform multi-response surface optimizations. To apply the desirability function approach, each estimated response is transformed into a scale-free value within the range $0 \leq d_i \leq 1$. This is known as desirability (d_i). The overall desirability function D (Equation (2)) is defined as the geometric average of the individual desirability functions of each response d_i (Y_i), where m is the number of responses. The optimal solutions are determined by maximizing D.

$$D = (d_1 \times d_2 \times \ldots d_m)^{1/m}, \quad (2)$$

The software application Statgraphics Centurion version XVI (Warrenton, VA, USA) was used to statistically analyze the results obtained by means of each separately and by the multi-response optimization design. Given that according to Shapiro–Wilk test, the resulting values follow a normal distribution (p-value < 0.05), and that according to Levene's test, they present the same variance (p-value < 0.05). An ANOVA test was carried out to detect any statistically significant differences (5% level of significance) between the means obtained by each one of the two separate methods and the combined method.

3. Results and Discussion

3.1. Developing a UAE Method for Total Phenolic Compounds by Means of a Box–Behnken Design

After the experimental matrix was completed (Table 2), an ANOVA was applied to evaluate the effect of the factors and the possible interactions between them. The results from the ANOVA are shown in Table 3. Based on these results, it can be confirmed that the analysis explains 82.50% of the total variability. In addition, to demonstrate the validity of the polynomial model, the ANOVA indicates the coefficients of the different parameters in the quadratic polynomial equation and their significance (p-values). According to such significance, the factors and/or interactions with a more relevant influence on the response can be determined. Thus, only those factors and/or interactions with p-values lower than 0.05 were considered to have a relevant influence on the response at the established level of significance (95%).

Regarding total phenolic compounds (Table 3), the linear terms were not significant (p-value > 0.05). With respect to quadratic interaction, the amplitude–amplitude (X_3X_2, p-value < 0.0001) and temperature–temperature (X_2X_2, p-value 0.01) interactions showed a relevant effect on the response. Finally, the following factor interactions were also determined as significant with p-values lower than 0.05: percentage methanol–amplitude (X_1X_3, p-value 0.00), percentage methanol–cycle (X_1X_4, p-value 0.00), percentage methanol–temperature (X_1X_2, p-value 0.01), temperature–ratio (X_2X_6, p-value 0.02), and amplitude–cycle (X_3X_4, p-value 0.02). Based on these results, it can be concluded that, except for pH, the rest of the variables have a relevant influence on the extraction of phenolic compounds from red onions samples. This highlights the importance of carrying out an experimental design when intending to extract bioactive compounds from natural matrices.

With regard to the quadratic interaction, amplitude–amplitude showed a positive effect ($b_3{}^2 = 0.88$) on the response variable. Amplitude is an important variable regarding extraction, as the energy provided by the ultrasounds is necessary to release the target compounds from the matrix [40]. The interaction of temperature–temperature also showed a positive effect ($b_2{}^2 = 0.35$) on the response variable. The temperature is an important variable; for a successful extraction, the temperature must be sufficient to favor the solubility, diffusion, and transfer of the compounds of interest in the solvent, but not so high as to produce degradation [44].

For a better understanding, a Pareto chart (Figure 3) was included to illustrate the influence from each factor and combination of factors on the response. The effect from each factor or factor interaction is graphically represented by bars arranged in decreasing order of influence on the response.

Table 3. ANOVA of the quadratic model adjusted to the extraction of total phenolic compounds from red onion.

Source	Source	Coefficient	Sum of Squares	df	Mean Square	F-Value	p-Value
Model		2.57	22.39	27	0.83	4.54	0
A—MeOH	X_1	0.11	0.31	1	0.31	1.72	0.2
B—Temperature	X_2	−0.11	0.31	1	0.31	1.67	0.21
C—Amplitude	X_3	−0.14	0.48	1	0.48	2.62	0.12
D—Cycle	X_4	−0.02	0.01	1	0.01	0.07	0.8
E—pH	X_5	−0.05	0.05	1	0.05	0.27	0.61
F—Ratio	X_6	−0.17	0.72	1	0.72	3.94	0.06
AB	$X_1 X_2$	−0.41	1.32	1	1.32	7.24	0.01
AC	$X_1 X_3$	0.65	3.34	1	3.34	18.27	0
AD	$X_1 X_4$	0.34	1.88	1	1.88	10.28	0
AE	$X_1 X_5$	0.26	0.56	1	0.56	3.07	0.09
AF	$X_1 X_6$	0.1	0.08	1	0.08	0.42	0.52
BC	$X_2 X_3$	0.15	0.18	1	0.18	0.97	0.33
BD	$X_2 X_4$	−0.01	0	1	0	0.01	0.92
BE	$X_2 X_5$	0.05	0.04	1	0.04	0.22	0.64
BF	$X_2 X_6$	−0.38	1.16	1	1.16	6.35	0.02
CD	$X_3 X_4$	0.36	1.04	1	1.04	5.67	0.02
CE	$X_3 X_5$	−0.00	0	1	0	0	0.98
CF	$X_3 X_6$	0.02	0	1	0	0.02	0.89
DE	$X_4 X_5$	0.09	0.06	1	0.06	0.32	0.58
DF	$X_4 X_6$	0.03	0.01	1	0.01	0.04	0.84
EF	$X_5 X_6$	0.2	0.33	1	0.33	1.82	0.19
A^2	X_1^2	−0.04	0.02	1	0.02	0.12	0.73
B^2	X_2^2	0.35	1.25	1	1.25	6.85	0.01
C^2	X_3^2	0.88	7.97	1	7.97	43.61	<0.0001
D^2	X_4^2	−0.04	0.01	1	0.01	0.08	0.78
E^2	X_5^2	−0.26	0.72	1	0.72	3.92	0.06
F^2	X_6^2	−0.24	0.58	1	0.58	3.18	0.09
Residual			4.75	26	0.18		
Lack of Fit			4.65	21	0.22	10.93	0.01
Pure Error			0.1	5	0.02		
Cor Total			27.14	53			

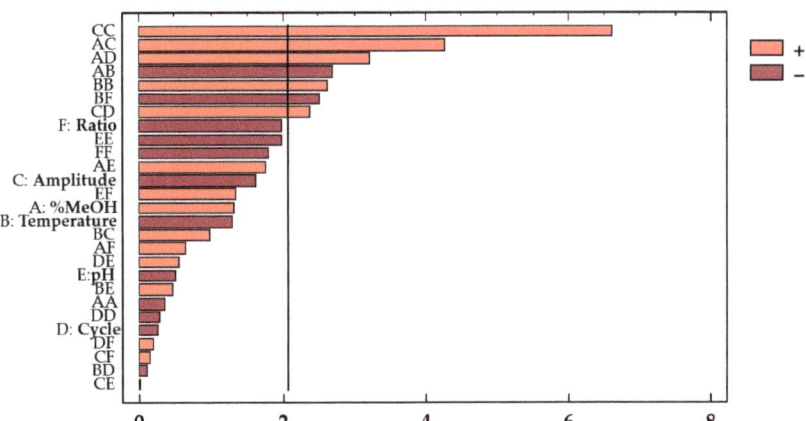

Figure 3. Standardized Pareto chart of the total phenolic compounds from red onion extracts.

Finally, based on the coefficients of the factors and interactive effects (Table 3), a polynomial equation can be obtained to predict total phenolic compounds' content (response variable, Y_{TPC}) as a function of the independent variables (Equation (3)). The full equation could be reduced by considering just the significant factors and interactions (p-value < 0.05). A reduced equation is represented in (Equation (4)).

$$Y_{TPC} \text{ (mg g}^{-1}) = 2.57 + 0.11 \cdot X_1 - 0.11 \cdot X_2 - 0.14 \cdot X_3 - 0.023 \cdot X_4 - 0.045 \cdot X_5 - 0.17 \cdot X_6 \\ - 0.046 \cdot X_1^2 - 0.41 \cdot X_1 X_2 + 0.65 \cdot X_1 X_3 + 0.34 \cdot X_1 X_4 + 0.26 \cdot X_1 X_5 + 0.098 \cdot X_1 X_6 + 0.35 \cdot X_2^2 + \\ 0.15 \cdot X_2 X_3 - 0.015 \cdot X_2 X_4 + 0.050 \cdot X_2 X_5 - 0.38 \cdot X_2 X_6 + 0.88 \cdot X_3^2 + 0.36 \cdot X_3 X_4 - 0.0031 \cdot X_3 X_5 \\ + 0.016 \cdot X_3 X_6 - 0.037 \cdot X_4^2 + 0.086 \cdot X_4 X_5 + 0.031 \cdot X_4 X_6 - 0.26 \cdot X_5^2 + 0.20 \cdot X_5 X_6 - 0.24 \cdot X_6^2, \quad (3)$$

$$Y_{TPC} \text{ (mg g}^{-1}) = 2.57 - 0.41 \cdot X_1 X_2 + 0.65 \cdot X_1 X_3 + 0.34 \cdot X_1 X_4 + 0.35 \cdot X_2^2 - 0.38 \cdot X_2 X_6 \\ + 0.88 \cdot X_3^2 + 0.36 \cdot X_3 X_4. \quad (4)$$

According to the fitted model based on the trends outlined above, a three-dimensional (3D) surface can be plotted. This 3D (Figure 4) graph facilitates the comprehension of the effect from the interactions of the most influential parameters, i.e., %MeOH–amplitude, %MeOH–cycle, %MeOH–temperature, or temperature–ratio on the total phenolic compounds' recovery.

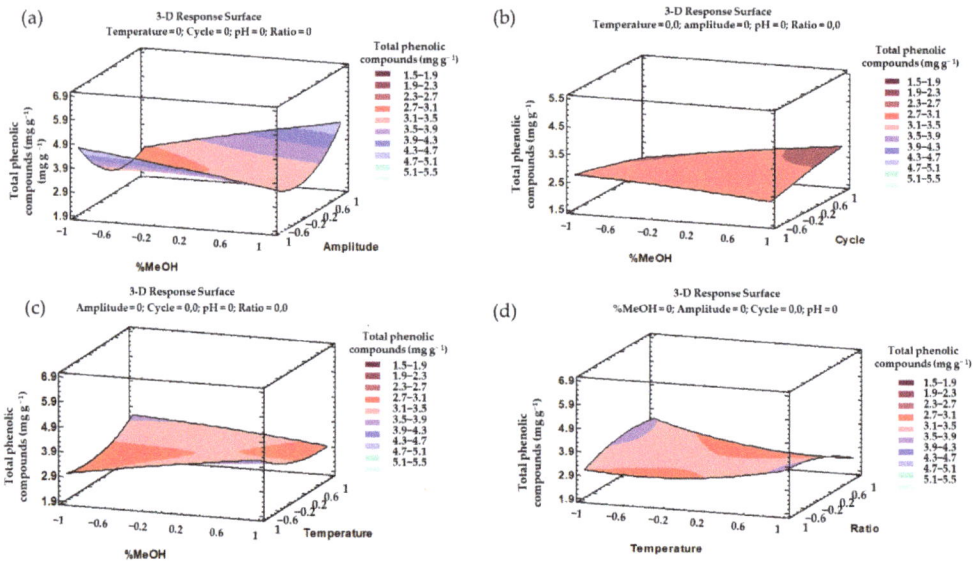

Figure 4. 3D surface graphs of the Box–Behnken design according to the polynomial equations representing the effects from the different interactions on the total phenolic compounds extractions: (**a**) %MeOH and amplitude; (**b**) %MeOH and cycle; (**c**) %MeOH and temperature; and (**d**) temperature and ratio.

3.2. Developing a UAE Method for Total Anthocyanins Using a Box–Behnken Design

As for total phenolic compounds, an ANOVA was applied to the anthocyanin data matrix (Table 2), showing that the analysis explains 94.50% of the total variability. The rest of the results obtained are shown in Table 4.

Regarding total anthocyanins (Table 4), both the linear term percentage of methanol (X_1, p-value < 0.0001) as well as the linear term pH (X_5, p-value < 0.0001) exhibited a relevant effect on the response. Concerning the quadratic interactions, the percentage of methanol–percentage of methanol ($X_1 X_2$, p-value < 0.0001), the pH–pH ($X_5 X_2$, p-value 0.00), and the cycle–cycle ($X_4 X_2$, p-value 0.03) interactions also showed a relevant effect on the response. Finally, the following factor interactions were also significant with p-values lower than 0.05: percentage methanol–temperature ($X_1 X_2$, p-value 0.00) and percentage methanol–pH ($X_1 X_4$, p-value 0.01).

Based on these results, it can be concluded that methanol and pH were the most significant variables regarding their influence on the anthocyanins' extraction. In fact, both variables showed a negative effect ($b_1 = -0.47$ and $b_5 = -0.23$, respectively) on the response variable. The solvent composition plays an important role in the extraction of the bioactive compounds because a similar polarity between anthocyanins and solvent is required for a successful extraction. Owing to the negative effect of this factor ($b_1 = -0.47$), when the solvent takes low polarity values within the range of the study, the variable shows the opposite effect, and a greater efficiency of the extraction is achieved. In this case, as the range goes from 50 to 100%, the results confirm that hydroalcoholic mixtures are more efficient than pure solvents (100% MeOH) for the extraction of amphiphilic or other moderately polar molecules, such as polyphenols. This had already been reported by other authors [12]. Likewise, pH also plays an important role in the bioactive compounds' extraction process, and particularly in the extraction of anthocyanins, as they are more stable when pH remains within the range of 1 to 3 [41]. The acids in the solvents contribute to breaking down the cell membranes, thus improving the release and solubilization of the different compounds, such as anthocyanins [41]. This is clearly visualized in the Pareto chart (Figure 5).

Table 4. ANOVA of the quadratic model adjusted for the extraction of total anthocyanins from red onion.

Source	Source	Coefficient	Sum of Squares	df	Mean Square	F-Value	p-Value
Model		1.72	14.37	27	0.53	16.54	<0.0001
A—MeOH	X_1	−0.47	5.38	1	5.38	167.14	<0.0001
B—Temperature	X_2	−0.02	0.01	1	0.01	0.2	0.66
C—Amplitude	X_3	0	0	1	0	0	0.95
D—Cycle	X_4	0	0	1	0	0	0.98
E—pH	X_5	−0.23	1.25	1	1.25	38.83	<0.0001
F—Ratio	X_6	0	0	1	0	0	0.99
AB	$X_1 X_2$	−0.20	0.34	1	0.34	10.44	0
AC	$X_1 X_3$	−0.02	0	1	0	0.09	0.77
AD	$X_1 X_4$	0.02	0.01	1	0.01	0.19	0.67
AE	$X_1 X_5$	0.18	0.25	1	0.25	7.62	0.01
AF	$X_1 X_6$	−0.01	0	1	0	0.05	0.83
BC	$X_2 X_3$	0	0	1	0	0	0.95
BD	$X_2 X_4$	−0.01	0	1	0	0.01	0.91
BE	$X_2 X_5$	0	0	1	0	0.01	0.93
BF	$X_2 X_6$	−0.03	0.01	1	0.01	0.26	0.61
CD	$X_3 X_4$	0	0	1	0	0	0.99
CE	$X_3 X_5$	−0.07	0.03	1	0.03	1.09	0.31
CF	$X_3 X_6$	0.03	0.02	1	0.02	0.53	0.48
DE	$X_4 X_5$	0.03	0.01	1	0.01	0.21	0.65
DF	$X_4 X_6$	0.01	0	1	0	0.01	0.91
EF	$X_5 X_6$	0.04	0.01	1	0.01	0.45	0.51
A^2	X_1^2	−0.63	4.02	1	4.02	124.94	<0.0001
B^2	X_2^2	−0.02	0.01	1	0.01	0.19	0.66
C^2	X_3^2	−0.02	0.01	1	0.01	0.17	0.68
D^2	X_4^2	−0.13	0.18	1	0.18	5.53	0.03
E^2	X_5^2	0.18	0.32	1	0.32	9.81	0
F^2	X_6^2	−0.10	0.11	1	0.11	3.49	0.07
Residual			0.84	26	0.03		
Lack of Fit			0.82	21	0.04	11.79	0.01
Pure Error			0.02	5	0		
Cor Total			15.21	53			

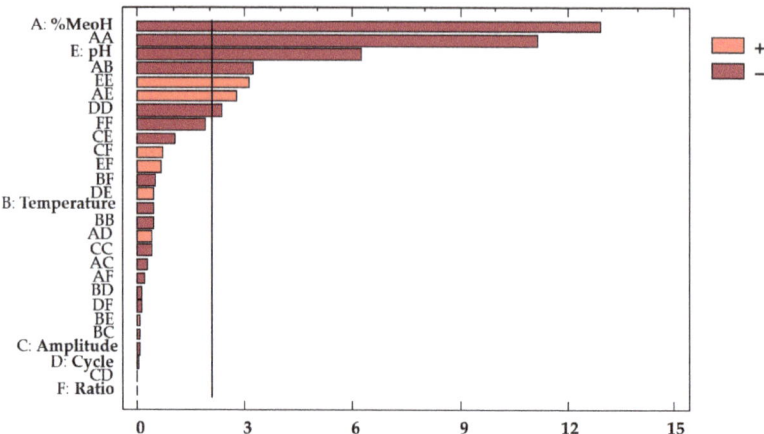

Figure 5. Standardized Pareto chart of total anthocyanins from red onion extracts.

Finally, the full polynomial equation (Equation (5)) and the reduced polynomial equation (Equation (6)) to predict the content of total anthocyanins (response variable, Y_{TA}) as a function of the independent variables are included below.

$$Y_{TA} \text{ (mg g}^{-1}) = 1.72 - 0.47 \cdot X_1 - 0.016 \cdot X_2 - 0.0024 \cdot X_3 - 0.00075 \cdot X_4 - 0.23 \cdot X_5 + \\ 0.00027 \cdot X_6 - 0.63 \cdot X_1^2 - 0.20 \cdot X_1 X_2 - 0.019 \cdot X_1 X_3 + 0.019 \cdot X_1 X_4 + 0.17 \cdot X_1 X_5 - \\ 0.014 \cdot X_1 X_6 - 0.025 \cdot X_2^2 - 0.0043 \cdot X_2 X_3 - 0.0076 \cdot X_2 X_4 + 0.0040 \cdot X_2 X_5 - 0.032 \cdot X_2 X_6 - \\ 0.023 \cdot X_3^2 + 0.00056 \cdot X_3 X_4 - 0.0660788 \cdot X_3 X_5 + 0.032 \cdot X_3 X_6 - 0.13 \cdot X_4^2 + 0.029 \cdot X_4 X_5 + \\ 0.0073 \cdot X_4 X_6 - 0.17 \cdot X_5^2 + 0.043 \cdot X_5 X_6 - 0.10 \cdot X_6^2, \quad (5)$$

$$Y_{TA} \text{ (mg g}^{-1}) = 1.72 - 0.47 \cdot X_1 - 0.23 \cdot X_5 - 0.62 \cdot X_1^2 - 0.20 \cdot X_1 X_2 + 0.17 \cdot X_1 X_5 \\ - 0.13 \cdot X_4^2 - 0.17 \cdot X_5^2, \quad (6)$$

All the trends outlined above can be graphically represented in their corresponding 3D surface graphs (Figure 6). These 3D representations illustrate the effect from the most influential interactions, i.e., %MeOH–temperature and %MeOH–pH, regarding total anthocyanins' recovery.

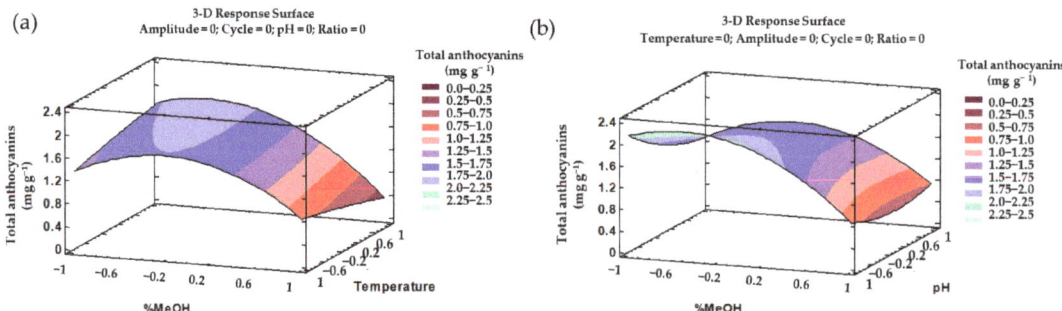

Figure 6. 3D surface graphs of the Box–Behnken design according to the polynomial equations representing the effects from the different interactions on the total anthocyanin extractions: (**a**) %MeOH and temperature; (**b**) %MeOH and pH.

3.3. Optimal Conditions, Extraction Time, and Precision of the Two Developed Methods

Based on the Box–Behnken design, the optimum values that maximize the response variables, both the yields of total phenolic compounds and the yields of total anthocyanins, can be construed. Such optimal conditions that maximize the two responses separately and simultaneously are presented in Table 5.

Table 5. Separate and simultaneous optimal conditions for the extraction of total phenolic compounds and total anthocyanins.

Factor	Total Phenolic Compounds	Total Anthocyanins	Multi-Response
%MeOH	53	57	50
Temperature (°C)	60	60	53
Amplitude (%)	30.1	89.9	30
Cycle (s)	0.43	0.64	0.4
pH	2.6	2	2
Ratio (g mL^{-1})	0.2:11	0.2:14.9	0.2:14
Result (mg g^{-1}) ± SD (n = 3)	7.30 ± 0.015	2.49 ± 0.053	7.23 ± 0.034 (TPC); 2.280 ± 0.081 (TA)

According to the results obtained for both variables, 60 °C was selected as the optimal temperature. Higher temperatures were not considered as relevant degradation was observed in the stability study [40]. Regarding the solvent, an approximate concentration of 1:1 MeOH/H$_2$O with acidic pH was selected as optimal for both variables. Each one of the components in a solvent mixture may play a specific role in the extraction. Thus, methanol increases the solubility of the bioactive compounds, while water contributes to the desorption of the solute from the sample. Therefore, a mixed solvent including both of these compounds should favor the extraction of phenolic compounds and anthocyanins [45]. Regarding pH, the extraction of the bioactive compounds is also promoted by acid solvents, as acid substances can break down cell membranes, which enhances the release and solubilization of the phenolic compounds and anthocyanins [46,47]. Amplitude was the factor that presented the greatest difference between both variables. The optimal amplitude for the extraction of phenolic compounds (30.1%) was close to the lower limit of the studied range, while for the extraction of anthocyanins, the optimal amplitude was near the upper limit (89.9%). This and other minor differences between each variable's optimal values mean that not all the phenolic compounds extracted would be anthocyanins. In fact, other phenolic compounds present in onion matrices, such as quercetin derivatives, which is a flavonoid, can be found at high concentrations in onion bulbs. This has already been reported by our research group [9].

In addition to establishing the optimal values for the most relevant parameters, the optimal time for the extraction of the phenolic compounds and anthocyanins in onion bulbs was also determined. For this purpose, several extractions were carried out at different times, keeping the rest of the extraction parameters under the optimal conditions previously determined (Table 5). The times studied were as follows: 2, 5, 10, 15, 20, and 25 min. Each time was studied three times, and the average results (n = 3) for phenolic compounds and anthocyanins are displayed in Figure 7.

It can be concluded, from Figure 7a, that 10 min provides the best extraction of total phenolic compounds. In fact, an extraction time of 10 min allows 7.30 ± 0.01 mg g^{-1} phenolic compounds to be extracted, while shorter times (2 or 5 min) achieve lower yields, probably because the extraction process cannot be completed in such a short time. Similarly, longer times (>10 min) also yield less phenolic compounds, but in this case, it might be due to the degradation suffered by the phenolic compounds when they are subjected to ultrasounds for so long. With respect to total anthocyanins (Figure 7b), only 2 min is required to obtain the best yields (2.62 ± 0.034 mg g^{-1}). In fact, after 5 min, the amount of anthocyanins extracted is practically the same. Therefore, 2 min was selected as the optimal extraction time, which represents a greater economic and time saving. Any time longer than 10 min resulted in lower extractions, probably owing to the already mentioned degradation of the phenolic compounds. Furthermore, individual anthocyanins were also represented, in order to know if there are opposite tendencies depending on each anthocyanin studied. All studied anthocyanins suffered a considerable decrease in their content when extraction times were greater than 10 min. Only the anthocyanins delphinidin 3,5-O-diglucoside (0.047 ± 0.004 mg g^{-1}) and delphinidin 3-O-glucoside (0.044 ± 0.002 mg g^{-1}) were unchanged with increasing time, showing no differences in their content as time increased.

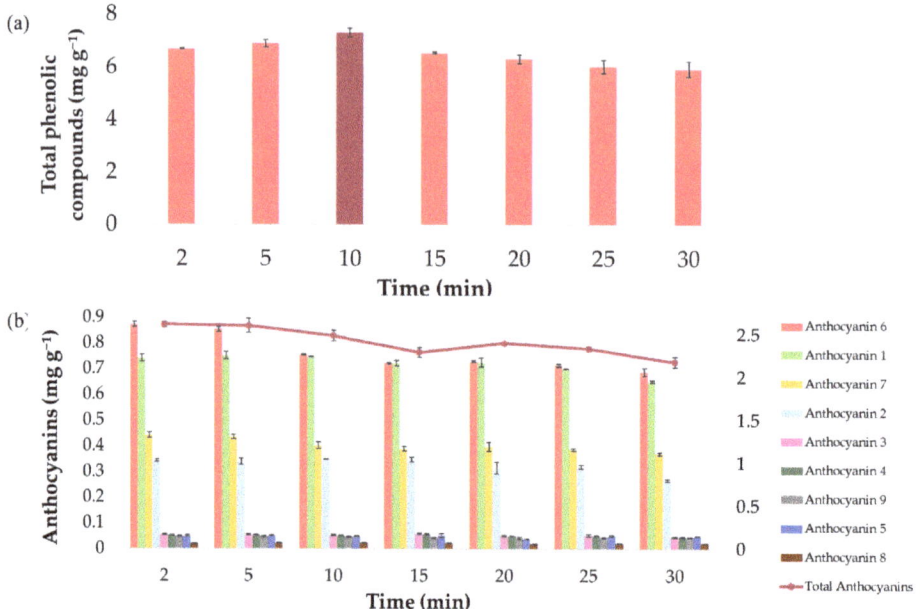

Figure 7. Average extraction yields (n = 3) resulting from different extraction times: (**a**) total phenolic compounds and (**b**) individual and total anthocyanins.

Finally, the precision of the developed methods was evaluated in terms of repeatability and intermediate precision. For the assessment of their intermediate precision, 10 experiments were conducted on 3 consecutive days (a total of 30 experiments). Then, their intermediate precision was determined according to the coefficient of variation (CV) of the 30 experiments. Repeatability was determined by calculating the coefficients of variation of each of the 10 experiments completed on a single day. This method to determine the precision of a particular process has been typically employed in other works on similar natural matrices [18,42]. The percentages of repeatability (3.01% for TPC and 2.86% for TA) and intermediate precision (4.12% for TPC and 3.56% for TA) obtained were lower than 5%. Therefore, the UAE methods for the extraction of total phenolic compounds and total anthocyanins can both be considered to have good repeatability and intermediate precision, as 5% is the generally accepted variation limit [48].

3.4. Multi-Response Optimization and Application to Different Onion Varieties

In addition to individually optimizing each response variable, a multi-response optimization was carried out. The multi-response optimization was evaluated to determine the optimum balanced conditions to successfully obtain extracts with a high amount of both anthocyanins and phenolic compounds. The optimal conditions that simultaneously maximize both responses are presented in Table 5. The yields obtained when applying the optimal values from the multi-response study using an intermediate time of 10 min were 7.23 ± 0.034 mg g^{-1} for total phenolic compounds and 2.280 ± 0.081 mg g^{-1} for total anthocyanins. These results were slightly lower than those obtained when the specific optimal values for each method were applied also using an intermediate time of 10 min (7.30 ± 0.015 mg g^{-1} for total phenolic compounds and 2.49 ± 0.053 mg g^{-1} for total anthocyanins). Specifically, and assuming that the values follow a normal distribution and that there is no difference between the variances, it can be stated that there is a statistically significant difference between the means of the two variables with a significance level of 5%, as the p-value of the F-test is less than 0.05.

Despite this statistical difference, the yields obtained are not exceedingly different. Therefore, this could represent a practicable set of conditions for those cases where time and cost savings (using a single solvent, for example) are a priority. This method could probably be applied by quality control analytical laboratories, where time and costs must be minimized [21]. Consequently, this combined method was applied to a number of onion varieties in order to verify the efficacy of the multi-response method when applied to onions of diverse chemical composition.

For this purpose, a total of 24 types of onions were purchased from different supermarkets and greengrocers. The onions were of different colors, varieties, or origin. All of them were subjected to the same pretreatment in order to obtain a fine powder, and then the samples of such powder were extracted in triplicate using the multi-response optimized method. The extracts resulting from the different varieties were analyzed to determine their phenolic compounds and anthocyanins contents as well as their antioxidant activity. Table 6 includes the results obtained from these extractions as the mean of the three replicates ± the standard deviation.

As expected, the only extracts that contain anthocyanins are those from red/purple onion varieties, given that these compounds are largely responsible for their reddish coloration [49]. Although yellow and white onions do not contain anthocyanins in their matrices, they do contain total phenolic compounds with antioxidant activity. As previously published by our research group [9], these onion varieties contain in their matrices other phenolic compounds such as quercetin derivatives, which largely confer their antioxidant properties. Even so, it is logical that, in general, the content of total phenolic compounds and the antioxidant activity is higher in the red varieties (6.60 ± 1.45 mg g^{-1} and 6.28 ± 0.80 mg g^{-1}, respectively) than in the yellow and white ones (4.51 ± 1.24 mg g^{-1} and 4.68 ± 1.22 mg g^{-1}, respectively), as these onions present both anthocyanins as quercetin derivatives. The fact that practically only the red onions have been reported to have a significant anthocyanin content may open new fields of applications for these varieties (perfume, cosmetics, food industries, medicine, and so on).

Finally, it should be highlighted how the use of advanced extraction techniques, like the UAE method when properly optimized, leads to valuable improvements. For this purpose, the results obtained from this study against those reported by other authors (see the bibliographic review in Table 1), who employed traditional extraction techniques, were compared. Even the most recent publications [25–27] are related to traditional techniques such as homogenization, centrifugation, or sonication. Not only do these techniques require longer extraction times and a greater use of solvents, but also they give rise to lower yields than those actually achieved in this work. Specifically, R. Metrani et al. with 0.103 ± 2.206 mg g^{-1} anthocyanins yields, A. D. Front et al. with 0.056 mg g^{-1} anthocyanins yields, and M. J. Park et al. with 0.02 ± 0.01–0.12 ± 0.01 mg g^{-1} anthocyanins yields from similar onion matrices did not reach the amounts of anthocyanins obtained by applying the method developed in this study. It can, therefore, be concluded that UAE, when properly optimized by means of an experimental design and coupled to a chromatographic analysis technique, represents a valuable tool for the extraction of bioactive compounds from onions. To the best of our knowledge, this UAE method had not been developed until present. Suitable analytical techniques that allow to identify and quantify the compounds of interest that are present in the final product and, thereby, its quality are extremely interesting.

Table 6. Quantification of total phenolic compounds, total anthocyanins, and antioxidant activity ($n = 3$) extracted by the multi-response developed UAE method to an assortment of onion varieties.

Onion Type	Peak 1 (mg g^{-1})	Peak 2 (mg g^{-1})	Peak 3 (mg g^{-1})	Peak 4 (mg g^{-1})	Peak 5 (mg g^{-1})	Peak 6 (mg g^{-1})	Peak 7 (mg g^{-1})	Peak 8 (mg g^{-1})	Peak 9 (mg g^{-1})	TA (mg g^{-1})	CFT (mg g^{-1})	Antioxidant Activity (mg g^{-1})
Spring white onion I	-	-	-	-	-	-	-	-	-	-	6.01 ± 0.09	4.64 ± 0.13
French white onion	-	-	-	-	-	-	-	-	-	-	6.71 ± 0.18	4.46 ± 0.15
Sweet white onion I	-	-	-	-	-	-	-	-	-	-	2.98 ± 0.04	3.49 ± 0.06
Spring white onion II	-	-	-	-	-	-	-	-	-	-	4.34 ± 0.02	3.26 ± 0.76
Sweet white onion II	-	-	-	-	-	-	-	-	-	-	2.74 ± 0.00	3.72 ± 0.16
CYO white onion	-	-	-	-	-	-	-	-	-	-	4.84 ± 0.04	6.92 ± 0.12
Sweet white onion III	-	-	-	-	-	-	-	-	-	-	5.40 ± 0.12	4.88 ± 0.98
White onion	-	-	-	-	-	-	-	-	-	-	3.31 ± 0.07	5.63 ± 0.31
Babosa white onion	-	-	-	-	-	-	-	-	-	-	5.78 ± 0.17	5.91 ± 0.10
Sweet white onion IV	-	-	-	-	-	-	-	-	-	-	3.30 ± 0.02	2.93 ± 0.03
Fuentes white onion	-	-	-	-	-	-	-	-	-	-	3.64 ± 0.01	4.07 ± 0.32
Yellow onion I	-	-	-	-	-	-	-	-	-	-	5.41 ± 0.17	5.41 ± 0.34
Yellow onion II	-	-	-	-	-	-	-	-	-	-	3.56 ± 0.03	6.14 ± 0.63
Yellow onion III	-	-	-	-	-	-	-	-	-	-	3.53 ± 0.02	6.15 ± 0.07
Yellow onion IV	-	-	-	-	-	-	-	-	-	-	5.00 ± 0.00	3.93 ± 0.10
Yellow onion V	-	-	-	-	-	-	-	-	-	-	5.61 ± 0.11	3.41 ± 0.16
Purple onion	0.43 ± 0.00	0.19 ± 0.00	0.04 ± 0.00	0.02 ± 0.00	0.04 ± 0.00	0.98 ± 0.02	0.27 ± 0.06	0.09 ± 0.00	0.02 ± 0.00	2.28 ± 0.05	7.97 ± 0.03	5.96 ± 0.01
Red onion I	0.20 ± 0.00	0.19 ± 0.00	0.03 ± 0.00	0.02 ± 0.00	0.03 ± 0.00	0.33 ± 0.01	0.23 ± 0.01	0.09 ± 0.00	0.04 ± 0.00	1.36 ± 0.03	6.70 ± 0.08	5.14 ± 0.16
Red label onion	0.24 ± 0.00	0.20 ± 0.01	0.04 ± 0.00	0.02 ± 0.00	0.03 ± 0.01	0.34 ± 0.00	0.24 ± 0.02	0.12 ± 0.01	0.05 ± 0.00	1.46 ± 0.04	8.01 ± 0.12	6.03 ± 0.01
Red onion II	0.53 ± 0.02	0.27 ± 0.01	0.02 ± 0.00	0.05 ± 0.0	0.06 ± 0.01	0.66 ± 0.07	0.34 ± 0.03	0.17 ± 0.01	0.06 ± 0.00	2.39 ± 0.05	7.23 ± 0.08	5.35 ± 0.00
Red onion III	0.31 ± 0.02	0.16 ± 0.01	0.05 ± 0.00	0.03 ± 0.0	0.04 ± 0.00	0.54 ± 0.03	0.23 ± 0.01	0.22 ± 0.00	0.07 ± 0.00	1.64 ± 0.05	6.52 ± 0.08	7.03 ± 0.67
Red onion IV	0.10 ± 0.01	0.09 ± 0.01	0.04 ± 0.00	0.03 ± 0.00	0.23 ± 0.00	0.07 ± 0.00	0.12 ± 0.00	0.03 ± 0.00	0.02 ± 0.00	0.93 ± 0.01	4.54 ± 0.05	7.25 ± 0.12
Purple onion II	0.10 ± 0.00	0.17 ± 0.00	0.04 ± 0.00	0.03 ± 0.00	0.04 ± 0.00	0.46 ± 0.01	0.38 ± 0.00	0.07 ± 0.00	0.03 ± 0.00	1.54 ± 0.02	4.30 ± 0.00	6.43 ± 0.01
Figueres Onion	0.02 ± 00	0.03 ± 0.0	0.02 ± 0.00	0.01 ± 0.0	0.05 ± 0.02	0.02 ± 0.01	0.06 ± 0.02	0.03 ± 0.01	0.02 ± 0.00	0.59 ± 0.00	7.50 ± 0.29	7.10 ± 0.53

4. Conclusions

Two ultrasound-assisted extraction methods were developed and optimized to extract total phenolic compounds and anthocyanins from red onion samples. A Box–Behnken design was used to optimize the relevant process parameters and the following values were established for the extraction of phenolic compounds: extraction solvent 53% MeOH, pH 2.6, 60 °C temperature, 30.1% amplitude, 0.43 s cycle, and 0.2:11 g onion sample/mL solvent ratio. The optimal values for the extraction of anthocyanins were established as follows: extraction solvent 57% MeOH, pH 2, 60 °C, amplitude of 90%, cycle of 0.64 s, and ratio of 0.2:15 g sample/mL solvent. It was found that most of the studied variables influence the extraction of phenolic compounds and anthocyanins from red onions samples. This highlights how important it is to support any method for the extraction of compounds from natural matrices on an appropriate experimental design. The two methods developed were confirmed to present high repeatability and intermediate precision (RSD < 5%), as well as to require rather short extraction times to achieve good yields. Finally, a multi-response optimization of the two responses, TA and TPC, was carried out, and the resulting UAE method was successfully applied to an assortment of onion varieties. The extraction method was proven to be adequate for the production of extracts from a number of onion varieties with disparate chemical composition. The different extracts were analyzed and high phenolic compounds and anthocyanins contents, as well as good antioxidant activity, were detected.

Supplementary Materials: The following are available online at https://www.mdpi.com/article/10.3390/antiox10111755/s1, Figure S1: Information about anthocyanins identified in red onion. Figure S2: MS spectra and structure of the nine anthocyanins identified in onion bulb: (**a**) cyanidin 3-*O*-glucoside; (**b**) cyanidin 3-O-laminaribioside; (**c**) cyanidin 3-*O*-(3″-malonylglucoside); (**d**) peonidin 3-*O*-glucoside; (**e**) delphinidin 3,5-*O*-diglucoside; (**f**) cyanidin 3-*O*-(6″-malonylglucoside); (**g**) cyanidin 3-*O*-(6″-malonyl-laminaribioside); (**h**) peonidin 3-*O*-malonylglucoside; and (**i**) delphinidin 3-*O*-glucoside. The molecular ion is framed in black and the fragments in different colors. Table S1: Mass spectra information of the nine anthocyanins present in onion bulb.

Author Contributions: Conceptualization, M.F.-G. and G.F.B.; methodology, A.V.G.-d.-P., M.V.-E. and C.C.; software, M.F.-G.; validation, A.V.G.-d.-P. and M.V.-E.; formal analysis, A.V.G.-d.-P., M.V.-E. and C.C.; investigation, A.V.G.-d.-P. and M.V.-E.; resources, M.P. and G.F.B.; data curation, E.E.-B., M.F.-G. and G.F.B.; writing—original draft preparation, A.V.G.-d.-P. and M.V.-E.; writing—review and editing, G.F.B. and E.E.-B.; visualization, G.F.B.; supervision, M.F.-G. and G.F.B.; project administration, G.F.B.; funding acquisition, G.F.B. and M.P. All authors have read and agreed to the published version of the manuscript.

Funding: This work has been supported by the project "EQC2018-005135-P" (Equipment for liquid chromatography using mass spectrometry and ion chromatography), of the State Subprogram of Research Infrastructures and Technical Scientific Equipment. This research was funded by the University of Cadiz, by the INIA (National Institute for Agronomic Research), and FEDER (European Regional Development Fund, within the framework of the Operational Programme under the Investment for Growth 2014–2020) who provided financial support (Project RTA2014-00083-C03-03), and by the Ministry of Science and Innovation of Spain (FPU grant AP-2018-03811 to Ana Velasco González de Peredo).

Institutional Review Board Statement: Not applicable.

Informed Consent Statement: Not applicable.

Data Availability Statement: The data presented in this study are contained within the article or Supplementary Material.

Acknowledgments: The authors express their acknowledgements to the Instituto de Investigación Vitivinícola y Agroalimentaria (IVAGRO) for providing the necessary facilities to carry out the research. A special remark goes to Carmelo García Barroso (in memoriam) for his contribution to the scientific community in the area of phenolic compounds and oenology and his important input to this research.

Conflicts of Interest: The authors declare no conflict of interest.

References

1. Puizina, J.; Javornik, B.; Bohanec, B.; Schweizer, D.; Maluszynska, J.; Papeš, D. Random amplified polymorphic DNA analysis, genome size, and genomic in situ hybridization of triploid viviparous onions. *Genome* **1999**, *42*, 1208–1216. [CrossRef] [PubMed]
2. Fredotovic, Z.; Sprung, M.; Soldo, B.; Ljubenkov, I.; Budic-Leto, I.; Bilusic, T.; Cikes-Culic, V.; Puizina, J. Chemical composition and biological activity of *Allium cepa* L. and *Allium* × *cornutum* (Clementi ex Visiani 1842) methanolic extracts. *Molecules* **2017**, *22*, 448. [CrossRef] [PubMed]
3. Tedesco, I.; Carbone, V.; Spagnuolo, C.; Minasi, P.; Russo, G.L. Identification and quantification of flavonoids from two southern italian cultivars of *Allium cepa* L., Tropea (Red Onion) and Montoro (Copper Onion), and their capacity to protect human erythrocytes from oxidative stress. *J. Agric. Food Chem.* **2015**, *63*, 5229–5238. [CrossRef] [PubMed]
4. Corzo-Martínez, M.; Corzo, N.; Villamiel, M. Biological properties of onions and garlic. *Trends Food Sci. Technol.* **2007**, *18*, 609–625. [CrossRef]
5. Lanzotti, V. The analysis of onion and garlic. *J. Chromatogr. A* **2006**, *1112*, 3–22. [CrossRef]
6. Kendler, B.S. Garlic (*Allium sativum*) and onion (*Allium cepa*): A review of their relationship to cardiovascular disease. *Prev. Med.* **1987**, *16*, 670–685. [CrossRef]
7. Nicastro, H.L.; Ross, S.A.; Milner, J.A. Garlic and onions: Their cancer prevention properties. *Cancer Prev. Res.* **2015**, *8*, 181–189. [CrossRef]
8. Yang, E.J.; Kim, G.S.; Kim, J.; Song, K.S. Protective effects of onion-derived quercetin on glutamate-mediated hippocampal neuronal cell death. *Pharmacogn. Mag.* **2013**, *9*, 302–308. [CrossRef]
9. González-de-Peredo, A.V.; Vázquez-Espinosa, M.; Espada-Bellido, E.; Carrera, C.; Ferreiro-González, M.; Barbero, G.F.; Palma, M. Flavonol composition and antioxidant activity of onions (*Allium cepa* L.) based on the development of new analytical ultrasound-assisted extraction methods. *Antioxidants* **2021**, *10*, 273. [CrossRef]
10. Carrera, C.; Aliaño-González, M.J.; Valaityte, M.; Ferreiro-González, M.; Barbero, G.F.; Palma, M. A Novel Ultrasound-Assisted Extraction Method for the Analysis of Anthocyanins in Potatoes (*Solanum tuberosum* L.). *Antioxidants* **2021**, *10*, 1375. [CrossRef]
11. Li, H.; Deng, Z.; Zhu, H.; Hu, C.; Liu, R.; Young, J.C.; Tsao, R. Highly pigmented vegetables: Anthocyanin compositions and their role in antioxidant activities. *Food Res. Int.* **2021**, *46*, 250–259. [CrossRef]
12. del Pilar Garcia-Mendoza, M.; Espinosa-Pardo, F.A.; Baseggio, A.M.; Barbero, G.F.; Junior, M.R.M.; Rostagno, M.A.; Martínez, J. Extraction of phenolic compounds and anthocyanins from juçara (*Euterpe edulis* Mart.) residues using pressurized liquids and supercritical fluids. *J. Supercrit. Fluids* **2017**, *119*, 9–16. [CrossRef]
13. Ren, F.; Reilly, K.; Kerry, J.P.; Gaffney, M.; Hossain, M.; Rai, D.K. Higher Antioxidant Activity, Total Flavonols, and Specific Quercetin Glucosides in Two Different Onion (*Allium cepa* L.) Varieties Grown under Organic Production: Results from a 6-Year Field Study. *J. Agric. Food Chem.* **2017**, *65*, 5122–5132. [CrossRef]
14. Pérez-Gregorio, R.M.; García-Falcón, M.S.; Simal-Gándara, J.; Rodrigues, A.S.; Almeida, D.P.F. Identification and quantification of flavonoids in traditional cultivars of red and white onions at harvest. *J. Food Compos. Anal.* **2010**, *23*, 592–598. [CrossRef]
15. Wu, X.; Prior, R.L. Identification and characterization of anthocyanins by high-performance liquid chromatography-electrospray ionization-tandem mass spectrometry in common foods in the United States: Vegetables, nuts, and grains. *J. Agric. Food Chem.* **2005**, *53*, 3101–3113. [CrossRef]
16. Gennaro, L.; Leonardi, C.; Esposito, F.; Salucci, M.; Maiani, G.; Quaglia, G.; Fogliano, V. Flavonoid and carbohydrate contents in tropea red onions: Effects of homelike peeling and storage. *J. Agric. Food Chem.* **2002**, *50*, 1904–1910. [CrossRef]
17. Petersson, E.V.; Puerta, A.; Bergquist, J.; Turner, C. Analysis of anthocyanins in red onion using capillary electrophoresis-time of flight-mass spectrometry. *Electrophoresis* **2008**, *29*, 2723–2730. [CrossRef]
18. Liu, J.; Sandahl, M.; Sjöberg, P.J.R.; Turner, C. Pressurised hot water extraction in continuous flow mode for thermolabile compounds: Extraction of polyphenols in red onions. *Anal. Bioanal. Chem.* **2014**, *406*, 441–445. [CrossRef]
19. Chaves, J.O.; De Souza, M.C.; Da Silva, L.C.; Lachos-Perez, D.; Torres-Mayanga, P.C.; da Fonseca Machado, A.P.; Forster-Carneiro, T.; Vázquez-Espinosa, M.; González-de-Peredo, A.V.; Barbero, G.F.; et al. Extraction of Flavonoids From Natural Sources Using Modern Techniques. *Front. Chem.* **2020**, *8*, 507887. [CrossRef]
20. Chen, S.; Zeng, Z.; Hu, N.; Bai, B.; Wang, H.; Suo, Y. Simultaneous optimization of the ultrasound-assisted extraction for phenolic compounds content and antioxidant activity of *Lycium ruthenicum* Murr. fruit using response surface methodology. *Food Chem.* **2018**, *242*, 1–8. [CrossRef]
21. Vázquez-Espinosa, M.; González-de-Peredo, A.V.; Espada-Bellido, E.; Ferreiro-González, M.; Toledo-Domínguez, J.J.; Carrera, C.; Palma, M.; Barbero, G.F. Ultrasound-Assisted Extraction of Two Types of Antioxidant Compounds (TPC and TA) from Black Chokeberry (*Aronia melanocarpa* L.): Optimization of the Individual and Simultaneous Extraction Methods. *Agronomy* **2019**, *9*, 456. [CrossRef]
22. Espada-Bellido, E.; Ferreiro-González, M.; Carrera, C.; Palma, M.; Barroso, C.G.; Barbero, G.F. Optimization of the ultrasound-assisted extraction of anthocyanins and total phenolic compounds in mulberry (*Morus nigra*) pulp. *Food Chem.* **2017**, *219*, 23–32. [CrossRef]

23. V González de Peredo, A.; Vázquez-Espinosa, M.; Espada-Bellido, E.; Ferreiro-González, M.; Amores-Arrocha, A.; Palma, M.; Barbero, G.F.; Jiménez-Cantizano, A. Alternative Ultrasound-Assisted Method for the Extraction of the Bioactive Compounds Present in Myrtle (*Myrtus communis* L.). *Molecules* **2019**, *24*, 882. [CrossRef]
24. Rao, J.S.; Kumar, B. 3D Blade root shape optimization. In Proceedings of the 10th International Conference on Vibrations in Rotating Machinery, London, UK, 11–13 September 2012; Woodhead Publishing: Sawston, UK, 2012; pp. 173–188. [CrossRef]
25. Metrani, R.; Singh, J.; Acharya, P.; Jayaprakasha, G.K.; Patil, B.S. Comparative Metabolomics Profiling of Polyphenols, Nutrients and Antioxidant Activities of Two Red Onion (*Allium cepa* L.) Cultivars. *Plants* **2020**, *9*, 1077. [CrossRef] [PubMed]
26. Frond, A.D.; Luhas, C.I.; Stirbu, I.; Leopold, L.; Socaci, S.; Andreea, S.; Ayvaz, H.; Andreea, S.; Mihai, S.; Diaconeasa, Z.; et al. Phytochemical characterization of five edible purple-reddish vegetables: Anthocyanins, flavonoids, and phenolic acid derivatives. *Molecules* **2019**, *24*, 1536. [CrossRef] [PubMed]
27. Park, M.J.; Ryu, D.H.; Cho, J.Y.; Ha, I.J.; Moon, J.S.; Kang, Y.H. Comparison of the antioxidant properties and flavonols in various parts of Korean red onions by multivariate data analysis. *Hortic. Environ. Biotechnol.* **2018**, *59*, 919–927. [CrossRef]
28. Saptarini, N.M.; Herawati, I.E. Extraction methods and varieties affect total anthocyanins content in acidified extract of papery skin of onion (*Allium cepa* L.). *Drug Invent. Today* **2018**, *10*, 471–474.
29. Zhang, C.; Li, X.; Zhan, Z.; Cao, L.; Zeng, A.; Chang, G.; Liang, Y. Transcriptome Sequencing and Metabolism Analysis Reveals the role of Cyanidin Metabolism in Dark-red Onion (*Allium cepa* L.) Bulbs. *Sci. Rep.* **2018**, *8*, 14109. [CrossRef]
30. Oancea, S.; Drăghici, O. pH and thermal stability of anthocyanin-based optimised extracts of romanian red onion cultivars. *Czech J. Food Sci.* **2013**, *283*–291. [CrossRef]
31. Rodrigues, A.S.; Pérez-Gregorio, M.R.; García-Falcón, M.S.; Simal-Gándara, J.; Almeida, D.P.F. Effect of meteorological conditions on antioxidant flavonoids in Portuguese cultivars of white and red onions. *Food Chem.* **2011**, *124*, 303–308. [CrossRef]
32. Pérez-Gregorio, M.R.R.; García-Falcón, M.S.S.; Simal-Gándara, J. Flavonoids changes in fresh-cut onions during storage in different packaging systems. *Food Chem.* **2011**, *124*, 652–658. [CrossRef]
33. Singleton, V.L.; Orthofer, R.; Lamuela-Raventós, R.M. Analysis of total phenols and other oxidation substrates and antioxidants by means of folin-ciocalteu reagent. *Methods Enzymol.* **1999**, *299*, 152–178. [CrossRef]
34. González-de-Peredo, A.V.; Vázquez-Espinosa, M.; Espada-Bellido, E.; Jiménez-Cantizano, A.; Ferreiro-González, M.; Amores-Arrocha, A.; Palma, M.; Barroso, C.G.; Barbero, G.F. Development of New Analytical Microwave-Assisted Extraction Methods for Bioactive Compounds from Myrtle (*Myrtus communis* L.). *Molecules* **2018**, *23*, 2992. [CrossRef]
35. Brand-Williams, W.; Cuvelier, M.E.; Berset, C. Use of a free radical method to evaluate antioxidant activity. *LWT-Food Sci. Technol.* **1995**, *28*, 25–30. [CrossRef]
36. Miliauskas, G.; Venskutonis, P.R.; Van Beek, T.A. Screening of radical scavenging activity of some medicinal and aromatic plant extracts. *Food Chem.* **2004**, *85*, 231–237. [CrossRef]
37. Razali, M.A.A.N.; Sanusi, N.; Ismail, H.; Othman, N.; Ariffin, A. Application of response surface methodology (RSM) for optimization of cassava starch grafted polyDADMAC synthesis for cationic properties. *Starch Stärke* **2012**, *64*, 935–943. [CrossRef]
38. Ferreira, S.L.C.; Bruns, R.E.; Ferreira, H.S.; Matos, G.D.; David, J.M.; Brandao, G.C.; da Silva, E.G.P. Box-Behnken design: An alternative for the optimization of analytical methods. *Anal. Chim. Acta* **2007**, *597*, 179–186. [CrossRef]
39. Pasquel Reátegui, J.L.; Machado, A.P.D.F.; Barbero, G.F.; Rezende, C.A.; Martínez, J. Extraction of antioxidant compounds from blackberry (*Rubus* sp.) bagasse using supercritical CO2 assisted by ultrasound. *J. Supercrit. Fluids* **2014**, *94*, 223–233. [CrossRef]
40. Carrera, C.; Ruiz-Rodríguez, A.; Palma, M.; Barroso, C.G. Ultrasound assisted extraction of phenolic compounds from grapes. *Anal. Chim. Acta* **2012**, *732*, 100–104. [CrossRef]
41. Pereira, D.T.V.; Tarone, A.G.; Cazarin, C.B.B.; Barbero, G.F.; Martínez, J. Pressurized liquid extraction of bioactive compounds from grape marc. *J. Food Eng.* **2018**, *240*, 105–113. [CrossRef]
42. Nooraziah, A.; Tiagrajah, V.J. A study on regression model using response surface methodology. *Appl. Mech. Mater.* **2014**, *666*, 235–239. [CrossRef]
43. Jadhav, S.B.; Chougule, A.S.; Shah, D.P.; Pereira, C.S.; Jadhav, J.P. Application of response surface methodology for the optimization of textile effluent biodecolorization and its toxicity perspectives using plant toxicity, plasmid nicking assays. *Clean Technol. Environ. Policy* **2015**, *17*, 709–720. [CrossRef]
44. Espada-Bellido, E.; Ferreiro-González, M.; Barbero, G.F.; Carrra, C.; Palma, M.; Barroso, C.G. Alternative Extraction Method of Bioactive Compounds from Mulberry (*Morus nigra* L.) Pulp Using Pressurized-Liquid Extraction. *Food Anal. Methods* **2018**, *11*, 2384–2395. [CrossRef]
45. Mustafa, A.; Turner, C. Pressurized liquid extraction as a green approach in food and herbal plants extraction: A review. *Anal. Chim. Acta* **2011**, *703*, 8–18. [CrossRef] [PubMed]
46. Ivanovic, J.; Tadic, V.; Dimitrijevic, S.; Stamenic, M.; Petrovic, S.; Zizovic, I. Antioxidant properties of the anthocyanin-containing ultrasonic extract from blackberry cultivar "Čačanska Bestrna". *Ind. Crops Prod.* **2014**, *53*, 274–281. [CrossRef]
47. Ju, Z.Y.; Howard, L.R. Effects of Solvent and Temperature on Pressurized Liquid Extraction of Anthocyanins and Total Phenolics from Dried Red Grape Skin. *J. Agric. Food Chem.* **2003**, *51*, 5207–5213. [CrossRef] [PubMed]
48. Association of Official Agricultural Chemists. Peer Verified Methods Advisory Committee. In *AOAC Peer Verified Methods Program*; AOAC International: Gaithersburg, MD, USA, 1998; pp. 1–35.
49. Ko, E.Y.; Nile, S.H.; Jung, Y.S.; Keum, Y.S. Antioxidant and antiplatelet potential of different methanol fractions and flavonols extracted from onion (*Allium cepa* L.). *3 Biotech* **2018**, *8*, 155. [CrossRef]

Article

Potential of Persimmon Dietary Fiber Obtained from Byproducts as Antioxidant, Prebiotic and Modulating Agent of the Intestinal Epithelial Barrier Function

Julio Salazar-Bermeo [1], Bryan Moreno-Chamba [1], María Concepción Martínez-Madrid [2], Domingo Saura [1], Manuel Valero [1,*] and Nuria Martí [1]

[1] Instituto de Investigación, Desarrollo e Innovación en Biotecnología Sanitaria de Elche (IDiBE), Universidad Miguel Hernández de Elche, 03202 Alicante, Spain; julio.salazar@goumh.umh.es (J.S.-B.); bryan.morenoc@umh.es (B.M.-C.); dsaura@umh.es (D.S.); nmarti@umh.es (N.M.)

[2] Departamento de Agroquímica y Medio Ambiente, Universidad Miguel Hernández de Elche, 03312 Alicante, Spain; c.martinez@umh.es

* Correspondence: m.valero@umh.es; Tel.: +34-96-522-2524

Abstract: Appropriate nutrition targets decrease the risk of incidence of preventable diseases in addition to providing physiological benefits. Dietary fiber, despite being available and necessary in balanced nutrition, are consumed at below daily requirements. Food byproducts high in dietary fiber and free and bonded bioactive compounds are often discarded. Herein, persimmon byproducts are presented as an interesting source of fiber and bioactive compounds. The solvent extraction effects of dietary fiber from persimmon byproducts on its techno- and physio-functional properties, and on the Caco-2 cell model after being subjected to in vitro gastrointestinal digestion and probiotic bacterial fermentation, were evaluated. The total, soluble, and insoluble dietary fiber, total phenolic, carotenoid, flavonoid contents, and antioxidant activity were determined. After in vitro digestion, low quantities of bonded phenolic compounds were detected in all fiber fractions. Moreover, total phenolic and carotenoid contents, as well as antioxidant activity, decreased depending on the extraction solvent, whereas short chain fatty acids production increased. Covalently bonded compounds in persimmon fiber mainly consisted of hydroxycinnamic acids and flavanols. After probiotic bacterial fermentation, few phenolic compounds were determined in all fiber fractions. Results suggest that persimmon's dietary fiber functional properties are dependent on the extraction process used, which may promote a strong probiotic response and modulate the epithelial barrier function.

Keywords: *Diospyros kaki*; antioxidant activity; in vitro digestion; probiotic bacterial fermentation; bioactive compounds

1. Introduction

Appropriate nutrition targets decrease the risk of incidence of preventable diseases in addition to providing physiological benefits [1–3]. The development of food products containing physiologically bioactive molecules capable of maintaining and/or improving beneficial long-term effects may contribute to achieve these objectives. For instance, persimmon fruit (*Diospyros kaki* Thunb.), a widespread cultivar in the south of Spain and China, has been found to provide a significant amount of bioactive compounds with physiological benefits [4,5].

The main compounds in persimmon have been reported to be polyphenols (gallic acid, coumaric acid, epicatechin, kaempferol, and ellagic acid), carotenoids (neoxanthin, antheraxanthin, lutein, zeaxanthin, β-carotene, and lycopene), and polysaccharides (pectin, cellulose, hemicellulose) [6–8]. Studies have shown the hypocholesterolemic, hypolipidemic, anti-atherogenic, anti-obesity, antidiabetic, antioxidant and antiviral effects of persimmon fruits and leaves in in vitro and animal models [9–11]. Due to its fast ripening, persimmon fruits are rapidly processed and generate a high amount of byproducts. The abundance of dietary fiber (DF) in fruit byproducts makes them attractive for second generation bio-refining and promoting the valorization of agricultural byproducts that are not part of value chain of the industry [12]. Obtained from byproducts, DF may help target disease prevention and the reduction of risks, such as atherosclerosis, cardiovascular disease, and colorectal cancer [13].

Fiber intake in Western populations reaches fifty percent of the daily recommended value of DF [13]. The beneficial effects of DF are directly influenced by their mechanical properties, known as physio-functional and techno-functional properties. Due to gastrointestinal degradation resistance, DF is determinant for gut microbiota ecology, diversity, and function. Metabolites produced from beneficial gut microbiota (e.g., Firmicutes and Bacteroidetes) consist of short chain fatty acids (SCFAs), such as acetate, propionate, and butyrate. SCFAs are metabolized by epithelial cells and increase the production of anti-inflammatory cytokines, influence cellular metabolism in colonocytes, fibroblasts, and adipocytes [14–16]. Diets low in fermentable substrates result in a thinner mucus layer lining the gut lumen, increasing the susceptibility to the infection of intestinal epithelial cells [17]. To our knowledge, this is the first study that evaluates the extraction effects of persimmon DF after gastrointestinal digestion and fermentation in human cell lines.

The enrichment of fiber content in food matrices throughout untreated fruit byproducts could have adverse effects on the glycemic index of some enriched foods due to the sugar content. Moreover, it could alter expected food sensory profiles. The treatment of byproducts with appropriate solvents may provide a DF with valuable bioactive compounds, while extracting other compounds of interest, such as carotenoids and phenolics [7]. Studies have reported that byproducts still retained a substantial amount of covalently bonded bioactive compounds, and their antioxidant activities, such as radical scavenging activity [18]. These remnants, after gastrointestinal digestion and when available, may be a key point for beneficial and pathogenic bacteria, health, and well-being. However, byproducts generated by food manufacturers may not be appropriate for immediate upcycling without previous treatments. The aim of this research was to evaluate the solvent extraction effects of DF from persimmon byproducts on its functional properties and safety on human epithelial cells after being subjected to in vitro gastrointestinal digestion and beneficial gut bacterial fermentation.

2. Materials and Methods

2.1. Chemicals and Reagents

Ethanol (99.5%), methanol (99.9%), acetone (99.9%), sulfuric acid (96%), petroleum ether (40–60 °C), acetic acid glacial (99.8%), acetonitrile (99.9%), hydrochloric acid (37%) and sodium hydroxide (40%) were obtained from PanReac (Barcelona, Spain). α-Amylase, pepsin, pancreatin, porcine bile extract, electrolytes ($CaCl_2$, KCl, KH_2PO_4, $NaHCO_3$, $MgCl_2$ and $(NH_4)_2CO_3$), Folin Ciocalteu's reagent, crystal violet staining, 3-(4,5-dimethylthiazol-2-yl)-2,5-diphenyltetrazolium bromide, dimethyl sulfoxide, phosphate buffered saline solution and reference reagents for identification of phenolics and SCFAs were purchased from Sigma-Aldrich (Madrid, Spain). Microbial culture media was obtained from Scharlab (Barcelona, Spain), while pure culture probiotic strains were purchased from Spanish Type Culture Collection (CECT) (Valencia, Spain). Cell culture medium and reagents were obtained from Fisher Scientific (Madrid, Spain).

2.2. Plant Material

Diospyros kaki Thumb. from the 'Rojo Brillante' variety were selected based on uniformity from a local market (Alicante, Spain). The fruits were in the orange ripening stage at $15 \pm 2°$ Brix, grouped into batches, washed, disinfected, the stem was separated, and the fruits were cut and processed at pilot scale; the juice was filtered, and byproducts made up by pulp and peels were collected and stored at -18 °C.

2.3. Solvent Assisted Extraction (SAE)

The byproduct was mixed with 70% (v/v) solvent:water solution at a 5:1 (v/w) ratio, then the mixture was heated at 60 °C and stirred for 15 min at 3600 rpm. Based on the solvent applied for the assisted extraction, the fractions obtained were: Persimmon Fiber Aqueous Extraction (PFAE), Persimmon Fiber Acetonic Extraction (PFAC), and Persimmon Fiber Ethanolic Extraction (PFEE). Final mixtures were filtered and freeze-dried before use.

2.4. Physicochemical Analysis

Total DF (TDF), insoluble DF (IDF), soluble DF (SDF), moisture, ash, and crude protein content were determined according to the Association of Official Analytical Chemist official enzymatic-gravimetric method 991.43 [19]. All analyses were carried out in triplicate.

2.5. Techno-Functional Properties

The water absorption activity (WAA) of the DF fractions was measured as described by [20] and expressed as the volume of water held by DF fractions after centrifugation. The water-holding activity (WHA) was expressed as the weight of water held by the weight corresponding to the DF fractions [21]. The swelling activity (SA) of the DF fractions was assessed according to [22] and expressed as milliliters of DF per gram of the DF samples. The oil-holding activity (OHA) of the DF fractions was evaluated and expressed as the weight of oil held by the weight of the DF samples (g/g) [21]. The emulsifying activity (EA) and emulsion stability (ES) were expressed as the volume of emulsion formed by the DF samples and the percentage (%) of emulsified and stable fraction, respectively [23]. The gel formation activity (GFA) was determined according to [24]. The DF solutions were expressed as the minimum percentage (w/v) of DF with GFA.

2.6. Physio-Functional Properties

The bile-holding activity (BHA) of the DF fractions was measured as the weight of porcine bile held by the DF fractions [25], while the fat/oil binding (FOB) capacity was measured as the adsorption capacity of fats on the DF matrix after simulated conditions of digestion. The FOB capacity of each fraction was expressed as grams of oil held by grams of DF (g/g).

2.7. In Vitro Gastrointestinal Digestion

The in vitro gastrointestinal digestion of the three treatments was simulated following the INFOGEST methodology described by [26] adapted for DF matrices. Simulated digestion fluids were prepared and sterilized prior digestion. A sample of 0.5 g of each extracted fraction and a control were subjected to three phases: oral, gastric, and intestinal at 37 °C. The pH, time, and simulated digestion fluids were adjusted for each phase. Afterwards, digested fractions were stored at −80 °C until further use.

2.8. Probiotic Fermentation Process (PFP)

To test the biological potential of extracted fiber and the effects of the digestion process, fermentation was performed before and after the in vitro gastrointestinal digestion on selected beneficial host microorganisms according to the methodologies established by [27–30]. A 10 mL volume of each homogenized fraction and control were centrifuged at $948\times g$, 10 min at 4 °C, and the supernatants and pellets were separated. Then, 100 mg of the pellet and 50 µL of the supernatant were combined and mixed with a 150 µL inoculum of four human host beneficial bacteria strains.

The strains were selected based on their microbiome diversity, health implications, and to test the production of SCFAs without the interference of other metabolites. Bacterial suspensions of *Bifidobacterium bifidum* CECT 870, *Lactobacillus casei* CECT 475, *Lactococcus lactis* subsp. *lactis* CECT 185, and *Streptococcus salivarius* subsp. *thermophilus* CECT 7207 in 5 mL sterilized distilled water at a concentration of 10^7 CFU/mL were used. Homogenized mixtures were incubated at 37 °C for 48 h in aerobiosis (*L. casei* and *S. salivarius*) and anaerobiosis (*B. bifidum* and *Lc. lactis*). Afterwards, fermented samples were centrifuged (pellet and soluble fraction) and stored at −80 °C until further analysis.

2.9. Cell Culture

Human epithelial colorectal adenocarcinoma cell line (Caco-2; American Type Culture Collection, HTB-37) was used in this study as a human intestinal barrier model. The cell line was grown and maintained in Dulbecco's Modified Eagle Medium (DMEM) supplemented with 10% heat-inactivated fetal bovine serum (SBF), 1% penicillin/streptomycin, 1% of nonessential amino acids, and N-2-hydroxyethylpiperazine-N-2-ethane sulfonic acid (HEPES) 1 M solution. Cells were maintained (37 °C and 5% CO_2 atmosphere) between 15–20 passages before assays. At every passage (70–80% confluence), the cells were rinsed with phosphate buffered saline, pH 7.2, supplemented with 1 mM 2,2′,2″,2‴-(Ethane-1,2-diyldinitrilo)tetra-acetic acid (PBS-EDTA solution), trypsinized with 0.25% trypsin, and trypsin-neutralized with new completed DMEM before being diluted.

2.9.1. Cell Viability

After the fermentation of the digested fractions, the obtained product was tested on the Caco-2 cells monolayer. Aliquots of 200 µL of DMEM with 1.5×10^4 Caco-2 cells (15–20 passages) were seeded in 96-well plates. After monolayer formation, plates were incubated (37 °C, 5% CO_2) for one week. The culture media was changed every other day with a new complete medium. An 8-day model was used for the viability assessment. Before the assay, the media was discarded from plates and media containing post digestion and fermentation metabolites was added. 200 µL of pure fermented fractions were added into the first row; then, they were two-fold diluted in the completed medium. The plates were incubated (37 °C, 5% CO_2) for 24 h. Two methods for cell viability were performed.

The crystal violet staining (CVS) assay was used to determine viable adhered cells [31]. The medium was discarded and 100 µL of DMEM with 0.5% CV was added to every well for 20 min at 37 °C. The CV solution was discarded from the plates, rinsed with pure water, and dried at room temperature for 2 h. 100 µL of pure methanol was added to the wells. Optical density (OD) at 590 nm was recorded by the microplate reader Cytation™ 3 Cell Imaging Multi-Mode (BioTek, Winooski, VT, USA).

The (3-(4,5-Dimethylthiazol-2-yl)-2,5-diphenyltetrazolium bromide (MTT) assay was also performed [32]. Culture media from 96-well plates was discarded. Then, 100 µL of DMEM with 5 mg/mL of MTT solution was added. The plates were incubated (37 °C, 5% CO_2) for 3 h. Then, medium was discarded, and plates were dried at room temperature for 2 h. A volume of 100 µL of pure dimethyl sulfoxide (DMSO) was added. The formazan production of viable cells was recorded at 550 nm using the Cytation 3 microplate reader. The percentage of viability of Caco-2 cells was determined comparing the viability of treated with untreated cells.

2.9.2. Trans Epithelial Electrical Resistance (TEER) Response of Caco-2 Cell Monolayer

The effect of fermented products of DF on intestinal epithelial barrier function was also tested using Caco-2 cells [33,34]. 1×10^5 cells (15–20 passages) were seeded into inserts of 0.4 µm pore size in 6-well plate. The culture media (DMEM) was changed every other day throughout 21 days of incubation. A 21-day model was used to study the TEER response during 8 h incubation with the fermented samples. The 21-day Caco-2 monolayers were rinsed twice with Hank's Balanced Salt Solution (HBSS) with 1 M HEPES (pH 7.4).

TEER was measured using a Millicell ERS-2 for 8 h of incubation (37 °C and 54 rpm of shaking). Caco-2 cells were incubated with media from fermentation samples diluted in HBSS at a 1:1 ratio (2 mL in apical chamber and 3 mL in basolateral chamber of different plates). TEER was also measured in a blank (insert with HBSS with no cellular monolayer) and a control (monolayer with no fermented samples added). Monolayers with TEER values above 350 Ω cm^2 were used for the assay. Fermented samples were diluted in HBSS to address their effect in Caco-2 monolayers without the influence of other nutrients. The assay was carried out by triplicate.

2.10. Determination of Biocompounds

2.10.1. Sample Preparation

To hydrolyze bonded bio-compounds from DF fractions, 0.25–0.50 g from each undigested, digested and fermented DF fraction were separately mixed with 5 mL of NaOH (2 M) for 18 h at room temperature, samples were then acidified with HCl (resulting pH <2) and extracted with methanol (80%, v/v) three times. The extracted fractions were filtered through a 0.45 μm filter; vacuum dried and stored at −80 °C until further use. To assess extracted and metabolized compounds after the digestion and fermentation process, samples from the supernatant formed after in vitro digestion and bacterial fermentation were not hydrolyzed, as control.

2.10.2. The Folin-Ciocalteu Reagent Assay

To determine the total phenolic content (TPC) of each fraction the Folin–Ciocalteu's reagent was used [35,36], acknowledging its limitations as a reducing capacity assay [37,38]. 0.125 mL of the sample was mixed with 0.5 mL of distilled water and 0.125 mL of Folin-Ciocalteu reagent. After 6 min in darkness, 1.25 mL of (Na_2CO_3) was added and 1 mL of distilled water. After 1 h, the absorbance values were measured on the Cytation 3 microplate reader at 760 nm and compared to a standard curve of Gallic acid.

2.10.3. Total Carotenoid Content (TCC)

The TCC was measured according to the method described by [39], with modifications. Approximately 5 g or 5 mL of each fraction was homogenized with 5 mL of petroleum ether, 2.5 mL of acetone and 2.5 mL of ethanol. The suspension was stirred for 30 min at 4 °C and centrifuged at $6000\times g$ for 10 min at 4 °C. The supernatants were pooled, and 10 mL of water was added. Absorbance values were measured on the Cytation 3 microplate reader at 450 nm and compared to a standard curve of β-carotene.

2.10.4. Total Flavonoid Content (TFC)

The TFC was measured according to [40]. In total, 1000 μL of a diluted sample (1:20 v/v) was mixed with 1000 μL of aluminum chloride (2%, w/v in methanol) the mixture was allowed to react for 10 min. Absorbance values were measured on the Cytation 3 microplate reader at 368 nm and compared to a standard curve of quercetin.

2.11. Recovery and Bioaccessibility Index of Free and Bonded Compounds

The recovery index and bio-accessibility index were identified for the analysis of the digestion process effect in the bio-compounds content [41]. The recovery index shows the quantity of phenolic compounds and carotenoids available in the fiber matrix after the in vitro digestion, comparing it with the total bio-compounds (free and bonded) present in each undigested fraction measured in the fiber matrix. The bio-accessibility index compares the total amount of bioactive compounds found after the digestion process in the intestinal phase with the amount in the supernatant from the digestion and fermentation process.

2.12. Antioxidant Activity

2.12.1. The 2,2′-Azino-Bis(3-Ethylbenzothiazoline-6-Sulfonic Acid) (ABTS$^{•+}$) Radical Cation-Based Decolorization Assay

The ABTS$^{•+}$ radical scavenging activity was determined as described by [42] with some modifications. The ABTS$^{•+}$ solution (4 mM) was prepared with potassium persulfate (2.45 mM) and diluted to an absorbance of 720 ± 20 at 734 nm 24 h beforehand. The reactions were performed by adding 200 μL of ABTS$^{•+}$ solution to 20 μL of each extract solution. Absorbance values were measured on the Cytation 3 microplate reader at 734 nm after 6 min of incubation at room temperature and compared to a standard curve of Trolox (6-hydroxy-2,5,7,8-tetramethylchroman-2-carboxylic acid).

2.12.2. The 2,2-Diphenyl-1-Picrylhydrazyl (DPPH) Radical-Based Assay

The DPPH free radical scavenging activity was determined as described by [43], with some modifications. A DPPH solution (0.06 mM) in methanol was prepared. The reactions were performed by adding 180 µL of DPPH solution to 20 µL of each extract solution. Absorbance values were measured on the Cytation 3 microplate reader at 515 nm after 20 min of incubation at room temperature and compared to a standard curve of Trolox.

2.13. High Performance Liquid Chromatography Analysis (HPLC-DAD)

Polyphenolic quantification of the most abundant compounds found in undigested, digested and fermented DF fractions was determined by HPLC-DAD. Briefly, A HPLC Agilent (Santa Clara, CA, USA) series 1200 instrument, equipped with a HPLC column Poroshell 120 SB-C18, 2.7 µm, 4.6 × 150 mm was used.

Phenolic compounds were analyzed with a flow rate elution of 0.7 mL/min. The mobile phases used were acetic acid in Milli-Q® water (0.5:99.5, v/v) as solvent A, and acetonitrile as solvent B. The chromatograms were recorded at full range UV/vis spectrum. Quantification was executed by comparing UV absorption spectra and retention times of each compound based on linear curves of authentic standards injected in the same conditions.

SCFAs production from the fermentation process was determined by HPLC-DAD following the methodology described by [44]. A HPLC Agilent series 1100 instrument, equipped with a HPLC column Supelcogel C610H 30 cm × 7.8 mm was used. Organic acids were analyzed, in standard and sample solutions, with a flow rate elution of 0.5 mL/min. The mobile phase used was phosphoric acid in Milli-Q® water (0.1:99.9 v/v). The chromatograms were recorded at 210 nm. Quantification of organic acids was executed by comparing UV absorption spectra and retention times of each compound based on linear curves of authentic standards injected in the same conditions.

2.14. HPLC Coupled to Electro-Spray Ion Trap Mass Spectrometry (HPLC-DAD-ESI-IT-MS^n)

The solvent assisted extracted fractions were analytically characterized by HPLC-DAD-ESI-IT-MS^n. A 1100 HPLC system with a G1315B diode array detector (Agilent, Waldbronn, Germany) coupled on-line to an Esquire 3000+ ion trap mass spectrometer (Bruker Daltonik, Bremen, Germany) with an atmospheric electro spray ionization (ESI-API) source.

MS-parameters were adjusted, mass spectra were recorded in negative polarity mode at a scan range of m/z 50–1100. at a scan rate of 13,000 Th/s (peak width = 0.6 Th, fwhm). Nitrogen was used as both drying and nebulizing gas at a flow rate of 9 L/min and a pressure of 45 psi, respectively. Nebulizer temperature was set at 365 °C, and a potential of −4500 V was applied on the capillary. Collision gas for induced dissociation was helium at a pressure of 4.9×10^{-6} mbar; mass spectra were obtained with an isolation width of 4.0 m/z for precursor ions and a fragmentation amplitude of 1.0 V.

Control of the system and data evaluation was achieved with ChemStation for LC version A.00.03 (Agilent) and Esquire software version 5.1 (Bruker), respectively. Column and HPLC settings were as detailed below. Identification of phenolics was accomplished by comparison of UV-vis absorption spectra, retention times, and mass spectra with those of authentic standards.

When standards were unavailable, pigments were tentatively identified by comparing their UV-vis absorption spectra and mass spectral behavior with in-lab spectral library, data published previously and databases available [8,45].

2.15. Statistical Analysis

All experiments were carried out in triplicate and the results were expressed as mean values ± standard error (SE). Data obtained for each test was analyzed by means of a one-way ANOVA test. Tukey's and Dunnett's post hoc tests were applied for comparisons of means; differences were considered significant at $p < 0.05$. Statistical analyses were carried out using the statistical package GraphPad Prism 8.0.2. Correlation analysis was performed between physicochemical, techno-functional, and physio-functional properties using Pearson correlation analysis.

3. Results

3.1. Physicochemical Analysis

Physicochemical parameters studied in all treatments (Table 1) showed discreet differences influenced by the solvent applied ($p < 0.05$). In the case of protein content, all fractions analyzed presented low quantities with values ranging between 0.0002 to 0.0021 g/g of sample, which was probably related to the SAE treatment and the low nitrogen content in persimmon fruits.

Table 1. Solvent effects on the physicochemical characteristics of extracted dietary fiber from persimmon byproduct.

Treatment	PFAE	PFEE	PFAC
Protein	0.0006 ± 0.0001 [a]	0.0014 ± 0.0000 [a, b]	0.0021 ± 0.0001 [b]
Ashes	0.08 ± 0.00 [a]	0.16 ± 0.00 [b]	0.04 ± 0.00 [a]
pH	5.50 ± 0.02 [a]	6.72 ± 0.01 [b]	6.83 ± 0.02 [c]
TSS *	1.17 ± 0.29 [a]	20.33 ± 0.29 [b]	21.17 ± 0.29 [b]
IDF	0.62 ± 0.13 [a]	0.68 ± 0.05 [a]	0.64 ± 0.08 [a]
SDF	0.19 ± 0.02 [a]	0.14 ± 0.03 [a]	0.30 ± 0.16 [b]
TDF	0.82 ± 0.11 [a]	0.81 ± 0.08 [a]	0.94 ± 0.08 [b]

Values expressed as g/g of sample. Significant differences in physicochemical characteristics were determined in persimmon fiber (PF) obtained by aqueous extraction (PFAE), ethanolic extraction (PFEE) and acetonic extraction (PFAC) ($p < 0.05$, ANOVA. Different letters near values in the same row indicate significative differences according to Tukey's post hoc test). TSS *, total soluble solids (°Brix); IDF insoluble dietary fiber; SDF soluble dietary fiber, TDF, total dietary fiber.

PFAC fractions (Figure 1A) showed the highest TDF content (0.94 ± 0.08), while no significant differences were observed between PFAE and PFEE ($p < 0.05$). The IDF content of all treatments was not affected by the solvent applied during extraction, as a result, no significant differences in IDF content were observed ($p < 0.05$). Treatments exhibited a higher IDF than SDF. On the other hand, SDF content was significantly affected ($p < 0.05$) where PFAC showed the highest yield in soluble polysaccharides (0.30 ± 0.16). SDF/IDF ratios for PFAE, PFEE, PFAC were 1:4, 1:6, and 1:3, respectively.

3.2. Techno-Functional Properties

Figure 1B,C show the results obtained for hydration (WAA and SA) and holding properties (WHA and OHA) from each treatment. The PFAC fraction showed the highest WAA, SA, WHA and OHA. Absorption, holding, and swelling capacities varied significantly, were influenced by the treatment ($p < 0.05$), and strongly related to the TDF, SDF and IDF content. WHA showed a positive correlation with TDF ($r = 0.93$) and SDF ($r = 0.94$), as well as WAA with SDF ($r = 0.77$) and TDF ($r = 0.99$), similarly to SA with SDF ($r = 0.95$) and TDF ($r = 0.92$). As for OHA, no significant differences were observed between treatments ($p < 0.05$). However, PFAC showed the highest value with a modest ($r = 0.73$) correlation between OHA and IDF.

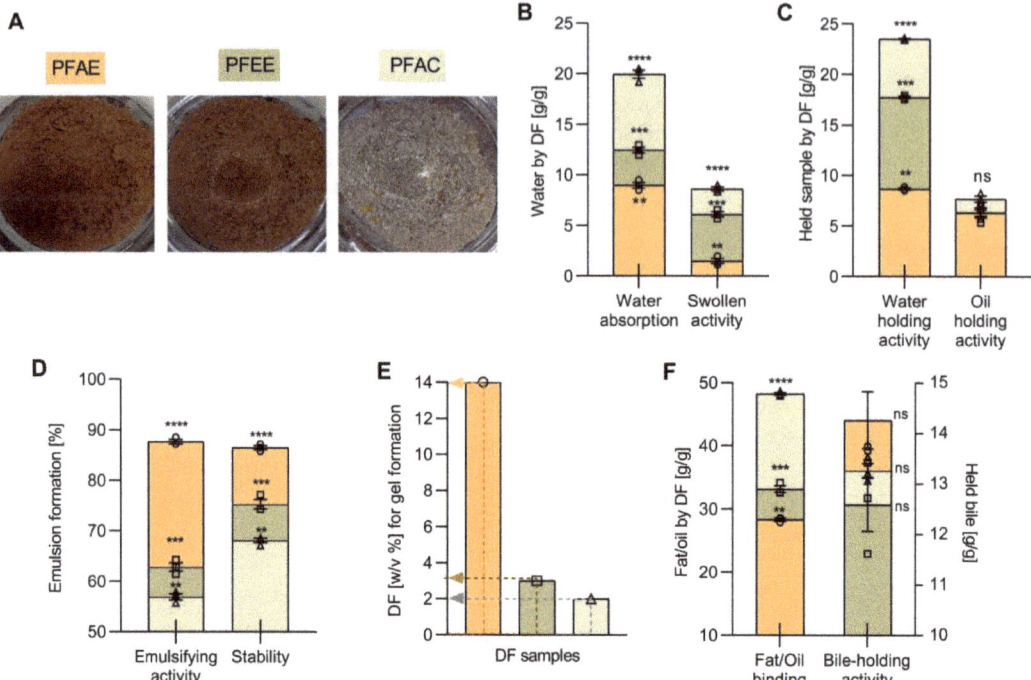

Figure 1. (**A**) Dietary fiber (DF) obtained from persimmon byproduct by aqueous extraction (PFAE, ○), ethanolic extraction (PFEE, □) and acetonic extraction (PFAC, △). Solvent assisted extraction (SAE) effect on techno-functional properties of persimmon fiber (PF); (**B**) hydration properties, (**C**) holding properties; (**D**) emulsifying properties, and (**E**) gel formation activity (GFA). (**F**) Solvent effect on physio-functional properties of PF (**** $p < 0.0001$, *** $p < 0.001$, ** $p < 0.01$, ns $p > 0.05$, ANOVA with Tukey's post hoc test).

Regarding emulsifying properties (Figure 1D), significant differences were obtained ($p < 0.05$), where PFAE showed the highest values for both emulsifying properties. The EA and ES were correlated with both SDF ($r = 0.99$), ($r = 0.93$) respectively, and TDF ($r = 0.83$), ($r = 0.94$) respectively; the decrease of these values influenced a lower percentage in PFAC and PFEE. Regarding the GFA (Figure 1E), the lowest amount of DF necessary to form a gel was recorded in PFAC which also reported the highest difference ($p < 0.05$); GFA was strongly related with the presence of SDF ($r = 0.99$).

3.3. Physio-Functional Properties

BHA (Figure 1F) was discreetly influenced by the treatments ($p < 0.05$), the highest values for PFAE and an inverse correlation with SDF ($r = -0.90$) were denoted. In addition, the FOB of extracted fiber fractions, which is an essential parameter in the characterization of functional DF, showcased significant differences observed among treatments ($p < 0.05$); PFAC showed the highest FOB value ($p < 0.05$) and a strong relation with the TDF content ($r = 0.99$).

3.4. Recovery and Bioaccessibility of Free and Bonded Biocompounds

Regarding to the recovery of phenolics, all fractions showed statistical differences after the digestion process; whereas PFAE (Figure 2A) and PFAC (Figure 2C) showed a similar reduction in their indexes, PFEE (Figure 2B) had a significant decrease on the TPC ($p < 0.05$). After the PFP, all fractions showed differences in the TPC recovery index. PFAE remained higher than the other fractions while PFEE and PFAC had a significant reduction after fermentation ($p < 0.05$) of the digested fraction. The bio-accessibility indexes of bonded phenolic compounds were relatively low (<10%) for all fractions.

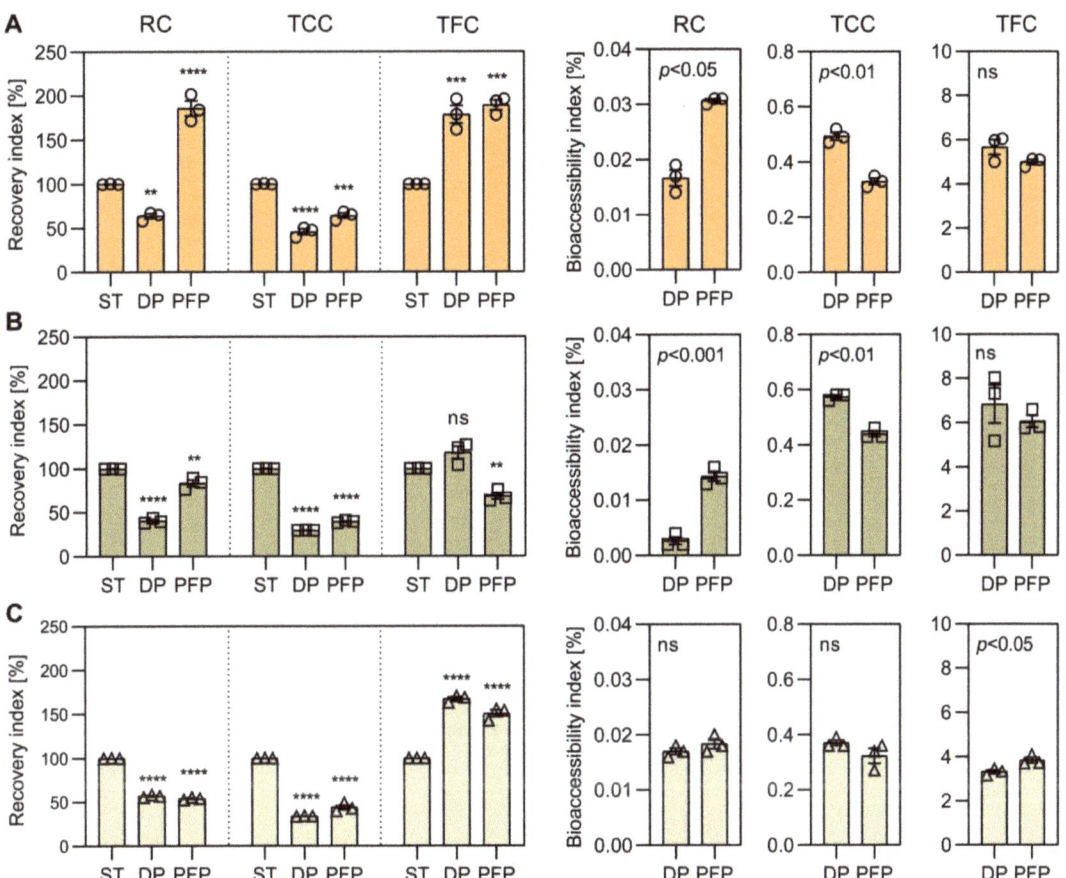

Figure 2. Recovery and bio-accessibility indexes of bonded total phenolic (TPC), total carotenoid (TCC) and total flavonoid content (TFC) in persimmon fiber (PF) after solvent treatment (ST), in vitro digestion (DP) and probiotic fermentation (PFP) processes. Recovery and bio-accessibility of bonded bio-compounds in PF obtained by (**A**) aqueous extraction (PFAE, ○), (**B**) ethanolic extraction (PFEE, □), and (**C**) acetonic extraction (PFAC, △) (recovery: **** $p < 0.0001$, *** $p < 0.001$, ** $p < 0.01$, ns $p > 0.05$, ANOVA with Dunnett's post hoc test; bio-accessibility: $p < 0.5$, ANOVA with t-test post hoc).

PFAE showed the highest bio-accessibility only after the fermentation process ($p < 0.05$). Of the phenolic group, flavonoid recovery was high in PFAE and PFAC after digestion but in lower concentrations than other phenolic compounds. Their indexes decreased significantly after fermentation in PFAC and PFEE fractions ($p < 0.05$).

The recovery and bio-accessibility of TCC and TFC showed a similar behavior than that observed for TPC after digestion (Figure 2). Fractions decreased more than 50% their values and PFAE fraction showed the highest index ($p < 0.05$). Both PFAE and PFAC fractions increased their recovery index over the digested fraction in the fermented fraction; both PFEE and PFAC showed a similar behavior and indexes ($p < 0.05$) after digestion and fermentation.

3.5. Antioxidant Activity

As regards to the antioxidant activity (Table 2) provided by the bio-compounds presented in the fiber matrix, PFAE fraction from undigested fiber showed the highest activity ($p < 0.05$) followed by the PFAC and PFEE highlighting the dependence on the solvent applied for the extraction process. However, the pellet formed after digestion showed a lower antioxidant activity than the soluble fraction, which displays a complex interaction between the extracted supernatant and the fiber matrix that formed the pellet.

Table 2. Effect of in vitro gastrointestinal digestion and probiotic fermentation processes in the antioxidant activity of persimmon fiber by colorimetric ABTS$^{\bullet+}$ and DPPH radical scavenging assays.

Samples	Treatments	ABTS$^{\bullet+}$ mg Trolox/g Sample	DPPH mg Trolox/g Sample
Extracted fiber	PFAE	2.81 ± 0.34 [a]	2.12 ± 0.13 [a]
	PFEE	2.71 ± 0.09 [b]	2.04 ± 0.09 [b]
	PFAC	2.71 ± 0.05 [b]	2.04 ± 0.04 [b]
Digested fiber	PFAE	1.21 ± 0.11 [f]	0.91 ± 0.11 [f]
	PFEE	1.43 ± 0.30 [e]	1.08 ± 0.09 [e]
	PFAC	1.05 ± 0.09 [g]	0.79 ± 0.19 [g]
Supernatant digested fiber	PFAE	1.68 ± 0.11 [d]	1.26 ± 0.11 [d]
	PFEE	1.68 ± 0.09 [c,d]	1.26 ± 0.09 [c,d]
	PFAC	1.68 ± 0.05 [c,d]	1.26 ± 0.05 [c,d]
Fermented fiber	PFAE	0.73 ± 0.12 [h]	0.55 ± 0.25 [h]
	PFEE	0.67 ± 0.32 [i]	0.51 ± 0.12 [i]
	PFAC	0.61 ± 0.13 [j]	0.46 ± 0.15 [j]
Supernatant fermented fiber	PFAE	1.68 ± 0.10 [c,d]	1.26 ± 0.10 [c,d]
	PFEE	1.68 ± 0.13 [c,d]	1.26 ± 0.13 [c,d]
	PFAC	1.68 ± 0.22 [c]	1.26 ± 0.22 [c]

The antioxidant activity of extracted persimmon fiber by aqueous extraction (PFAE), ethanolic extraction (PFEE) and acetonic extraction (PFAC) decreased after digestion and fermentation of persimmon fiber ($p < 0.05$, ANOVA. Different letters in the same column indicate significant differences among samples according to Tukey's post hoc test).

3.6. Phenolic Profile

In total, 23 phenolic compounds were identified (Table 3). Among the most abundant hydroxycinnamic acids, hydroxybenzoic acids, flavonols, flavanols, flavones, tannins, and stilbenes. Fiber bounded phenolic compounds found in persimmon belonged to the group of hydroxybenzoic acid derivatives. Retention time, MS, and UV-vis characteristics of compounds peaks were similar to those of the authentic standards and agreed with our previous results and reported elsewhere in persimmon [8,45,46]. Few differences in the phenolic profile were observed from treatments, while a greater profile of fiber bonded compounds was observed.

From the hydroxybenzoic acids subgroup, compound No. **1** displayed a fragmentation pattern at m/z 126 and eluted at 7.7 min, presented absorbance maxima at 280 nm in accordance with literature, and was reported as Gallic acid. This compound was reported in all PFAE, PFAC, and PFEE fractions before and after digestion and fermentation; however, it was not observed in high intensity in the supernatant after the digestion process. Compounds **3** and **5** were ellagic acid and salicylic acid at m/z 303 and 301, which corresponded with authentic standards and were observed in all fractions before the fermentation process.

Table 3. Identified compounds in treated persimmon fiber fractions by HPLC-DAD-ESI-IT- MS^n.

No.	RT (Min)	HPLC-DAD UV–vis Spectrum λmax (nm)	[M−H]− m/z	HPLC-DAD-ESI-IT-MS^n Experiments m/z	Compound Identity	Molecular Formula	Extracts Present
1	7.7	272/280	126.1 (100) 170.5 (57)	MS^2 [170.5] 126.1 (100) 168.4 (31) 124.1 (15.4)	Gallic acid *	$C_7H_6O_5$	U1, U2, U3, DP1, DP2, DP3, FP1, FP2, FP3
2	16.0	305/272	164.5 (100)	MS^2 [164.5] 120.1 (100)	p-Coumaric acid *	$C_9H_8O_3$	U1, U2, U3, DP1, DP2, DP3, DS3,
3	16.1	260/272	303.7 (100)	MS^2 [303.7] 301.6 (100) 259.2 (60) 186.5 (16.9)	Ellagic acid *	$C_{14}H_6O_8$	U1, U2, U3, DP1 DP2, DP3, DS1, DS2, DS3
4	16.6	320/272	209.8 (100) 225.0 (45)	MS^2 [209.8] 165.4 (100) 150.3 (60) 164.4 (22)	3,5-Dimethoxy-4-hydroxycinnamic acid *	$C_{11}H_{12}O_5$	U1, U2, U3, DP1, DP2, DP3, DS1, DS2, DS3
5	18.7	305/280	301.6 (100) 241.1 (94) 138.2 (76)	MS^2 [301.6] 138.2 (100) 139.2 (10) 94.0 (4)	Salicylic acid *	$C_7H_6O_3$	U1, U2, U3, DP1, DP2, DP3, DS1, DS2, DS3
6	20.8	270/332	514.3 (100)	MS2 [514.3] 170.9 (100) 342.1 (31)	3,5-Dicaffeoylquinic acid **	$C_{25}H_{24}O_{12}$	DP1, DP2, DP3
7	20.9	270/332	1021.3 (100) 948.1 (42) 194.7 (42)	MS^2 [948.6] 888.5 (100) 909.7 (47) 930.1 (42)	Spinacetin 3-O-(2″-p-coumaroylglucosyl)(1->6)-[apiosyl (1->2)]-glucoside **	$C_{43}H_{48}O_{24}$	U1, U2, U3, DP1, DP2, DP3, DS1, DS2, DS3, FP1, FP2, FP3
8	21.6	270/332	498.3 (100)	MS^2 [498.6] 350.6 (100) 412.9 (94)	6″-O-Malonyldaidzin **	$C_{24}H_{22}O_{12}$	DP1, DP2, DP3, DS1, DS2, DS3, FP1, FP2, FP3
9	22.4	270/332	331.2 (100)	MS^2 [331.3] 228.9 (100) 293 (96) 210.8 (90)	Galloyl-hexoside I **	$C_{13}H_{16}O_{10}$	DP1, DP2, DP3, FP1, FP2, FP3
10	22.6	270/332	334.4 (100)	MS^2 [334.4] 316.1 (100) 172.6 (22) 332.2 (16) 287.7 (2)	Galloyl-hexoside II **	$C_{13}H_{16}O_{10}$	U1, U2, U3, DP1, DP2, DP3, DS1, DS2, DS3, FP1, FP2, FP3
11	23.2	270/332	448.3 (100)	MS^2 [448.3] 384.1 (100) 402.2 (54)	Cyanidin 3-O-galactoside **	$C_{21}H_{21}O_{11}$	DP1, DP2, DP3, DS1, DS2, FP1, FP2, FP3
12	23.5	270/332	289.8 (100)	MS^2 [289.8] 271.6 (100) 142.2 (22) 130.2 (7)	Epicatechin *	$C_{15}H_{14}O_6$	U1, U2, U3, DP1, DP2, DP3, DS1, DS2, DS3, FP1, FP2, FP3
13	24.2	280/320	446.3 (100)	MS^2 [446.1] 249.2 (100)	Kaempferol-7-glucoside **	$C_{21}H_{19}O_{11}$	DP1, DP2, DP3, DS1, DS2, DS3, FP1

Table 3. Cont.

No.	RT (Min)	HPLC-DAD UV−vis Spectrum λmax (nm)	[M−H]− m/z	HPLC-DAD-ESI-IT-MS^n Experiments m/z	Compound Identity	Molecular Formula	Extracts Present
14	25.2	280/320	783.5 (100) 391.3 (27.7)	MS^2 [391.3] 343.1 (100) 170.9 (56)	Quercetin glucoside I **	$C_{33}H_{40}O_{21}$	DP1, DP2, DP3, FP1, FP2, FP3
15	25.9	272/280	389.5 (100)	MS^2 [389.6] 339 (100) 342.2 (90) 297.2 (43)	Resveratrol glucoside I **	$C_{20}H_{22}O_8$	DP1, DP2, FP1, FP1, FP2, FP3
16	26.1	320/360	316.2 (100)	MS^2 [316.5] 172.6 (100) 297.9 (75) 128.3 (7)	Methoxyluteolin **	$C_{16}H_{12}O_7$	U1, U2, U3, DP1, DP2, DP3, DS1, DS2, DS3, FP1, FP2, FP3
17	26.4	272/280	389.2 (100)	MS^2 [389.2] 369.1 (100) 352.9 (56) 296.1 (38) 343.1 (21)	Resveratrol glucoside II **	$C_{20}H_{22}O_8$	DP1, DP2, DP3, DS1, DS2, U3, FP1
18	27.2	280/320	783.6 (100) 391.3 (30)	MS^2 [391.3] 218.9 (100) 357.1 (92)	Quercetin glucoside II **	$C_{33}H_{40}O_{21}$	DP1, DP2, DP3
19	27.6	280/320	754.2 (100) 718 (95)	MS^2 [754.2] 718.4 (100) MS^3 [718.4] 661.4 (100)	Kaempferol 3-O-glucosyl-rhamnosyl-galactoside **	$C_{21}H_{20}O_{11}$	U1, U2, U3, DP1, DP2, DP3, DS1, DS2, DS3, FP1, FP2, FP3
20	28.5	280/320	267.6 (100)	MS^2 [267.6] 98.0 (100) 297.9 (75) 128.3 (7)	7-Hydroxy-4'-methoxyisoflavone **	$C_{16}H_{12}O_4$	U1, U2, U3, FP1, FP2
21	29.0	280/320	312.1 (100)	MS^2 [312.1] 98.0 (100) 310.0 (18) 124.1 (17)	5,4'-Dihydroxy-6,7-dimethoxyflavone **	$C_{17}H_{14}O_6$	U1, U2, U3, DP1, DS1, DS2, DS3, FP1, FP2, FP3
22	29.5	320/360	355.2 (100)	MS^2 [355.2] 265 (100) 291 (56) 234.8 (29)	Ferulic acid glucoside **	$C_{16}H_{20}O_9$	DP1, DP2, DP3
23	30.2	320/360	297.2 (100)	MS^2 [297.4] 277.1 (100) 234.5 (30)	p-Coumaroyl tartaric acid **	$C_{13}H_{12}O_8$	DP1, DP2, FP1

U: Undigested, D: Digested, F: Fermented, P: Pellet, S: Supernatant, 1: Persimmon Fiber Aqueous Extraction (PFAE), 2: Persimmon Fiber Ethanolic Extraction (PFEE), 3: Persimmon Fiber Acetonic Extraction (PFAC). * Authentic standards. ** Tentatively identified.

From the hydroxamic acid group, compounds **2, 4, 6, 22** and **23** corresponded to p-coumaric acid at m/z 164, sinapic acid at m/z 209, dicaffeoylquinic acid at m/z 514, ferulic acid glucoside at m/z 355, and p-coumaroyl tartaric acid at m/z 297. The product ions observed in the ESI(-)-MS2 experiment were in agreement with previous findings [8,47,48]. p-Coumaric acid was observed as bonded in the fractions before and after digestion and extractable in the PFAC fraction after the digestion process. Likewise, synaptic acid was also observed in the extractable fraction in the PFAE and PFEE, whereas dicaffeoylquinic acid and ferulic acid glucoside were only observed to be bonded and was detected in low intensity after the digestion process. Similarly, p-coumaroyl tartaric acid was found bonded to the fiber matrix and only detectable in hydrolyzed, digested, and fermented samples.

Flavonols detected at peaks 7, 13, 14, 18, and 19 with retention times of 20.9, 24.2, 25.2, 27.2, and 27.6 min were identified according to their main ions and fragmentation patterns. Spinacetin 3-O-(2″-p-coumaroylglucosyl) (1->6)-[apiosyl(1->2)]-glucoside at m/z 1021, was found in all fractions (PFAE, PFAC, PFEE) during the digestion and fermentation process;

kaempferol-7-glucoside at m/z 446, was detected bonded and released to the supernatant after digestion (PFAE, PFAC, PFEE) and fermentation process (PFAE). Quercetin glucoside isomers were identified at m/z 783 and found bonded and released to the supernatant after the digestion fermentation process in all fractions. Kaempferol 3-O-glucosyl-rhamnosyl-galactoside at m/z 754 was detected free and bonded in all extracted fractions.

Isoflavonoids, compounds **8** and **20** eluted at RT 21.6- and 28.5-min. Compound **8** at m/z 498 was identified as 6''-O-malonyldaidzin which was released after the digestion and fermentation process; whereas compound **20** at m/z 267 was identified as 7-hydroxy-4'-methoxyisoflavone and found bonded before the digestion process and released after the fermentation process in all PFAE, PFAC and PFEE fractions.

Regarding tannins, two galloyl-hexosides were found at 22.4 and 22.6 min at m/z 331 and 334, respectively; the first, compound **9**, was found bonded in all fractions after digestion and released after the fermentation process, while compound **10** was found free and bonded before and after the digestion process in all fractions. Also, stilbene isomers were detected at peaks 15 and 17 and eluted at 25,9 and 26.4 min respectively with m/z values of 389 which corresponded to resveratrol glucosides which were found bonded to the fiber and released to the supernatant after the digestion and fermentation process.

Flavones were also found bonded and released after the digestion and fermentation process in all fractions. Peak No. 16 which eluted at 26.1 min at m/z 316 was tentatively identified as 6-methoxyluteolin; peak No. 21 which eluted at 29.0 min at m/z 312 was identified as 5,4'-dihydroxy-6,7-dimethoxyflavone. Compounds **11** and **12** eluted at 23.2 and 23.5 min and were identified as cyanidin 3-O-galactoside at m/z 448 and epicatechin at m/z 290. While the flavanol, epicatechin, was detected in all fractions free and bonded, the anthocyanin cyanidin 3-O-galactoside was found bonded after digestion and released after the fermentation process in all PFAE, PFAC, and PFEE fractions.

3.7. Polyphenolic Quantification

The most abundant phenolic compounds found in extracted, digested and fermented DF fractions were quantified by HPLC-DAD (Limit of quantification, LOQ: 5 µg/mL) (Table 4).

After gastric digestion, a significant decrease ($p < 0.05$) was observed; the results varied from treatments and phenolic compounds. As a result of the alkaline hydrolysis, the most abundant phenolic bounded to the fiber matrix was gallic acid and its highest yield observed before digestion 11,471.5 ± 260 mg/100 g in PFAC and 9131.3 ± 250 mg/100 g in PFEE and after digestion 8042 ± 424 mg/100 g in PFAE; these values were higher than the reported for the Ichida-gaki variety [46].

Phenolic compounds quantified displayed a higher concentration in the undigested fraction, unlike ellagic acid which increased after digestion and the fermentation process in the PFAC fraction, this derived hydroxybenzoic acid was detected in higher concentrations due to the probiotic bacterial fermentation process and the hydrolysis process; previous studies have reported significant increase in yields of ellagic acid because of the solvent effect [47]. *p*-Coumaric acid was also higher in the PFAC fraction, before, after digestion and fermentation. Salicylic acid was found in all samples and fractions and showed a higher concentration in the PFAC.

From the compounds identified by HPLC-DAD-ESI-IT-MS^n, gallic acid made up to 92% of the total amount of phenolics found in fiber fractions. On the other hand, and in accordance with the recovery and bio-accessibility results, low concentration of bonded phenolics was found to have been released to the supernatant, as a result, gallic acid was quantified after the fermentation process in PFEE and PFAC.

Table 4. Quantification of the most abundant polyphenolic compounds found in treated persimmon fiber fractions.

Samples	Treatment	Phenolic Compounds (mg/g or mL Sample)				
		Gallic Acid	Sinapic Acid	p-Coumaric Acid	Salicylic Acid	Ellagic Acid
Extracted fiber	PFAE	58.63 ± 0.46 [c]	0.69 ± 0.04 [b]	0.57 ± 0.01 [d]	2.19 ± 0.18 [b]	1.26 ± 0.01 [c]
	PFEE	91.31 ± 2.45 [b]	0.69 ± 0.04 [b]	0.14 ± 0.01 [d]	2.18 ± 0.15 [b]	1.89 ± 0.19 [b]
	PFAC	114.72 ± 2.60 [a]	0.77 ± 0.08 [a]	3.73 ± 0.22 [b]	2.24 ± 0.09 [b]	2.05 ± 0.08 [b]
Digested fiber	PFAE	80.42 ± 4.24 [b]	0.46 ± 0.00 [d]	1.28 ± 0.08 [c]	3.57 ± 0.18 [a]	1.03 ± 0.05 [d]
	PFEE	64.94 ± 3.32 [c]	0.34 ± 0.03 [f]	0.08 ± 0.00 [e]	1.16 ± 0.09 [e]	1.90 ± 0.11 [b,c]
	PFAC	104.20 ± 9.38 [a]	0.67 ± 0.03 [b]	3.57 ± 0.18 [b]	1.35 ± 0.11 [e]	4.22 ± 0.13 [a]
Supernatant digested fiber	PFAE	<0.001	<0.001	<0.001	<0.001	<0.001
	PFEE	<0.001	<0.001	<0.001	<0.001	<0.001
	PFAC	<0.001	<0.001	<0.001	<0.001	<0.001
Fermented fiber	PFAE	33.06 ± 2.64 [d]	0.51 ± 0.01 [d]	<0.001	1.87 ± 0.15 [c]	1.40 ± 0.06 [c]
	PFEE	50.04 ± 0.97 [c]	0.47 ± 0.04 [d]	<0.001	1.71 ± 0.14 [d]	2.01 ± 0.04 [b]
	PFAC	57.01 ± 3.04 [c]	0.57 ± 0.01 [c]	4.56 ± 0.09 [a]	2.20 ± 0.11 [b]	4.59 ± 0.37 [a]
Supernatant fermented fiber	PFAE	<0.001	<0.001	<0.001	<0.001	<0.001
	PFEE	0.006 ± 0.00 [e]	<0.001	<0.001	<0.001	<0.001
	PFAC	0.004 ± 0.00 [e]	<0.001	<0.001	<0.001	<0.001

Phenolic compounds present in persimmon fiber by aqueous extraction (PFAE), ethanolic extraction (PFEE) and acetonic extraction (PFAC) decreased significantly in the fermented samples. Gallic acid was the most abundant phenolic in samples, especially PFAC ($p < 0.05$, ANOVA. Different letters in the same column indicate significant differences among samples according to Tukey's post hoc test).

3.8. SCFA Profile

SCFAs produced after PFP were quantified in all fractions, before and after the digestion process, and shown in Table 5. The digestion process increased the production of acetic acid; the PFAE fraction presented a higher production of acetic acid before digestion while the PFAC fraction showed a higher concentration of acetic acid when fermenting the fiber matrix after digestion ($p < 0.05$). Propionic acid was produced in lower concentrations before digestion, once digested, it increased its amount significantly ($p < 0.05$); fractions did not show differences in the fermentation production of propionic acid after digestion. Butyric acid was the highest SCFA produced by the probiotic population where the PFAE fraction and the digested fractions showed to be an optimal matrix to produce this compound and allowed the highest yield ($p < 0.05$).

Table 5. Short chain fatty acid (SCFA) profile of probiotic fermented supernatants from treated persimmon fiber fractions.

SCFA (mg/L)	Treatment	Undigested Fraction	Digested Fraction
Acetic acid	PFAE	0.65 ± 0.01 [a,b]	0.67 ± 0.01 [a,b]
	PFEE	0.65 ± 0.00 [a]	0.78 ± 0.01 [c]
	PFAC	0.68 ± 0.01 [b]	0.76 ± 0.01 [c]
Propionic acid	PFAE	0.09 ± 0,00 [a]	0.23 ± 0.01 [b]
	PFEE	n.d.	0.21 ± 0.01 [b]
	PFAC	0.20 ± 0.02 [b]	0.20 ± 0.01 [b]
Butyric acid	PFAE	0.95 ± 0.01 [a]	1.29 ± 0.02 [d]
	PFEE	1.03 ± 0.00 [b]	1.10 ± 0.00 [d]
	PFAC	1.06 ± 0.00 [c]	1.01 ± 0.00 [b]

Significant differences were determined among persimmon fiber samples treated by aqueous extraction (PFAE), ethanolic extraction (PFEE) and acetonic extraction (PFAC). In vitro gastrointestinal digestion process stimulated significant differences in propionic and butyric acids ($p < 0.05$, ANOVA. Different letters near values indicate significant differences among all samples according to Tukey's post hoc test).

3.9. Cytotoxicity Assays

The viability results of Caco-2 cells in interaction with probiotic fermented supernatants (PFSn) were displayed in Figure 3A. PFAE, PFEE and PFAC fractions were applied; these fractions contained 2.19 ± 0.53, 2.09 ± 0.45 and 1.97 ± 0.41 mg/mL of total SCFAs in the supernatants PFAE, PFEE and PFAC respectively, and were diluted at 50, 25, 12.5, 6.25 and 3.13% of their initial concentration in DMEM.

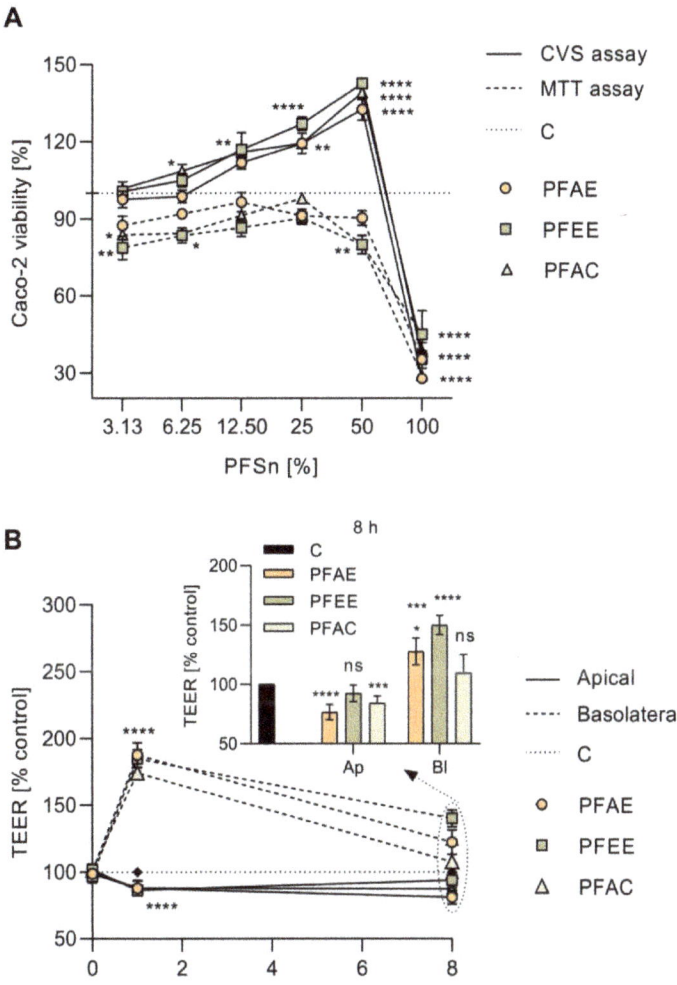

Figure 3. Effect of probiotic fermented supernatants (PFSn) of persimmon fiber (PF) obtained from aqueous extraction (PFAE), ethanolic extraction (PFEE) and acetonic extraction (PFAC) in Caco-2 cells. (**A**) Viability of Caco-2 cells exposed to different concentrations of PFSn of PFAE, PFEE and PFAC by crystal violet staining (CVS) and 3-(4,5-dimethylthiazol-2-yl)-2,5-diphenyltetrazolium bromide (MTT) assays. (**B**) Trans epithelial electronic resistance (TEER) response of Caco-2 cells measured from apical-basolateral (Ap) and basolateral-apical (Bl) directions during 8 h of incubation with PFSn in comparison to untreated monolayers (C) (**** $p < 0.0001$, *** $p < 0.001$, ** $p < 0.01$, * $p < 0.05$, ns $p > 0.05$, ANOVA with Dunnett's post hoc test).

After incubation, the MTT assay showed that the fermentation supernatant fractions did not exhibit cytotoxic effect in Caco-2 cells exposed from 6.25 to 50% of PFSn. However, a significant reduction of Caco-2 cellular viability ($p < 0.0001$) was recorded in cells exposed to pure fractions. Among them, PFAC supernatant stimulated the lowest Caco-2 cell viability (28.94 ± 6.02%), followed by PFAE (35.30 ± 8.08%) and PFEE (44.88 ± 1.73%), when compared to untreated cells. A 20% reduction of Caco-2 cell viability was also detected at 3.13% of PFAC ($p < 0.05$) and PFEE ($p < 0.01$) when compared to untreated cells.

CVS results (Figure 3A) also confirmed the absence of cytotoxic effects from PFSn in Caco-2 cells exposed from 3.13 to 50% of fermented fractions. Moreover, an increase of Caco-2 cell viability was recorded, especially at 50% of fermented samples ($p < 0.0001$). PFEE stimulated the highest Caco-2 cell proliferation, followed by PFAC and PFAE. Alike the MTT assay, when cells were exposed to pure supernatants, a significant decrease of viability was recorded ($p < 0.0001$).

3.10. TEER Response of Cell Culture

Regarding the TEER response of Caco-2 cell monolayers incubated with PFSn, it was found a significant decrease of TEER response of the monolayer when supernatants were added apically after 1 h of incubation ($p < 0.001$) (Figure 3B). After 8 h of incubation, TEER response of PFEE-treated cells was similar to untreated cells (93.89 ± 5.61%, $p > 0.05$) while a reduction of TEER response was recorded in PFAC (87.46 ± 4.71%, $p < 0.01$) and PFAE-treated cells (81.21 ± 5.71%, $p < 0.001$). Contrary, TEER values were significantly higher when PFSn were added in basolateral chamber ($p < 0.0001$), especially after 1 h of incubation. After 8 h of incubation, PFEE-treated cells presented the highest TEER response (140.16 ± 6.48%, $p < 0.0001$), followed by PFAE (122.32 ± 9.16%, $p < 0.001$) and PFAC (113.92 ± 0.93%, $p < 0.01$).

4. Discussion

Our results suggest that after appropriate treatment on persimmon fruit byproducts, the bonded metabolites and functional properties were significantly stimulated by the solvent applied; moreover, we found that the obtained DF fractions acted as a nutrient source for beneficial bacteria. Byproducts of this metabolization SCFAs were obtained and interacted with the epithelial cell barrier where they displayed a protective function.

Physicochemical parameters studied in all treatments showed slightly differences influenced by the solvent applied. The most significant TDF value was recorded in PFAC fractions, which were higher than values recorded in persimmon flours [48]. The 65% proportion of IDF in TDF in all extracted fractions of fiber from persimmon byproduct was alike other reports from persimmon byproducts [49], which suggests that the solvent applied did not promote a higher insoluble portion. While some studies have reported a higher SDF/IDF ratio to provide the appropriate physiological effects and bio-accessibility [50], others are reporting increasing evidence of the relationship between IDF and gut health [51]. Due to several logistic challenges, the maturity of processed fruits often fluctuates. Therefore, the physiochemical composition can vary widely; as a result, SDF and TDF content and ratio as well as other nutrients concentration may vary, mostly due to the ripening stage of the processed fruits. Over-ripening has been reported to polymerize tannins in persimmon fruits, and to increase TSS which implies a loss in astringency but also fruit firmness [52,53]. These effects may result in a lower DF yield but with an increased quantity of bonded compounds. These challenges, altogether with environmental conditions may also generate diverse outputs in the extraction and characterization of PF and should be acknowledged when processing byproducts.

After SAE, all fractions seemed to have a low risk for deterioration by microorganisms, enzymatic or physical reactions. Moreover, in adequate conditions, bonded bio-compounds may seem to act as antimicrobial and prebiotic agents themselves. However, byproducts generated by manufacturers tend to be spoiled or partly fermented which will also affect the chemical, microbial composition, and compromise safety of use for human consumption. These issues must be addressed through byproduct quality control and appropriate processing of obtained byproducts prior food upcycling.

Regarding techno-functional properties, WHA and SA are directly implied in health and nutrition as they may promote a satiety effect when swelling and holding water during digestion processes. This behavior in WAA and WHA has been reported for commercially available prebiotics; moreover, obtained results were above intervals reported for persimmon flours (12.19 g/g) and for other fruit byproducts such as lemon (14.4 g/g), orange (9.9 g/g), peach (14 g/g) or apple fiber (15.4 g/g) [45,48]; especially, values obtained from PFAC fractions. Both EA and ES results show the potential application of these fibers to the decrease in the interfacial tension among the hydrophobic and hydrophilic compounds in the food matrix and during the digestion and colonic fermentation process.

OHA was similar to treated persimmon flours obtained from high hydrostatic pressure [54]. Given the strong relation recorded between SDF and TDF with OHA (r = 0.99); DF fractions might derive from a higher availability of hydrophobic bonds within the fiber in all treatments. Moreover, OHA values in all fractions showed a potential function as an emulsifying or stabilizing agent ingredient in fatty matrices as they could help prevent an over greasy incorporation and reduce fat content. On the other hand, GFA, which affects texture and mouthfeel of food matrices, was similar to values for cassava flour (4%) [55]; given GFA's strong correlation with SDF content in DF fractions (r = 0.99), persimmon extracted fibers could, in addition to provide functional benefits, act as gel forming agents in matrices which require thickening or gelling.

The BHA has been strongly related with the presence of fiber bounded phenolic compounds and linked with the modulation of glucose blood levels, reduction of cholesterol and the conversion of cholesterol to bile acids in the liver to reduce glycemic and lipidemic levels [46,56,57]. In addition, the OHA of fiber fractions which is an essential parameter in the characterization of DF was relatively higher than values reported for persimmon flours [45,48]. Like OHA values, FOB might imply a higher availability of hydrophobic bonds within the PFAC fiber organizational structure. This property shows the capacity of fiber to adsorb or retain oil/fat in its matrix, simulating the conditions of food digestion.

TPC recovery index results showcased statistical differences after digestion process with PFEE fraction with a remarkable decrease. The significant reduction after fermentation of recovery index in PFEE and PFAC fractions are similar to the reports from other food matrices [58]. These results imply a high impact from the extraction treatment in the digestion and fermentation processes and in the bioactive compounds bonded to the fibers, as a result, the PFAE fraction was able to reduce the impact of the digestion process in the reduction, degradation or polymerization of phenolic compounds while allowing the availability of bio-compounds for gut bacteria consumption. On the other hand, results related to bio-accessibility indexes after fermentation are mainly implied with bacterial enzyme action on fiber polysaccharides and the release of bonded phenols [59–61].

The fractions of bioactive compounds released from the fiber matrices and found in the supernatant fractions are directly implied in the availability of these compounds during the digestion process and the gut fermentation process for its absorption into the bloodstream. The probiotic bacterial population used in this assay allowed a significant amount of phenolic compounds to be released from the fiber matrix compared to the index of phenolic compounds released after the digestion process.

Assessing TPC by the Folin–Ciocalteu reagent has been also acknowledged as nonspecific to phenolic compounds [37,38]. Even though fiber fractions may present low quantities of interfering compounds, the reagent can be reduced by other compounds present in in vitro digestion, and fermentation such as amino acids, peptides, and reduction sugars. Despite limitations of the TPC assay with the Folin–Ciocalteu reagent, it is an accessible, simple, and reproducible tool when exploring phenolic antioxidants and its reducing capacity. More insightful analyzes through enzymatic reactions, chromatographic and spectrometric quantitation assays should be performed to avoid bias and confirm results.

Regarding to recovery and bio-accessibility indexes of carotenoids, results imply a lower proportion, and concentration of carotenoids released from the fiber matrix, a low stability of both carotenoids and phenolics after the digestion process. Despite a lower recovery index, bio-accessibility index in carotenoids was higher than the phenolic fraction; however, it remained low compared to the TCC bonded to the fibers. Studies have reported persimmon byproducts as a remarkable source of carotenoids, specially β-carotene, lycopene and β-cryptoxanthin. Bonded bio-compounds act as a source of nutrients for gut bacteria, where the fiber matrix acts as a protective agent against the digestion process and allows bonded material to be released by the gut bacteria [60].

The selected bacterial population used in this assay did not allow a significant carotenoid amount to be released from the fiber matrix (<0.6%), in fact, results imply these populations might have consumed some of the carotenoids present in the matrix. It is important to address the high concentration of phenolic compounds and carotenoids bonded to the fiber matrix, while many approaches have been made to integrate food byproducts into food matrixes, taking into consideration the high quantity of these bio-compounds, few have assessed the extractability or availability of them after digestion and fermentation process. Hence, we provide information about the low extractability of these compounds' despite of previous treatment. As a result, in order to increase the release of these bonded compounds from the fiber matrices, multiple approaches and technologies have to be implemented takin into account the functional properties of DF, the stability of bio-compounds, and the biological interaction of the outcomes.

As regards to the antioxidant activity provided by the bio-compounds presented in fiber matrix, after digestion, the pellet formed showed a lower antioxidant activity than the supernatant fraction, which displays a complex interaction between the soluble fraction and the fiber matrix that formed the pellet. Similar values were reported for the antioxidant activity in soluble fractions and pellets after fermentation. This behavior in the antioxidant activity has been reported in ABTS$^{\bullet+}$ assays in other food matrices [58].

Many studies have reported a similar distribution of antioxidants and bio-compounds in both pellets and soluble fractions, while others have reported contradictory results [48]. Probably, the variability of these results is based on the fiber matrix composition, and the sample preparation methodology, while some authors only use organic solvents for the extraction of bio-compounds, others also modify conditions such as pH, temperature and pressure in order to obtain highest yield of bonded bio-compounds present in the food matrix.

After hydrolysis, the major polyphenolic compound detected and quantified was gallic acid. This hydroxybenzoic acid has been reported to be bonded to the fiber matrix. However, our results suggest that it was released from the galloylated tannins reported and detected in persimmon's non-extractable fractions [46]. Tannins present in persimmon contain gallic acid residues linked with glucose via glycosidic bonds. The hydroxyl group of both glucose and gallic acid can be considered as the potential interacting sites for the formation of hydrogen bonds with cellulose and hemicellulose. Moreover, tannins can interact with carbohydrates non-covalently or covalently which influences in the extractability of phenolic compounds [62] in persimmon DF. These effects yielded a high concentration of gallic acid after hydrolysis elucidates the complex configuration of persimmon DF.

Soluble tannins have been assessed in variety 'Rojo Brillante' due to the formation of salivary protein complexes resulting in astringent sensations [8]. This must be accounted when introducing persimmon DF from byproducts in food matrixes. Additionally, flavonoids coupled with hydrolysable gallic acid moiety through carbon-carbon linkage were also substantial phytochemical moieties bonded in the persimmon DF matrix after SAE. Few differences were assessed in the composition of bonded polyphenols as a result of the solvent applied; whereas the digestion and fermentation processes displayed and released a higher variety of bioactive compounds.

Chemical characteristics of phenolic compounds, such as solubility, hydrophobicity, molecular weight, or configuration are evidently affected by the course of the digestion [63] and fermentation processes. Effects and variability displayed on each phenolic compound are related to its configuration through the fiber matrix and its bond with other carbohydrates and in agreement with the data shown for the bio-accessibility indexes, similarly to other studies reported [64]. Results show the release of these phenolic compounds to the colon where they may undergo metabolism and transformation by bacterial populations into absorbable and beneficial metabolites like SCFAs.

Beneficial microflora plays an important role in metabolization of non-digestible carbohydrates and polyphenols. Both have an important role in the protection of intestinal tract because they keep their antioxidant activity, generate SCFAs and are subsequently set off. From SCFAs, acetic acid, propionic acid, and butyric acid have been among the most documented metabolites because of their health implications [16,65]; PFAE, PFEE and PFAC fractions after digestion were tested as a potential prebiotic matrix with beneficial bacteria that has been reported for the SCFAs metabolization after fermentation. The results varied from each fraction before and after digestion, the SCFA production followed the order propionate > acetate > butyrate, persimmon fibers showed a higher production of SCFAs than the reported for orange, mango, and pear byproducts [66].

Results imply the utilization of the fiber matrix to produce SCFAS is dependent of the digestive process and the application of solvent treatments in the fiber present in the food matrix. We acknowledge the complex composition and outcomes of the gut microbiota fermentation; for these reasons, the main gut beneficial bacteria were tested. However, these results must be confirmed with entire gut microbiome in vitro and in vivo.

PFP is known to exert health benefits in colonic epithelium such as an enhancement of barrier function [67]. Therefore, the interaction of PFSn was determined. Pure fermented fractions of PFAE, PFAC, PFEE decreased the viability of Caco-2 cells according to CVS and MTT assays, suggesting that metabolites produced by PFP contained substances that might either directly inhibit cell-proliferation or inactive it due to alteration of microenvironment, lowering pH or scavenging reactive oxygen intermediates [68], especially as it was the only bio-accessible nutrient in the microenvironment. This effect has also been recorded in PFSn by lactic acid bacteria or Bifidobacterium spp., with cytotoxic effects in HT-29, SW-480 and Caco-2 cell lines [67,69].

Regarding cells incubated with fermented fractions and DMEM, the CVS assay showed no cytotoxic effect of these samples in cell viability; in fact, Caco-2 cell viability increased in the presence of DMEM, dose-dependent. Interestingly, a low decrease of Caco-2 cell viability was observed by MTT assay. Although not significant ($p > 0.05$), fermented samples from 6.25 to 50% of purity may interfere with MTT or succinate dehydrogenase activity, with a loss of viability as an outcome. In addition, it has been reported a direct reduction of MTT to formazan by the interference of phenolics which were detected in PFSn from fiber fractions [70]. The outcome is related to a slightly decrease in Caco-2 viability which may explain the obtained results. Even so, the viability recorded by MTT correlated with those obtained by CVS, indicating that the PFSn from fiber fractions were no cytotoxic and promoted Caco-2 cell viability when compared to untreated cells.

Moreover, it was determined that fermented fractions generated a significant increase of TEER values when added basolateral-apical after 1 and 8 h of incubation ($p < 0.0001$), with the highest TEER value recorded by PCAE and PFEE supernatants. The TEER response of Caco-2 cell monolayer to fermented samples may be due to the presence of SCFAs which have been identified in fermented fractions, especially butyrate in PFAE (1.29 ± 0.02 mg/mL), PFEE (1.10 ± 0.00 mg/mL) and PFAC (1.01 ± 0.00 mg/mL). It has been reported that butyrate increases TEER values of Caco-2 cell monolayers [33] which suggests that PFAE and PFEE (samples with the highest butyrate concentrations) improved the intestinal barrier function of Caco-2 cell monolayers when compared to untreated monolayers, after 8 h of incubation. On the other hand, lower TEER values were recorded by fermented samples when added apical-basolateral, especially by PFAE supernatants after 8 h of incubation. When added apical-basolateral, PFSn interacted directly with intracellular junctional complexes (tight junctions, gap junctions, adherence junctions and desmosome) [71]. The modulation of SCFAs in Caco-2 cell monolayers has been reported to produce lower TEER values [72–74], which may increase the permeability of another compounds. Overall, PFSn especially from PFEE and PFAE, showed the potential to improve barrier function in Caco-2 cell monolayers, which has been related to restrictions of the channel from the lumen and into the systemic circulation (abluminal) of larger potentially toxic compounds, as well as allowing the absorption of nutrients, electrolytes and bio-compounds [73].

5. Conclusions

This study focused on the solvent extraction effect of DF from persimmon byproducts on its physiological, technological, and prebiotic features. The in vitro gastrointestinal digestion and probiotic bacterial fermentation decreased TPC and TCC and therefore the antioxidant activity of DF. Hydrolysis of covalently bonded compounds in persimmon fiber yielded a high amount of gallic acid. Moreover, metabolites produced by bacterial fermentation were no cytotoxic for human epithelial cells. Overall results show the biological potential of persimmon's DF is dependent on the SAE process and may promote a strong probiotic response and modulate the epithelial barrier function in a Caco-2 cell model. The underlying mechanism will be further investigated in the future works. These findings contribute to existing knowledge of persimmon byproducts as a DF and bound phenolics source and provide a new insight to its suitability.

Author Contributions: Conceptualization, M.C.M.-M., D.S., M.V. and N.M.; Data curation, J.S.-B., B.M.-C. and D.S.; Formal analysis, J.S.-B. and B.M.-C.; Funding acquisition, D.S. and M.V.; Investigation, J.S.-B. and B.M.-C.; Methodology, J.S.-B. and B.M.-C.; Project administration, D.S. and M.V.; Resources, M.C.M.-M. and N.M.; Supervision, M.C.M.-M. and N.M.; Validation, M.C.M.-M., D.S., M.V. and N.M.; Visualization, M.C.M.-M. and N.M.; Writing—original draft, J.S.-B. and B.M.-C.; Writing—review & editing, M.V. All authors have read and agreed to the published version of the manuscript.

Funding: This study was supported by the Ministerio de Ciencia, Innovación y Universidades through the funded project 'Simbiosis industrial en el aprovechamiento integral del caqui (*Diospyros kaki*); Ejemplo de bioeconomía' (CTM2017-88978-R).

Institutional Review Board Statement: Not applicable.

Informed Consent Statement: Not applicable.

Data Availability Statement: Data is contained within the article.

Acknowledgments: The authors wish to thank Mitra Sol Technologies S.L. for the given technical assistance.

Conflicts of Interest: The authors declare no conflict of interest.

References

1. Kearney, J. Food consumption trends and drivers. *Philos. Trans. R. Soc. B Biol. Sci.* **2010**, *365*, 2793–2807. [CrossRef]
2. Liu, R.H. Health benefits of fruit and vegetables are from additive and synergistic combinations of phytochemicals. *Am. J. Clin. Nutr.* **2003**, *78*, 517S–520S. [CrossRef]
3. Yao, L.H.; Jiang, Y.M.; Shi, J.; Tomas-Barberan, F.A.; Datta, N.; Singanusong, R.; Chen, S.S. Flavonoids in food and their health benefits. *Plant Foods Hum. Nutr.* **2004**, *59*, 113–122. [CrossRef]
4. Matheus, J.R.V.; Andrade, C.J.D.; Miyahira, R.F.; Fai, A.E.C. Persimmon (*Diospyros kaki* L.): Chemical properties, bioactive compounds and potential use in the development of new products—A review. *Food Rev. Int.* **2020**, 1–18. [CrossRef]
5. Matsumura, Y.; Ito, T.; Yano, H.; Kita, E.; Mikasa, K.; Okada, M.; Furutani, A.; Murono, Y.; Shibata, M.; Nishii, Y.; et al. Antioxidant potential in non-extractable fractions of dried persimmon (*Diospyros kaki* Thunb.). *Food Chem.* **2016**, *202*, 99–103. [CrossRef] [PubMed]
6. Cano, M.P.; Gómez-Maqueo, A.; Fernández-López, R.; Welti-Chanes, J.; García-Cayuela, T. Impact of high hydrostatic pressure and thermal treatment on the stability and bioaccessibility of carotenoid and carotenoid esters in astringent persimmon (*Diospyros kaki* Thunb, var. Rojo Brillante). *Food Res. Int.* **2019**, *123*, 538–549. [CrossRef] [PubMed]
7. Gea-Botella, S.; Agulló, L.; Martí, N.; Martínez-Madrid, M.C.; Lizama, V.; Martín-Bermudo, F.; Berná, G.; Saura, D.; Valero, M. Carotenoids from persimmon juice processing. *Food Res. Int.* **2021**, *141*, 109882. [CrossRef]

8. Jiménez-Sánchez, C.; Lozano-Sánchez, J.; Marti, N.; Saura, D.; Valero, M.; Segura-Carretero, A.; Fernández-Gutiérrez, A. Characterization of polyphenols, sugars, and other polar compounds in persimmon juices produced under different technologies and their assessment in terms of compositional variations. *Food Chem.* **2015**, *182*, 282–291. [CrossRef] [PubMed]
9. Kim, G.N.; Shin, M.R.; Shin, S.H.; Lee, A.R.; Lee, J.Y.; Seo, B.I.; Kim, M.Y.; Kim, T.H.; Noh, J.S.; Rhee, M.H.; et al. Study of Antiobesity Effect through Inhibition of Pancreatic Lipase Activity of *Diospyros kaki* Fruit and *Citrus unshiu* Peel. *BioMed Res. Int.* **2016**, *2016*, 1723042. [CrossRef]
10. Son, J.E.; Hwang, M.K.; Lee, E.; Seo, S.G.; Kim, J.E.; Jung, S.K.; Kim, J.R.; Ahn, G.H.; Lee, K.W.; Lee, H.J. Persimmon peel extract attenuates PDGF-BB-induced human aortic smooth muscle cell migration and invasion through inhibition of c-Src activity. *Food Chem.* **2013**, *141*, 3309–3316. [CrossRef]
11. Ueda, K.; Kawabata, R.; Irie, T.; Nakai, Y.; Tohya, Y.; Sakaguchi, T. Inactivation of pathogenic viruses by plant-derived tannins: Strong effects of extracts from persimmon (*Diospyros kaki*) on a broad range of viruses. *PLoS ONE* **2013**, *8*, e55343. [CrossRef]
12. Orejuela-Escobar, L.M.; Landázuri, A.C.; Goodell, B. Second generation biorefining in Ecuador: Circular bioeconomy, zero waste technology, environment and sustainable development: The nexus. *J. Bioresour. Bioprod.* **2021**, *6*, 83–107. [CrossRef]
13. Soliman, G.A. Dietary Fiber, Atherosclerosis, and Cardiovascular Disease. *Nutrients* **2019**, *11*, 1155. [CrossRef]
14. Blad, C.C.; Tang, C.; Offermanns, S. G protein-coupled receptors for energy metabolites as new therapeutic targets. *Nat. Rev. Drug Discov.* **2012**, *11*, 603–619. [CrossRef] [PubMed]
15. den Besten, G.; van Eunen, K.; Groen, A.K.; Venema, K.; Reijngoud, D.J.; Bakker, B.M. The role of short-chain fatty acids in the interplay between diet, gut microbiota, and host energy metabolism. *J. Lipid Res.* **2013**, *54*, 2325–2340. [CrossRef] [PubMed]
16. Park, J.; Kim, M.; Kang, S.G.; Jannasch, A.H.; Cooper, B.; Patterson, J.; Kim, C.H. Short-chain fatty acids induce both effector and regulatory T cells by suppression of histone deacetylases and regulation of the mTOR–S6K pathway. *Mucosal Immunol.* **2015**, *8*, 80–93. [CrossRef] [PubMed]
17. Cani, P.D. Human gut microbiome: Hopes, threats and promises. *Gut* **2018**, *67*, 1716–1725. [CrossRef] [PubMed]
18. Akter, S.; Ahmed, M.; Eun, J.B. Effect of blanchig and drying temperatures on the physicochemical characteristics, dietary fiber and antioxidant-related parameters of dried persimmons peel powder. *Int. J. Food Sci. Nutr.* **2010**, *61*, 11. [CrossRef]
19. AOAC Association of Official Analytical Chemist. *Official Methods of Analysis*; Association of Official Analytical Chemist: Washington, DC, USA, 2012.
20. Raghavendra, S.N.; Rastogi, N.K.; Raghavarao, K.S.M.S.; Tharanathan, R.N. Dietary fiber from coconut residue: Effects of different treatments and particle size on the hydration properties. *Eur. Food Res. Technol.* **2004**, *218*, 563–567. [CrossRef]
21. Robertson, J.A.; de Monredon, F.D.; Dysseler, P.; Guillon, F.; Amado, R.; Thibault, J.F. Hydration properties of dietary fibre and resistant starch: A European collaborative study. *LWT-Food Sci. Technol.* **2000**, *33*, 72–79. [CrossRef]
22. Gómez-Ordóñez, E.; Jiménez-Escrig, A.; Rupérez, P. Dietary fibre and physicochemical properties of several edible seaweeds from the northwestern Spanish coast. *Food Res. Int.* **2010**, *43*, 2289–2294. [CrossRef]
23. Yasumatsu, K.; Sawada, K.; Moritaka, S.; Misaki, M.; Toda, J.; Wada, T.; Ishii, K. Whipping and emulsifying properties of soybean products. *Agric. Biol. Chem.* **1972**, *36*, 719–727. [CrossRef]
24. Chau, C.F.; Cheung, P.C.K. Functional properties of flours prepared from three Chinese indigenous legume seeds. *Food Chem.* **1998**, *61*, 429–433. [CrossRef]
25. Eastwood, M.A.; Kirkpatrick, J.R.; Mitchell, W.D.; Bone, A.; Hamilton, T. Effects of dietary supplements of wheat bran and cellulose on faeces and bowel function. *Britrish Med. J.* **1973**, *4*, 392–394. [CrossRef] [PubMed]
26. Minekus, M.; Alminger, M.; Alvito, P.; Ballance, S.; Bohn, T.; Bourlieu, C.; Carrière, F.; Boutrou, R.; Corredig, M.; Dupont, D.; et al. A standardised static in vitro digestion method suitable for food–An international consensus. *Food Funct.* **2014**, *5*, 1113–1124. [CrossRef]
27. Jiménez-Zamora, A.; Pastoriza, S.; Rufián-Henares, J. Revalorization of coffee by-products. Prebiotic, antimicrobial and antioxidant properties. *LWT-Food Sci. Technol.* **2015**, *61*, 12–18. [CrossRef]
28. Kwon, H.J.; Kim, M.Y.; Seo, K.H.; Lee, H.G.; Kim, H. In-vitro prebiotic activity of grape seed fluor highly rich in flavonoid and dietary fiber. *J. Food Nutr. Res.* **2018**, *6*, 621–625. [CrossRef]
29. Praveen, M.A.; Parvathy, K.K.; Jayabalan, R.; Balasubramanian, P. Dietary fiber from Indian edible seaweeds and its in-vitro prebiotic effect on the gut microbiota. *Food Hydrocoll.* **2019**, *96*, 343–353. [CrossRef]
30. Zarinah, Z.; Anis, A.A.; Napisah, H.; Shazila, S. Prebiotic activity score of breadfruit resistant starch (*Artocarpus altilis*), breadfruit flour, and inulin during *in-vitro* fermentation by pure cultures (*Lactobacillus plantarum* and *Bifidobacterium bifidum*). *J. Agrobiotechnology* **2018**, *9*, 122–131.
31. Saotome, K.; Morita, H.; Umeda, M. Cytotoxicity test with simplified crystal violet staining method using microtitre plates and its application to injection drugs. *Toxicol. Vitr.* **1989**, *3*, 317–321. [CrossRef]
32. Peng, L.; He, Z.; Chen, W.; Holzman, I.R.; Lin, J. Effects of butyrate on intestinal barrier function in a Caco-2 cell monolayer model of intestinal barrier. *Pediatric Res.* **2007**, *61*, 37–41. [CrossRef] [PubMed]
33. Chen, T.; Kim, C.Y.; Kaur, A.; Lamothe, L.; Shaikh, M.; Keshavarzian, A.; Hamaker, B.R. Dietary fibre-based SCFA mixtures promote both protection and repair of intestinal epithelial barrier function in a Caco-2 cell model. *Food Funct.* **2017**, *8*, 1166–1173. [CrossRef] [PubMed]
34. Hubatsch, I.; Ragnarsson, E.G.; Artursson, P. Determination of drug permeability and prediction of drug absorption in Caco-2 monolayers. *Nat. Protoc.* **2007**, *2*, 2111–2119. [CrossRef] [PubMed]

35. Singleton, V.L.; Orthofer, R.; Lamuela-Raventós, R.M. Analysis of total phenols and other oxidation substrates and antioxidants by means of folin-ciocalteu reagent. In *Methods in Enzymology*; Academic Press: Cambridge, MA, USA, 1999; Volume 299, pp. 152–178.
36. Wolfe, K.; Wu, X.; Liu, R.H. Antioxidant activity of apple peels. *J. Agric. Food Chem.* **2003**, *51*, 609–614. [CrossRef] [PubMed]
37. Huang, D.; Ou, B.; Prior, R.L. The chemistry behind antioxidant capacity assays. *J. Agric. Food Chem.* **2005**, *53*, 1841–1856. [CrossRef]
38. Górnaś, P.; Dwiecki, K.; Siger, A.; Tomaszewska-Gras, J.; Michalak, M.; Polewski, K. Contribution of phenolic acids isolated from green and roasted boiled-type coffee brews to total coffee antioxidant capacity. *Eur. Food Res. Technol.* **2015**, *242*, 641–653. [CrossRef]
39. Andrade-Cuvi, M.J.; Moreno, C.; Zaro, M.J.; Vicente, A.R.; Concellón, A. Improvement of the antioxidant properties and postharvest life of three exotic Andean fruits by UV-C treatment. *J. Food Qual.* **2017**, *2017*. [CrossRef]
40. Luximon-Ramma, A.; Bahorun, T.; Soobrattee, M.A.; Aruoma, O.I. Antioxidant activities of phenolic, proanthocyanidin, and flavonoid components in extracts of *Cassia fistula*. *J. Agric. Food Chem.* **2002**, *50*, 5042–5047. [CrossRef]
41. Ortega, N.; Macià, A.; Romero, M.P.; Reguant, J.; Motilva, M.J. Matrix composition effect on the digestibility of carob flour phenols by an in-vitro digestion model. *Food Chem.* **2011**, *124*, 65–71. [CrossRef]
42. Re, R.; Pellegrini, N.; Proteggente, A.; Pannala, A.; Yang, M.; Rice-Evans, C. Antioxidant activity applying an improved ABTS radical cation decolorization assay. *Free Radic. Biol. Med.* **1999**, *26*, 1231–1237. [CrossRef]
43. Brand-Williams, W.; Cuvelier, M.E.; Berset, C. Use of a free radical method to evaluate antioxidant activity. *LWT Food Sci. Technol.* **1995**, *28*, 25–30. [CrossRef]
44. Calero, R.R.; Lagoa-Costa, B.; Fernandez-Feal, M.M.D.C.; Kennes, C.; Veiga, M.C. Volatile fatty acids production from cheese whey: Influence of pH, solid retention time and organic loading rate. *J. Chem. Technol. Biotechnol.* **2018**, *93*, 1742–1747. [CrossRef]
45. Martínez-Las Heras, R.; Landines, E.F.; Heredia, A.; Castelló, M.L.; Andrés, A. Influence of drying process and particle size of persimmon fibre on its physicochemical, antioxidant, hydration and emulsifying properties. *J. Food Sci. Technol.* **2017**, *54*, 2902–2912. [CrossRef] [PubMed]
46. Hamauzu, Y.; Suwannachot, J. Non-extractable polyphenols and in vitro bile acid-binding capacity of dried persimmon (*Diospyros kaki*) fruit. *Food Chem.* **2019**, *293*, 127–133. [CrossRef] [PubMed]
47. Maroun, G.R.; Rajha, H.N.; El Darra, N.; El Kantar, S.; Chacar, S.; Debs, E.; Louka, N. Emerging technologies for the extraction of polyphenols from natural sources. In *Polyphenols: Properties, Recovery, and Applications*; Galanakis, C.M., Ed.; Woodhead Publishing: Vienna, Austria, 2018; pp. 265–293.
48. Lucas-González, R.; Viuda-Martos, M.; Pérez-Álvarez, J.; Fernández-López, J. Evaluation of Particle Size Influence on Proximate Composition, Physicochemical, Techno-Functional and Physio-Functional Properties of Flours Obtained from Persimmon (*Diospyros kaki* Trumb.) Coproducts. *Plant Foods Hum. Nutr.* **2017**, *72*, 67–73. [CrossRef]
49. Park, Y.-S.; Jung, S.-T.; Kang, S.-G.; Delgado-Licon, E.; Leticia Martinez Ayala, A.; Tapia, M.S.; Martín-Belloso, O.; Trakhtenberg, S.; Gorinstein, S. Drying of persimmons (Diospyros kaki L.) and the following changes in the studied bioactive compounds and the total radical scavenging activities. *LWT Food Sci. Technol.* **2006**, *39*, 748–755. [CrossRef]
50. Vitaglione, P.; Napolitano, A.; Fogliano, V. Cereal dietary fibre: A natural functional ingredient to deliver phenolic compounds into the gut. *Trends Food Sci. Technol.* **2008**, *19*, 451–463. [CrossRef]
51. Yang, L.; Zhao, Y.; Huang, J.; Zhang, H.; Lin, Q.; Han, L.; Liu, J.; Wang, J.; Liu, H. Insoluble dietary fiber from soy hulls regulates the gut microbiota in vitro and increases the abundance of bifidobacteriales and lactobacillales. *J. Food Sci. Technol.* **2020**, *57*, 152–162. [CrossRef]
52. Tessmer, M.A.; Besada, C.; Hernando, I.; Appezzato-da-Gloria, B.; Aquiles, A.; Salvador, A. Microstructural changes while persimmon fruits mature and ripen. Comparison between astringent and non-astringent cultivars. *Postharvest Biol. Technol.* **2016**, *120*, 52–60. [CrossRef]
53. Salvador, A.; Arnal, L.; Besada, C.; Larrea, V.; Quiles, A.; Pérez-Munuera, I. Physiological and structural changes during ripening and deastringency treatment of persimmon fruit cv. 'Rojo Brillante'. *Postharvest Biol. Technol.* **2007**, *46*, 181–188. [CrossRef]
54. Rodríguez-Garayar, M.; Martín-Cabrejas, M.A.; Esteban, R.M. High hydrostatic pressure in astringent and non-astringent persimmons to obtain fiber-enriched ingredients with improved functionality. *Food Bioprocess Technol.* **2017**, *10*, 854–865. [CrossRef]
55. Akubor, P.I.; Ukwuru, M.U. Functional properties and biscuit making potential of soybean and cassava flour blends. *Plant Foods Hum. Nutr.* **2003**, *58*, 1–12. [CrossRef]
56. Houten, S.M.; Watanabe, M.; Auwerx, J. Endocrine functions of bile acids. *EMBO J.* **2006**, *25*, 1419–1425. [CrossRef]
57. Niu, Y.; Xie, Z.; Zhang, H.; Sheng, Y.; Yu, L.L. Effects of structural modifications on physicochemical and bile acid-binding properties of psyllium. *J. Agric. Food Chem.* **2013**, *61*, 596–601. [CrossRef]
58. Bouayed, J.; Hoffmann, L.; Bohn, T. Total phenolics, flavonoids, anthocyanins and antioxidant activity following simulated gastro-intestinal digestion and dialysis of apple varieties: Bioaccessibility and potential uptake. *Food Chem.* **2011**, *128*, 14–21. [CrossRef]
59. Holscher, H.D. Dietary fiber and prebiotics and the gastrointestinal microbiota. *Gut Microbes* **2017**, *8*, 172–184. [CrossRef]
60. Selma, M.V.; Espín, J.C.; Tomás-Barberán, F.A. Interaction between phenolics and gut microbiota: Role in human health. *J. Agric. Food Chem.* **2009**, *57*, 6485–6501. [CrossRef]

61. Slavin, J. Fiber and prebiotics: Mechanisms and health benefits. *Nutrients* **2013**, *5*, 1417–1435. [CrossRef] [PubMed]
62. Mamet, T.; Ge, Z.Z.; Zhang, Y.; Li, C.M. Interactions between highly galloylated persimmon tannins and pectins. *Int. J. Biol. Macromol.* **2018**, *106*, 410–417. [CrossRef] [PubMed]
63. Barba, F.J.; Mariutti, L.R.; Bragagnolo, N.; Mercadante, A.Z.; Barbosa-Canovas, G.V.; Orlien, V. Bioaccessibility of bioactive compounds from fruits and vegetables after thermal and nonthermal processing. *Trends Food Sci. Technol.* **2017**, *67*, 195–206. [CrossRef]
64. Padayachee, A.; Netzel, G.; Netzel, M.; Day, L.; Mikkelsen, D.; Gidley, M.J. Lack of release of bound anthocyanins and phenolic acids from carrot plant cell walls and model composites during simulated gastric and small intestinal digestion. *Food Funct.* **2013**, *4*, 906–916. [CrossRef]
65. Trompette, A.; Gollwitzer, E.S.; Pattaroni, C.; Lopez-Mejia, I.C.; Riva, E.; Pernot, J.; Ubags, N.; Fajas, L.; Nicod, L.P.; Marsland, B.J. Dietary Fiber Confers Protection against Flu by Shaping Ly6c. *Immunity* **2018**, *48*, 992–1005. [CrossRef] [PubMed]
66. Tejada-Ortigoza, V.; Garcia-Amezquita, L.E.; Kazem, A.E.; Campanella, O.H.; Cano, M.P.; Hamaker, B.R.; Serna-Saldívar, S.O.; Welti-Chanes, J. In vitro fecal fermentation of high pressure-treated fruit peels used as dietary fiber sources. *Molecules* **2019**, *24*, 697. [CrossRef]
67. Sharma, M.; Chandel, D.; Shukla, G. Antigenotoxicity and cytotoxic potentials of metabiotics extracted from isolated probiotic, *Lactobacillus rhamnosus* MD 14 on Caco-2 and HT-29 human colon cancer cells. *Nutr. Cancer* **2020**, *72*, 110–119. [CrossRef]
68. Raman, M.; Ambalam, P.; Kondepudi, K.K.; Pithva, S.; Kothari, C.; Patel, A.T.; Purama, R.K.; Dave, J.M.; Vyas, B.R. Potential of probiotics, prebiotics and synbiotics for management of colorectal cancer. *Gut Microbes* **2013**, *4*, 181–192. [CrossRef]
69. Arun, K.B.; Madhavan, A.; Reshmitha, T.R.; Sithara, T.; Nisha, P. Short chain fatty acids enriched fermentation metabolites of soluble dietary fibre from *Musa paradisiaca* drives HT29 colon cancer cells to apoptosis. *PLoS ONE* **2019**, *14*, e0216604. [CrossRef]
70. Wang, P.; Henning, S.M.; Heber, D. Limitations of MTT and MTS-based assays for measurement of antiproliferative activity of green tea polyphenols. *PLoS ONE* **2010**, *5*, e10202. [CrossRef] [PubMed]
71. Hsieh, C.Y.; Osaka, T.; Moriyama, E.; Date, Y.; Kikuchi, J.; Tsuneda, S. Strengthening of the intestinal epithelial tight junction by *Bifidobacterium bifidum*. *Physiol. Rep.* **2015**, *3*, e12327. [CrossRef]
72. Castro-Herrera, V.M.; Rasmussen, C.; Wellejus, A.; Miles, E.A.; Calder, P.C. In Vitro Effects of Live and Heat-Inactivated *Bifidobacterium animalis* subsp. *lactis*, BB-12 and *Lactobacillus rhamnosus* GG on Caco-2 Cells. *Nutrients* **2020**, *12*, 1719. [CrossRef]
73. Ohata, A.; Usami, M.; Miyoshi, M. Short-chain fatty acids alter tight junction permeability in intestinal monolayer cells via lipoxygenase activation. *Nutrition* **2005**, *21*, 838–847. [CrossRef]
74. Srinivasan, B.; Kolli, A.R.; Esch, M.B.; Abaci, H.E.; Shuler, M.L.; Hickman, J.J. TEER measurement techniques for in vitro barrier model systems. *J. Lab. Autom.* **2015**, *20*, 107–126. [CrossRef] [PubMed]

Article

A Novel Ultrasound-Assisted Extraction Method for the Analysis of Anthocyanins in Potatoes (*Solanum tuberosum* L.)

Ceferino Carrera [1], María José Aliaño-González [1], Monika Valaityte [2], Marta Ferreiro-González [1], Gerardo F. Barbero [1,*] and Miguel Palma [1]

[1] Department of Analytical Chemistry, Faculty of Sciences, Agrifood Campus of International Excellence (ceiA3), IVAGRO, University of Cadiz, 11510 Puerto Real, Spain; ceferino.carrera@uca.es (C.C.); mariajose.alianogonzalez@alum.uca.es (M.J.A.-G.); marta.ferreiro@uca.es (M.F.-G.); miguel.palma@uca.es (M.P.)

[2] Department of Biology, Faculty of Marine and Environmental Sciences, University of Cadiz, 11510 Puerto Real, Spain; mvalaityte@yahoo.com

* Correspondence: gerardo.fernandez@uca.es; Tel.: +34-956-016755; Fax: +34-956-016460

Abstract: Purple potato is one of the least known and consumed potato varieties. It is as rich in nutrients, amino acids and starches as the rest of the potato varieties, but it also exhibits a high content of anthocyanins, which confer it with some attractive health-related properties, such as antioxidant, pain-relieving, anti-inflammatory and other promising properties regarding the treatment of certain diseases. A novel methodology based on ultrasound-assisted extraction has been optimized to achieve greater yields of anthocyanins. Optimal extraction values have been established at 70 °C using 20 mL of a 60% MeOH:H$_2$O solution, with a pH of 2.90 and a 0.5 s^{-1} cycle length at 70% of the maximum amplitude for 15 min. The repeatability and intermediate precision of the extraction method have been proven by its relative standard deviation (RSD) below 5%. The method has been tested on Vitelotte, Double Fun, Highland and Violet Queen potatoes and has demonstrated its suitability for the extraction and quantification of the anthocyanins found in these potato varieties, which exhibit notable content differences. Finally, the antioxidant capacity of these potato varieties has been determined by means of 2,2-diphenyl-1-picrylhydrazyl (DPPH) radical scavenging and the values obtained were similar to those previously reported in the literature.

Keywords: anthocyanins; Box-Behnken design; purple potato; red potato; *Solanum tuberosum*; ultrasound-assisted extraction; optimization; UHPLC; antioxidant activity; DPPH

1. Introduction

Potato (*Solanum tuberosum* L.) is one of the most extensively consumed food products around the world, as it is an important source of nutrients and calories with a reduced content of undesirable compounds, such as fat or salt, and a high content of essential amino acids and starches [1–4].

There are a large number of potato varieties, each of them with specific chemical and physical profiles, the white, cream, or yellow flesh varieties being the most widely consumed [5,6]. However, other varieties such as red or purple flesh potatoes are not so well known, as consumers are not so familiar with their color, and they tend to be included in some gourmet dishes [7].

Different research studies have been carried out over the last few years reporting a number of important properties—antioxidant, anti-inflammatory or prebiotic—associated with purple flesh potatoes, which belong to the same species [8–10]. They also favor mitochondrial functions with significant effects against obesity, diabetes or other metabolic disorders, and cancer inhibitory effects have also been reported [11–17]. A deeper study on the composition of purple potatoes discovered a large concentration of dietary fiber, minerals, vitamins, flavonoids, anthocyanins and phenolic compounds that are associated with several of the above-mentioned properties exhibited by potatoes [18,19].

Anthocyanins are natural pigments from the phenolic compound family [20,21] that confer a red, purple or blue color to many vegetables or fruits such as blueberries, açai, grapes, onions, sloes or even purple potatoes [22–28]. Anthocyanins are chemical compounds of high interest because they can directly scavenge reactive oxygen species (ROS), modulate the synthesis or activity of some antioxidant enzymes or even have an influence on mitochondrial respiration to prevent ROS generation or a decisive effect on the cellular antioxidant system [29–31]. Additionally, anthocyanins have exhibited significant activities in the reduction of inflammation markers from illnesses such as cancer, cardiovascular diseases, obesity or neurodegenerative diseases [32–35], which represents an important advance in preventing and treating these disorders. Furthermore, anthocyanins have proved an important relationship with pain mitigation [36]. Therefore, counting with effective methodologies that allow the extraction of anthocyanins from purple potatoes efficiently would be of utmost importance for its critical application in medicine, cosmetics and pharmaceutical industries or even in the food industry as a natural dye [37,38].

Different techniques have been employed for the extraction and analysis of anthocyanins from purple potatoes, acetylated derivatives of malvidin, petunidin, peonidin and delphinidin being the main anthocyanins that have been identified up to date [39–41].

Ultrasound-assisted extraction (UAE) has been demonstrated to be an efficient method for extracting bioactive compounds from different natural sources, which includes blackberry, lavender, raspberry, pepper, walnut, onion or even purple potato [28,42–47]. This is due to its many advantages, such as shorter extraction times, less solvent consumption and lower production costs with higher yields, reaching even twice the extraction yield, in comparison to conventional extraction techniques [48–51]. In addition, it is quite easy to apply, cost-effective and does not require complex maintenance operations [52]. This extraction technique is based on the phenomenon of cavitation, a mechanical effect caused by ultrasounds, that instigate mass transfer and compound extraction [53,54] at low temperature and with short extraction times. Hence, the compounds of interest, such as anthocyanins, are not degraded because of the extraction procedure, and larger yields are obtained [55].

An efficient method for the extraction of anthocyanins from purple potato depends on a number of factors that exert a varying degree of influence on the process. These factors include solvent properties, extraction temperature and time, among others. In order to obtain extracts enriched with anthocyanins from purple potato that can be used for the previously mentioned purposes, it is important to know the optimal conditions for the extraction that guarantee the highest performance.

In the present research, a Box-Behnken design (BBD) combined with response surface methodology (RSM) was applied to investigate the extraction of anthocyanins from purple potatoes. BBD-RSM is a mathematical and statistical method that evaluates how a number of selected variables affect the efficiency of the anthocyanin extraction method so that it can be optimized. In addition, ultra-high-performance liquid chromatography (UHPLC) coupled to ultraviolet-visible spectroscopy was selected as the method to quantify anthocyanin extractions from purple potatoes.

In conclusion, this research intends to establish the optimal conditions to achieve the maximum possible extraction of anthocyanins from purple potato samples using UAE, a technique that presents multiple advantages in comparison to other conventional methods. The objective of the authors is to develop a novel methodology that can be applied to verify the quality not only of the raw material but also of the intermediate and final products that can be obtained from purple potatoes. The final method should be suitable for industries, as well as for their analysis by commercial laboratories that intend to verify the composition and quality of any of the main or derived products that can be obtained from purple potatoes.

2. Materials and Methods

2.1. Biological Material

The flesh from purple potatoes of the Vitelotte variety acquired from a market in Cadiz (province of Andalusia, Spain) was used to design and optimize the extraction and analysis methodologies. Seed potatoes of the Vitelotte, Double Fun, Violet Queen (purple potatoes) and Highland (red potato) varieties were purchased from Fitoagricola.net (Castellón, Spain) and grown at the same plot (Torrecera, Spain; N: 36°36′39.9″; W: 5°56′22.6″) from January until May 2021.

In all the cases, the skin and bulbs were removed, and the flesh was cut into small pieces with the aim of quantifying the anthocyanins content in this part of the potatoes. After that, the samples were lyophilized using an LYOALFA 10/15 freeze dryer (Telstar Technologies, S.L., Terrassa, Barcelona, Spain) until a constant weight was reached. Finally, the samples were crushed by means of a ZM200 knife mill (Retsch GmbH, Haan, Germany) and stored in a freezer at $-20\,°C$ until further analysis.

2.2. Chemical and Solvents

The solvents chosen for the extraction were methanol of HPLC purity (Fisher Scientific, Loughborough, UK) and Milli-Q water, acquired from a Milli-Q water purification system (Millipore, Bedford, MA, USA). The extraction solvents were prepared in liquid-liquid mixtures at different percentages, and the pH values were adjusted using HCl (1 M) and NaOH solution (0.5 M) (Panreac, Barcelona, Spain).

For the UHPLC analysis, methanol of HPLC grade (Fisher Scientific, Loughborough, UK) and formic acid of HPLC grade (Panreac, Barcelona, Spain) were used for chromatographic separation.

A commercial standard, namely cyanidin chloride ($\geq$95% purity, Sigma-Aldrich Chemical Co., St. Louis, MO, USA), was used for the quantification of the anthocyanins in the extracts.

Finally, the antioxidant activity of the yields was evaluated by 2,2-diphenyl-1-picrylhydrazyl (DPPH) radical scavenging and Tris base, both supplied by Sigma-Aldrich (St. Louis, MO, USA). The standard used for control was 6-hydroxy-2,5,7,8-tetramethylchroman-2-carboxylic acid (Trolox) from Sigma-Aldrich (Steinheim, Germany).

2.3. Ultrasound-Assisted Extraction

2.3.1. Ultrasound-Assisted Extraction Equipment

For the ultrasound-assisted extraction, a UP200S sonifier (200 W, 24 kHz) (Dr. Hielscher, GmbH, Berlin, Germany) was employed. This equipment is fitted with an amplitude controller, a cycle controller and a water bath equipped with a temperature controller (FRIGITERM-10, J.P. Selecta, S.A., Barcelona, Spain).

2.3.2. Ultrasound-Assisted Extraction Optimization

As previously mentioned, BBD-RSM was the method chosen for the optimization of the UAE extraction. In this type of design, three levels per factor are evaluated: (-1) a lower level, (0) an intermediate level and (1) a higher level. In addition, it does not present axial points so that it has a more spherical arrangement of the design points than other statistical designs. Consequently, not only is a lower total number of experiments required, but also any experiments to be run under extreme conditions and that might cause the degradation of the anthocyanins or that may represent an excessive financial expense are excluded [56].

Four independent factors were selected to optimize the performance of the equipment: (i) composition of the extraction solvents (%MeOH in Milli-Q water; 0, 25 and 50%), (ii) extraction temperature (10, 40 and 70 °C), (iii) pH of the solvent (2, 4.5 and 7) and (iv) sample-to-solvent ratio (1:100, 1:150 and 1:200 g/mL). The cycle was set up at $0.5\,s^{-1}$ and the amplitude at 70% of the ultrasound equipment's maximum amplitude (200 W). These values and ranges were assigned based on the research team's previous experience on similar matrices, as well as on the relevant bibliography [57–60]. The resulting design comprised a total of 27 extractions, including 3 repetitions at the center point, and all the experiments were performed at random.

Regarding the response variable, the area of the identified anthocyanins was measured and normalized according to the amount of sample weighted. The sum of these normalized areas was taken as the response variable. A summary of the experimental conditions for the BBD-RSM (Table 1).

The extracts obtained from each experiment were centrifuged twice for 5 min at 5985× g, filtered using a 0.20 µm nylon syringe filter (Membrane Solutions, Dallas, TX, USA) and analyzed under the conditions below.

Table 1. Box-Behnken design experiment to optimize the extraction of total anthocyanins from purple potato.

Experiment	% MeOH	Temperature	pH	Ratio	Relative Area (Measured)	Relative Area (Predicted)	Relative Error in the Prediction (%)
1	0	−1	0	−1	72,953.22	74,783.20	2.45
2	1	0	0	−1	188,615.38	173,546.00	8.68
3	1	0	0	1	205,339.62	197,649.00	3.89
4	0	0	0	0	98,233.05	114,049.00	13.87
5	0	1	0	−1	209,930.07	209,546.00	0.18
6	0	0	−1	1	24,3104.45	263,197.00	7.63
7	0	0	1	−1	142,070.86	143,175.00	0.77
8	1	0	1	0	197,433.36	182,238.00	8.34
9	1	−1	0	0	120,781.49	127,148.00	5.01
10	0	1	−1	0	324,735.63	330,203.00	1.66
11	0	0	1	1	97,182.43	109,982.00	11.64
12	−1	0	−1	0	64,754.72	72,322.10	10.46
13	0	1	1	0	255,505.62	247,362.00	3.29
14	−1	0	0	−1	33,589.36	32,288.70	4.03
15	−1	1	0	0	68,379.59	63,210.00	8.18
16	0	0	−1	−1	187,493.78	195,891.00	4.29
17	0	−1	−1	0	236,810.85	201,385.00	17.59
18	1	1	0	0	348,856.93	316,545.00	10.21
19	0	−1	0	1	74,903.66	77,659.60	3.55
20	0	−1	1	0	77,330.88	78,294.80	1.23
21	−1	0	1	0	376.90	347.63	8.42
22	1	0	−1	0	282,293.09	316,194.00	10.72
23	−1	−1	0	0	41,213.04	45,278.10	8.98
24	0	0	0	0	116,928.90	114,049.00	2.53
25	0	0	0	0	96,983.88	114,049.00	14.96
26	0	1	0	1	260,240.38	240,782.00	8.08
27	−1	0	0	1	26,220.55	22,278.80	17.69

2.4. Identification of Anthocyanins by UHPLC-PDA-QToF-MS

An ultra–high-performance liquid chromatography equipment coupled to a photodiode array detector and to a quadrupole time-of-flight mass spectrometry (UHPLC-PDA-QToF-MS) model Xevo G2 (Waters Corp., Milford, MA, USA) was employed for the identification of the anthocyanins from the purple and red potato samples. This system contains a 100 × 2.1 mm reverse-phase C18 analytical column (Acquity UPLC BEH C18, Waters) with a particle size of 1.7 µm. Phase A contained water and formic acid at 2%, and phase B was pure methanol. The flow rate was 0.4 mL/min. The injection gradient employed was 5% B at 0 min; 20% B at 3.30 min; 30% B at 3.86 min; 40% B at 5.05 min; 55% B at 5.35 min; 60% B at 5.64 min; 95% B at 5.94 min; 95% B at 7.50 min. In summary, a total time of 12 min was required for each analysis, including 4 min for re-equilibration. The electrospray was applied in positive ionization mode. The desolvation gas temperature was 500 °C with a flow of 700 L/h, and the capillary cone was set at 700 V. On the other hand, the cone gas flow was 10 L/h, the temperature of the source was 150 °C and the cone voltage was 20 V. Finally, the trap collision energy was 4 eV. Full-scan mode in a 100–1200 m/z range was employed for the identification of the anthocyanins

The major anthocyanins identified in the purple potato extracts were [M$^+$]: 2. Petunidin 3-p-coumaroylrutinoside-5-glucoside (m/z 933.2390); 4. Peonidin 3-p-coumaroylrutinoside-5-glucoside (m/z 917.2419); 5. Malvidin 3-p-coumaroylrutinoside-5-glucoside (m/z 947.2563); 7. Malvidin 3-feruoylrutinoside-5-glucoside (m/z 977.2499). The major anthocyanins identified in the red potato extracts were [m$^+$]: 1. Pelargonidin 3-p-coumaroylrutinoside-5-glucoside (m/z 887.2379); 3. Pelargonidin 3-p-coumaroylrutinoside-5-glucoside (m/z 887.2385); 6. Pelargonidin derivative (m/z 887.2387).

The identified anthocyanins are in agreement with those described by Gutierrez-Quequezana et al. [42] and by Kita et al. [61]. The mass chromatograms of the identified anthocyanins are included in the supplementary material (Figures S1 and S2).

2.5. Separations and Quantification of the Anthocyanins by UHPLC–UV–Vis

Once anthocyanins were identified, their separation and quantification in the purple and red potato samples was conducted by means of an Elite UHPLC LaChrom System (Hitachi, Tokyo, Japan). This system features an L-2200U autosampler, an L2300 column oven, which was set up at 50 °C, two L-2160U pumps and a UV–Vis detector L-2420U. The last one was set at 520 nm for identification purposes. A reverse–phase C18 (Phenomenex, Kinetex, CoreShell Technology, Torrance, CA, USA) 2.1 × 50 mm and 2.6 μm particle size column was employed. As mentioned above, phase A was water with 5% of formic acid, and phase B was pure methanol. In order to avoid impurities and possible bubbles, both solvents were filtered using a 0.22 μm filter (RephiLe Bioscience, Ltd., Shanghai, China) and degassed using an ultrasonic bath (Elma S300, Elmasonic, Singen, Germany). The flow rate was 0.7 mL/min.

A 0.22 μm nylon syringe filter (Membrane Solutions, Dallas, TX, USA) was used for filtering the extracts before their analysis. The injection volume was set at 15 μL, and the time and % of solvent B for the UHPLC separation were the following: 0.00 min, 2%; 1.50 min, 2%; 3.30 min, 15%; 4.80 min, 15%; 5.40 min, 35%; 6 min, 100%.

This method was chosen because it allows the separation of the seven major anthocyanins identified in the samples in less than 7 min. Short-time analysis represents an important advantage, particularly for quality control laboratories, where hundreds of analyses are to be completed on a daily basis.

Cyanidin chloride was selected as a reference standard; the calibration curve obtained (y = 300,568.88x − 28,462.43) exhibited a coefficient of regression $R^2 = 0.9999$. Furthermore, the limits of detection (LOD) and quantification (LOQ) were 0.198 mg L^{-1} and 0.662 mg L^{-1}, respectively. The normal distribution of residuals was tested by Shapiro-Wilk test, and a W value of 0.8514 (very close to 1) was obtained, as well as a p-value of 0.803 (above 0.05), which confirms hypothesis H$_0$. Because different anthocyanins show similar absorbance values and considering the individual molecular weights, this calibration curve was used to quantify the different anthocyanin contents in the potato extract samples. All the analyses were performed in duplicate.

2.6. DPPH Analysis

The antioxidant activity of the anthocyanins found in purple and red potato samples was evaluated using DPPH assays based on the procedure designed by Brand-Williams et al. [62] with the modifications implemented by Miliauskas et al. [63]. The α-diphenyl-β-picrylhydrazyl (DPPH; $C_{18}H_{12}N_5O_6$) molecule is characterized by the delocalization of the spare electron in the molecule, which is the reason why these molecules do not dimerize like most other free radicals do [64]. This specific characteristic makes DPPH a stable free radical that exhibits a violet color in the solution due to a strong absorption band at about 515 nm. When DPPH is in the presence of a substance that can donate a hydrogen atom (an antioxidant substance), the odd electron from the nitrogen atom in DPPH is reduced. This causes a change in the color of the solution from violet into pale yellow that can be measured at

515 nm and that can be used to determine the antioxidant activity of any compound that is mixed with it [59].

For this research, a 6×10^{-5} M DPPH solution in methanol was prepared. A 2 mL volume of this solution was mixed with 100 µL of purple or red potato extract and with 900 µL of Tris-HCl buffer (0.1 M at pH 7.4). The mixture was kept at room temperature in the absence of light for 40 min. After that, the absorbance of the solution was measured at 515 nm.

Trolox was selected as the standard, and a six-point linear regression model was calculated from 0 until 1.4 mM in triplicate. The regression equation obtained (y = 88.94x + 0.75) exhibited a determination coefficient $R^2 = 0.9959$. The antioxidant activity was expressed as mg of Trolox equivalents (TE) per gram of dry weight potato (mg TE g^{-1} DW).

2.7. Optimization Study

After completing the 27 extractions for the BBD-RSM and evaluating the extraction variables of the total anthocyanins extractions, a second-order polynomial equation, where all the variables were considered, was applied.

The polynomial equation is the following:

$$Y = \beta_0 + \beta_1 X_1 + \beta_2 X_2 + \beta_3 X_3 + \beta_4 X_4 + \beta_{12} X_1 X_2 + \beta_{13} X_1 X_3 + \beta_{14} X_1 X_4 + \beta_{23} X_2 X_3 + \beta_{24} X_2 X_4 + \beta_{34} X_3 X_4 + \beta_{11} X_1^2 + \beta_{22} X_2^2 + \beta_{33} X_3^2 + \beta_{44} X_4^2 \quad (1)$$

where Y is the aforesaid response, and β_0 corresponds to the ordinate, whereas X_1 (% MeOH in the solvent), X_2 (extraction temperature), X_3 (pH of the solvent) and X_4 (sample-to-solvent ratio) are independent variables. Finally, β_i are the linear coefficients, β_{ij} are the cross-product coefficients and β_{ii} are the quadratic coefficients.

The effect of each factor and of their interactions on the response variable, the second-order mathematical model, the surface graphs, the optimal levels of the significant variables and the variance analysis were calculated by means of Statgraphic Centurion software (version XVII) (Statgraphics Technologies, Inc., The Plains, VA, USA) and Minitab software (version X) (Minitab LLC, State College, PA, USA). After that, repeatability and intermediate precision tests under the established optimum conditions were conducted. For the repeatability analysis, eight extractions were performed on the same day, and for the intermediate precision, eight extractions were conducted on three days in a row (a total of 24 extractions). The coefficient of variation was the statistical parameter used as the reference to determine the repeatability and intermediate precision levels of the optimized method.

3. Results and Discussion

3.1. Optimization of the Method

Four variables were selected to optimize the UAE methodology in order to maximize anthocyanins recoveries from purple potato samples. The BBD-RSM was employed, and a total of 27 extracts were obtained under different conditions and analyzed by UHPLC-UV-vis. The measured and predicted values were correlated (Table 1), and an average 7.3% difference was obtained, ranging between 0.18% and 17.7%.

The model obtained showed an R-squared statistic of 0.89, and the p-value for Durbin-Watson in the ANOVA (0.257) was above 0.05, which means that there are no significant differences between the predicted and the observed values.

The t-test using Minitab software with a 95% confidence level was applied to obtain the p-values for each of the optimized variables. Consequently, the variables presenting values of $p < 0.05$ were selected as influential. These data can be seen in Table 2.

Table 2. Results from the BBD-RSM analysis on the anthocyanins obtained from purple potato samples.

Variable	Sum of Squares	F Value	p-Value
%MeOH	1.36×10^{11}	41.92	0.00
Temperature	6.66×10^{10}	20.52	0.00
pH	3.18×10^{10}	9.81	0.01
Ratio	8.73×10^{8}	0.27	0.61
%MeOH*%MeOH	6.57×10^{9}	2.02	0.18
%MeOH*Temperature	1.64×10^{9}	0.50	0.49
%MeOH*pH	9.60×10^{8}	0.30	0.60
%MeOH*Ratio	4.97×10^{7}	0.02	0.90
Temperature*Temperature	7.08×10^{9}	2.18	0.17
Temperature*pH	4.05×10^{8}	0.12	0.73
Temperature*Ratio	2.01×10^{8}	0.06	0.81
pH*pH	2.17×10^{10}	6.70	0.02
pH*Ratio	2.53×10^{9}	0.78	0.39
Ratio*Ratio	2.07×10^{5}	0.00	0.99
Error total	3.24×10^{9}		

As can be observed in Table 2, the percentage of methanol (p = 0.00), the solvent pH (p = 0.01) and the extraction temperature (p = 0.00) exhibited a clear influence on the normalized area of the anthocyanins. The ratio was the only independent variable that did not significantly influence anthocyanin extraction from purple potato samples. On the other hand, the quadratic interaction of the solvent pH (p = 0.02) was the only interaction with a significant influence as well. The rest of the interactions between the optimized variables did not have a significant influence on the extractions.

In order to determine the influence from each variable, a Pareto chart was elaborated (Figure 1), and a positive correlation with the percentage of solvent (methanol), the temperature and the quadratic interaction of the solvent pH was observed, while solvent pH had a negative effect on the recovery of the anthocyanins.

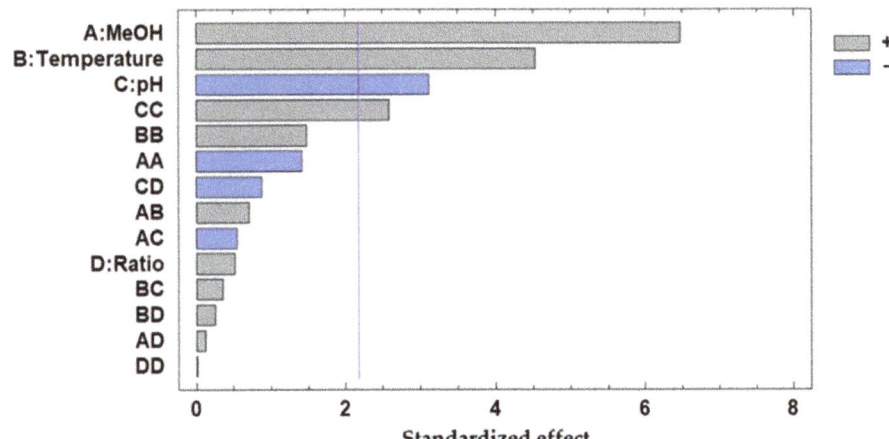

Figure 1. Pareto chart corresponding to the BBD-RSM analyses of the anthocyanins in the purple potato sample extracts. A: %MeOH in the solvent; B: extraction temperature; C: extraction solvent pH; D: sample-to-solvent ratio.

The second-order polynomial equation to calculate the total extraction of anthocyanins under the established optimal conditions according to the coefficients resulting from the BBD–RSM analyses is:

$$Y = 114{,}049.0 + 106{,}440.0X_1 + 74{,}471.3X_2 - 51{,}482.7X_3 + 8528.2X_4 + 20{,}227.2X_1X_2 - 15{,}495.5X_1X_3 + 3523.3X_1X_4 + 10{,}062.5X_2X_3 + 7089.9X_2X_4 - 25{,}124.8X_3X_4 - 35{,}089.0X_1^2 + 36{,}446.9X_2^2 + 63{,}815.8X_3^2 + 197.1X_4^2 \quad (2)$$

Extraction temperature and solvent pH were established as influential variables, and the interaction between them was evaluated by means of a surface response graph (Figure 2). The dark blue areas represent the lowest relative areas, whereas the highest relative area appears in light green color and corresponds to the conditions required for the largest possible anthocyanins recovery. In this case, the area representing the maximum anthocyanins recovery was obtained when the solvent pH and the temperature were close to the highest values in the range.

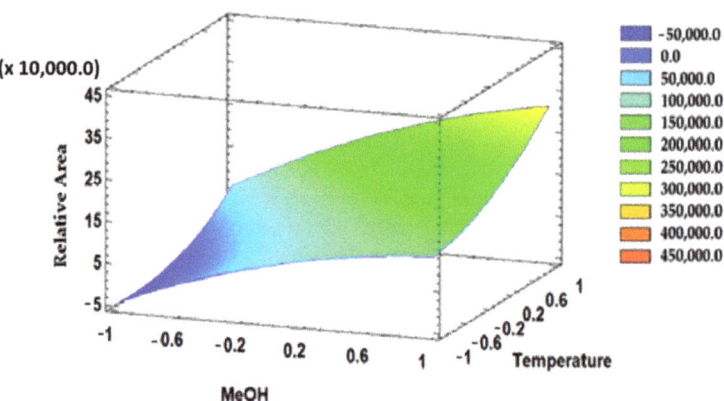

Figure 2. Surface response diagram representing the effect of temperature and percentage of methanol in the solvent with respect to the relative area of the anthocyanins extracted.

The optimal conditions for the maximum recovery of anthocyanins from purple potato were as follows: 0.5 g of sample extracted adding 20 mL of solvent at 50% methanol in water at pH 2.90 and at 70 °C. These conditions were in agreement with those obtained by other authors when extracting anthocyanins from other matrices [28,46]. The optimal temperature was at the highest end of the range studied (70 °C). Given that higher temperature levels could make the solvent boil and affect the final recovery, the authors decided not to run any tests beyond the established range.

The percentage of methanol in the solvent was at the highest value within the range as well. On the other hand, because some authors have observed that greater extraction of anthocyanins is achieved when methanol percentages are above 50%, especially from plant matrices [30,65,66], higher methanol concentrations were also tested (50–100%).

These extractions were carried out under the previously established optimal conditions (0.5 g of sample, 70 °C, 20 mL of solvent with a pH of 2.90), while the percentage of methanol was being modified. Each extraction was carried out in triplicate, and the total response (total anthocyanins) was registered as previously explained.

The relative area of the anthocyanins when different percentages of methanol were used is shown in Figure 3. Standard deviations of data were included as error bars. Thus, it was observed that the relative area reached its maximum level when the percentage of methanol was 60%, with significant differences to the rest of the percentages evaluated. At values above 60%, the relative area went down, with significant differences when methanol concentration was between 70% and 100%. Finally, the minimum relative area was registered at the maximum percentage of methanol concentration (100%). For this reason, it was determined that the optimum percentage of methanol in the solvent was 60%.

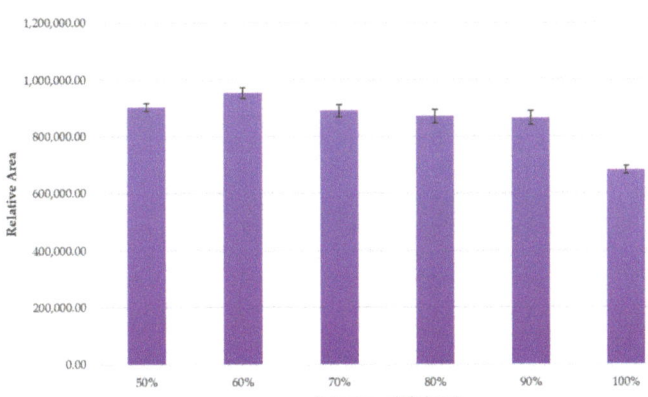

Figure 3. Anthocyanins' relative area according to different methanol percentages in the extraction solvent (n = 3).

3.2. Optimal Extraction Time of the Method

Once the optimal extraction conditions were established, the optimal extraction time was to be determined. For that purpose, 0.5 g of purple potato sample was extracted at 70 °C using 20 mL of solvent formed by 60% methanol in water with 2.90 pH and a 0.5 s^{-1} cycle at 70% of the ultrasound maximum amplitude. For these extraction runs, different times were studied from 5 up to 25 min. All the extractions were run in triplicate, and the relative area was the response variable used to determine optimal time. Results can be found in Figure 4. Standard deviations of data were included as error bars.

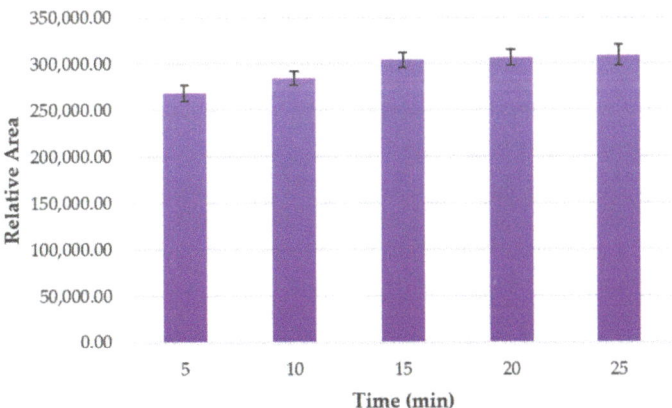

Figure 4. Anthocyanins' relative area according to the different extraction times (n = 3).

A notable increment in the relative area of the anthocyanins was observed when the time was increased, with significant differences between 5, 10 and 15 min, achieving the maximum relative area at 15 min of extraction. However, no significant differences could be noticed when longer times were used. Consequently, 15 min of extraction was established as the optimum time for extracting anthocyanins from purple potato samples.

3.3. Repeatability and Intermediate Precision

Once the method was optimized, the following step consisted of evaluating its repeatability and intermediate precision to ensure the accuracy and precision for the detection and quantification of anthocyanins for quality control purposes of the raw materials and other products derived from purple potatoes. Thus, for the repeatability evaluation, eight extractions were carried out on the same day by applying the established optimal conditions. Likewise, for the intermediate precision assessment, eight extractions were performed on two different days up to a total of 24 extractions (n = 8 + 8 + 8). The average relative area, together with the residual relative standard deviation (RSD) resulting from these extractions, is presented in Table 3.

Table 3. Repeatability and intermediate precision results of the developed method.

	Repeatability	Intermediate Precision
Average	304,245.41	297,867.39
SD *	8012.67	10,154.10
RSD **	2.63	3.41

Repeatability (n = 8); intermediate precision (n = 24); * standard deviation; ** relative standard deviation.

As can be observed, the RSD resulting from the repeatability and intermediate precision tests were 2.63% and 3.41%, respectively. Because both of them were below 5%, it was confirmed that the optimized method for the extraction of anthocyanins from purple potato exhibited a high precision level.

3.4. Re-Extraction Analysis

Once the extraction method was developed, in order to determine its effectiveness, a re-extraction study was carried out. To do so, the resulting residue was re-extracted after the double centrifugation of the extract. This residue was again subjected to an extraction process under the optimal conditions of the method. The extracts obtained were analyzed to determine the concentration of anthocyanins in this second extraction cycle. The test was carried out in triplicate. After analyzing the extracts, a concentration of 3.16% was obtained with respect to the amount of anthocyanins obtained in the first extraction cycle. This amount is less than 5%, so it is considered that one extraction cycle is sufficient for a qualitative extraction of the anthocyanins present in the potato by using the method developed by UAE.

3.5. Applying the Optimized Method to Different Potato Varieties

The anthocyanins contained in the four potato varieties cultivated (Vitelotte, Double Fun, Highland and Violet Queen) were extracted by applying the optimized method. For this purpose, 0.5 g of samples from each variety was subjected to the developed extraction method using 20 mL of solvent (60:40 MeOH:H$_2$O, pH 2.90) at 70 °C with a 0.5 s^{-1} cycle and at 70% of the maximum amplitude for 15 min. These extractions were performed in triplicate.

UHPLC–UV–vis was used for anthocyanin quantification and the DPPH methodology for the evaluation of the antioxidant activity of the extracts. The results are displayed in Table 4. It can be observed that the purple varieties exhibit a larger anthocyanin content than the red variety (Highland). In addition, their anthocyanin profile was also different, as the purple varieties contained mainly petunidin and malvidin derivatives, while the red variety content was mostly formed of pelargonidin derivatives. On the other hand, malvidin 3-*p*-coumaroylrutinoside-5-glucoside was the main anthocyanin present in the Vitelotte and Double Fun varieties, whereas petunidin 3-*p*-coumaroylrutinoside-5-glucoside was the most abundant anthocyanin in Violet Queen's samples (Figure S3). A one-way ANOVA was performed, and differences between data are represented in Table 4. As can be observed, significant differences were observed in petunidin 3-*p*-coumaroylrutinoside-5-glucoside between the Violet Queen variety and Vitelotte and Double Fun varieties. Significant differences were also detected in malvidin 3-*p*-coumaroylrutinoside-5-glucoside and malvidin 3-feruoylrutinoside-5-glucoside. The red variety showed the lowest anthocyanin concentration, but significant differences between the purple varieties were also detected according to the total anthocyanin content.

Table 4. Concentration of anthocyanins (mg g^{-1} DW) and antioxidant activity (DPPH, mg TE g^{-1} DW) of purple and red flesh potatoes. Different letters on the same line mean that the values are significantly different.

Compound (mg/g DW)/Variety	Vitelotte	Double Fun	Violet Queen	Highland
1	n.d.	n.d.	n.d.	0.14 ± 0.00
2	0.66 ± 0.05 [a]	0.59 ± 0.02 [a]	3.82 ± 0.10 [b]	n.d.
3	n.d.	n.d.	n.d.	1.12 ± 0.01
4	n.d.	n.d.	0.14 ± 0.01	n.d.
5	3.97 ± 0.14 [c]	1.98 ± 0.09 [b]	0.91 ± 0.09 [a]	n.d.
6	n.d.	n.d.	n.d.	0.22 ± 0.00
7	0.34 ± 0.08 [b]	0.21 ± 0.03 [a]	n.d.	n.d.
Total anthocyanins (mg/g)	4.97 ± 0.17 [c]	2.78 ± 0.10 [b]	4.87 ± 0.13 [c]	1.48 ± 0.01 [a]
DPPH (mg TE/g DW)	28.81	21.67	29.87	11.19
Flesh Color	Purple	Purple/White	Purple	Red

1: Pelargonidin 3-*p*-coumaroylrutinoside-5-glucoside; 2: petunidin 3-*p*-coumaroylrutinoside-5-glucoside; 3: pelargonidin 3-*p*-coumaroylrutinoside-5-glucoside; 4: peonidin 3-*p*-coumaroylrutinoside-5-glucoside; 5: malvidin 3-*p*-coumaroylrutinoside-5-glucoside; 6: pelargonidin derivative; 7: malvidin 3-feruoylrutinoside-5-glucoside. n.d.: not detected.

Regarding their antioxidant activity, it was observed that the purple flesh varieties exhibited a notably higher mg TE g^{-1} DW than the varieties with red flesh. This result agrees with those found in the literature [67,68]. In addition, the values obtained were within the same range as those reported by other authors. For example, the Vitelotte variety exhibited 28.81 mg TE g^{-1} DW, with 14.4–42.4 mg TE g^{-1} DW the usual range reported in the bibliography. On the other hand, the Highland variety showed antioxidant activity of 11.9 mg TE g^{-1} DW with a usual range between 8.0 and 27.8 mg TE g^{-1} DW [61,69].

4. Conclusions

The present study consists of the optimization of a novel method for the UAE extraction and analysis of anthocyanins from purple potatoes. The Vitelotte variety was selected for this purpose. A Box-Behnken design (BBD) combined with response surface methodology (RSM) was employed for the optimization of four relevant variables, and the total yield of the four anthocyanins previously identified was used as the response variable.

The optimal conditions were established as 0.5 g of sample extracted using 20 mL of 60% methanol solvent in water at pH 2.9 at 70 °C with a cycle and amplitude of 0.5 and 70%, respectively. Finally, 15 min was established as the optimum extraction time. The repeatability and intermediate precision of the optimized methodology were confirmed by an RSD below 5% in all cases. In addition, re-extraction analysis was performed, but the results proved that one cycle of extraction with the optimized methodology is enough to ensure the maximum extraction of anthocyanins from purple potato samples.

After that, the developed method was applied to three purple varieties (Vitelotte, Double Fun and Violet Queen) and a red variety (Highland). The results corroborated the suitability of the optimized method for the detection and quantification of anthocyanins from purple and red potato samples. In addition, significant differences regarding anthocyanins content were observed.

Finally, the antioxidant activity of the extracts was measured using the DPPH method, and highly significant variety-related differences were observed. Furthermore, the developed method was confirmed to preserve the antioxidant activity of the anthocyanins extracted from the potato samples.

As previously mentioned, the composition of either raw material or elaborated products from purple potatoes may be altered or degraded as a result of certain processing methods or for other causes. For this reason, it would be really interesting to count on a rapid, reliable, economic and straightforward methodology to control the quality of the anthocyanins in these products.

Supplementary Materials: The following materials are available online at https://www.mdpi.com/article/10.3390/antiox10091375/s1. Figure S1: Chromatograms obtained for the potato variety Violet Queen by UHPLC-PDA-QToF-MS. The m/z identified were m/z 933.2390 (Time: 6.29 min: Compound 2. Petunidin 3-p-coumaroylrutinoside-5-glucoside), m/z 917.2419 (Time: 6.53 min: Compound 4. Peonidin 3-p-coumaroylrutinoside-5-glucoside), m/z 947.2563 (Time: 6.55 min: Compound 5. Malvidin 3-p-coumaroylrutinoside-5-glucoside), and m/z 977.2499 (Time 6.62 min: Compound 7. Malvidin 3-feruoylrutinoside-5-glucoside). Figure S2: Chromatogram obtained for the potato variety Highland by UHPLC-PDA-QToF-MS. The m/z identified was m/z 887.2387, corresponding with (Time: 5.92 min: Compound 1. Pelargonidin 3-p-coumaroylrutinoside-5-glucoside; Time: 6.46 min: Compound 3. Pelargonidin 3-p-coumaroylrutinoside-5-glucoside; Time: 6.59 min: Compound 6. Pelargonidin derivative). Figure S3: Overlay of the UHPLC chromatogram of the major anthocyanins in purple-fleshed Vitelotte (blue) and Violet Queen (red) potato varieties at 520 nm. **2:** Petunidin 3-p-coumaroylrutinoside-5-glucoside; **5:** Malvidin 3-p-coumaroylrutinoside-5-glucoside.

Author Contributions: Conceptualization, C.C. and M.P.; methodology, G.F.B. and C.C.; software, M.J.A.-G., M.F.-G. and M.V.; validation, M.J.A.-G. and M.F.-G.; formal analysis, C.C., M.V. and M.J.A.-G.; investigation, M.F.-G. and C.C.; resources, M.P. and G.F.B.; data curation, C.C.; writing—original draft preparation, C.C. and M.J.A.-G.; writing—review and editing, M.P., M.F.-G. and G.F.B.; visualization, G.F.B. and M.V.; supervision, M.P., M.F.-G. and G.F.B.; project administration, G.F.B. and M.P.; funding acquisition, G.F.B. and M.P. All authors have read and agreed to the published version of the manuscript.

Funding: This work has been supported by the project "EQC2018-005135-P" (Equipment for liquid chromatography by means of mass spectrometry and ion chromatography) of the State Subprogram of Research Infrastructures and Technical Scientific Equipment.

Institutional Review Board Statement: Not applicable.

Informed Consent Statement: Not applicable.

Data Availability Statement: The data presented in this study are contained within the article and Supplementary Materials.

Acknowledgments: The authors are grateful to the Instituto de Investigación Vitivinícola y Agroalimentaria (IVAGRO) for providing the necessary facilities to carry out the research.

Conflicts of Interest: The authors declare no conflict of interest.

References

1. Pęksa, A.; Miedzianka, J.; Nemś, A.; Rytel, E. The free-amino-acid content in six potatoes cultivars through storage. *Molecules* **2021**, *26*, 1322. [CrossRef] [PubMed]
2. Xu, A.; Guo, K.; Liu, T.; Bian, X.; Zhang, L.; Wei, C. Effects of different isolation media on structural and functional properties of starches from root tubers of purple, yellow and white sweet potatoes. *Molecules* **2018**, *23*, 2135. [CrossRef] [PubMed]
3. Noguerol, A.T.; Igual, M.; Pagán-Moreno, M.J. Nutritional, physico-chemical and mechanical characterization of vegetable fibers to develop fiber-based gel foods. *Foods* **2021**, *10*, 1017. [CrossRef]
4. Gaudino, E.C.; Colletti, A.; Grillo, G.; Tabasso, S.; Cravotto, G. Emerging processing technologies for the recovery of valuable bioactive compounds from potato peels. *Foods* **2020**, *9*, 1598. [CrossRef] [PubMed]
5. Akyıldız, A.; Polat, S.; Agçam, E.; Fenercioglu, H. Potato and potato-processing technology. In *Handbook of Vegetable Preservation and Processing*; Hui, Y.H., Evranuz, E.Ö., Eds.; CRC Press: Boca Raton, FL, USA, 2015; pp. 591–594.
6. Ru, W.; Pang, Y.; Gan, Y.; Liu, Q.; Bao, J. Phenolic compounds and antioxidant activities of potato cultivars with white, yellow, red and purple flesh. *Antioxidants* **2019**, *8*, 419. [CrossRef] [PubMed]
7. Durham, C.A.; Wechsler, L.J.; Morrissey, M.T. Using a fractional model to measure the impact of antioxidant information, price, and liking on purchase intent for specialty potatoes. *Food Qual. Prefer.* **2015**, *46*, 66–78. [CrossRef]
8. Qiu, G.; Wang, D.; Song, X.; Deng, Y.; Zhao, Y. Degradation kinetics and antioxidant capacity of anthocyanins in air-impingement jet dried purple potato slices. *Food Res. Int.* **2018**, *105*, 121–128. [CrossRef] [PubMed]

9. Sun, J.; Chen, H.; Kan, J.; Gou, Y.; Liu, J.; Zhang, X.; Wu, X.; Tang, S.; Sun, R.; Qian, C.; et al. Anti-inflammatory properties and gut microbiota modulation of an alkali-soluble polysaccharide from purple sweet potato in DSS-induced colitis mice. *Int. J. Biol. Macromol.* **2020**, *153*, 708–722. [CrossRef]
10. Jiang, T.; Zhou, J.; Liu, W.; Tao, W.; He, J.; Jin, W.; Guo, H.; Yang, N.; Li, Y. The anti-inflammatory potential of protein-bound anthocyanin compounds from purple sweet potato in LPS-induced RAW264. 7 macrophages. *Food Res. Int.* **2020**, *137*, 109647. [CrossRef]
11. Torres, A.; Noriega, L.G.; Delgadillo-Puga, C.; Tovar, A.R.; Navarro-Ocaña, A. Caffeoylquinic acid derivatives of purple sweet potato as modulators of mitochondrial function in mouse primary hepatocytes. *Molecules* **2021**, *26*, 319. [CrossRef]
12. Strugała, P.; Dzydzan, O.; Brodyak, I.; Kucharska, A.Z.; Kuropka, P.; Liuta, M.; Kaleta-Kuratewicz, K.; Przewodowska, A.; Michałowska, D.; Gabrielska, J.; et al. Antidiabetic and antioxidative potential of the blue congo variety of purple potato extract in streptozotocin-induced diabetic rats. *Molecules* **2019**, *24*, 3126. [CrossRef] [PubMed]
13. Sun, Y.; Pan, Z.; Yang, C.; Jia, Z.; Guo, X. Comparative assessment of phenolic profiles, cellular antioxidant and antiproliferative activities in ten varieties of sweet potato (*Ipomoea Batatas*) storage roots. *Molecules* **2019**, *24*, 4476. [CrossRef] [PubMed]
14. Lee, S.G.; Chae, J.; Kim, D.S.; Lee, J.-B.; Kwon, G.-S.; Kwon, T.K.; Nam, J.-O. Enhancement of the antiobesity and antioxidant effect of purple sweet potato extracts and enhancement of the effects by fermentation. *Antioxidants* **2021**, *10*, 888. [CrossRef] [PubMed]
15. Petropoulos, S.A.; Sampaio, S.L.; Di Gioia, F.; Tzortzakis, N.; Rouphael, Y.; Kyriacou, M.C.; Ferreira, I. Grown to be blue—Antioxidant properties and health effects of colored vegetables. Part I: Root vegetables. *Antioxidants* **2019**, *8*, 617. [CrossRef]
16. Kostyn, K.; Boba, A.; Kostyn, A.; Kozak, B.; Starzycki, M.; Kulma, A.; Szopa, J. Expression of the tyrosine hydroxylase gene from rat leads to oxidative stress in potato plants. *Antioxidants* **2020**, *9*, 717. [CrossRef]
17. Im, Y.R.; Kim, I.; Lee, J. Phenolic composition and antioxidant activity of purple sweet potato (*Ipomoea Batatas* (L.) Lam.): Varietal comparisons and physical distribution. *Antioxidants* **2021**, *10*, 462. [CrossRef]
18. Frond, A.D.; Iuhas, C.I.; Stirbu, I.; Leopold, L.; Socaci, S.; Andreea, S.; Ayvaz, H.; Andreea, S.; Mihai, S.; Diaconeasa, Z.; et al. Phytochemical characterization of five edible purple-reddish vegetables: Anthocyanins, flavonoids, and phenolic acid derivatives. *Molecules* **2019**, *24*, 1536. [CrossRef]
19. Sun, Q.; Du, M.; Navarre, D.A.; Zhu, M. Effect of cooking methods on bioactivity of polyphenols in purple potatoes. *Antioxidants* **2021**, *10*, 1176. [CrossRef]
20. Fallah, A.A.; Sarmast, E.; Jafari, T. Effect of dietary anthocyanins on biomarkers of oxidative stress and antioxidative capacity: A systematic review and meta-analysis of randomized controlled trials. *J. Funct. Foods* **2020**, *68*, 103912. [CrossRef]
21. Belwal, T.; Singh, G.; Jeandet, P.; Pandey, A.; Giri, L.; Ramola, S.; Bhatt, I.D.; Venskutonis, P.R.; Georgiev, M.I.; Clément, C.; et al. Anthocyanins, multi-functional natural products of industrial relevance: Recent biotechnological advances. *Biotechnol. Adv.* **2020**, *43*, 107600. [CrossRef]
22. Li, G.; Lin, Z.; Zhang, H.; Liu, Z.; Xu, Y.; Xu, G.; Li, H.; Ji, R.; Luo, W.; Qiu, Y.; et al. Anthocyanin accumulation in the leaves of the purple sweet potato (*Ipomoea Batatas* L.) cultivars. *Molecules* **2019**, *24*, 3743. [CrossRef]
23. Aliaño-González, M.J.; Espada-Bellido, E.; Ferreiro-González, M.; Carrera, C.; Palma, M.; Ayuso, J.; Álvarez, J.Á.; Barbero, G.F. Extraction of anthocyanins and total phenolic compounds from Açai (*Euterpe Oleracea Mart.*) using an experimental design methodology. Part 2: Ultrasound-assisted extraction. *Agronomy* **2020**, *10*, 326. [CrossRef]
24. González-De-Peredo, A.V.; Vázquez-Espinosa, M.; Espada-Bellido, E.; Ferreiro-González, M.; Carrera, C.; Palma, M.; Álvarez, J.Á.; Barbero, G.F.; Ayuso, J. Optimization of analytical ultrasound-assisted methods for the extraction of total phenolic compounds and anthocyanins from sloes (*Prunus Spinosa* L.). *Agronomy* **2020**, *10*, 966. [CrossRef]
25. Aliaño-González, M.J.; Jarillo, J.A.; Carrera, C.; Ferreiro-González, M.; Álvarez, J.Á.; Palma, M.; Ayuso, J.; Barbero, G.F.; Espada-Bellido, E. Optimization of a novel method based on ultrasound-assisted extraction for the quantification of anthocyanins and total phenolic compounds in blueberry samples (*Vaccinium Corymbosum* L.). *Foods* **2020**, *9*, 1763. [CrossRef] [PubMed]
26. Tang, P.; Giusti, M.M. Metal chelates of petunidin derivatives exhibit enhanced color and stability. *Foods* **2020**, *9*, 1426. [CrossRef] [PubMed]
27. González-De-Peredo, A.V.; Vázquez-Espinosa, M.; Carrera, C.; Espada-Bellido, E.; Ferreiro-González, M.; Barbero, G.F.; Palma, M. Development of a rapid UHPLC-PDA method for the simultaneous quantification of flavonol contents in onions (*Allium Cepa* L.). *Pharmaceuticals* **2021**, *14*, 310. [CrossRef]
28. Xue, H.; Tan, J.; Li, Q.; Tang, J.; Cai, X. Optimization ultrasound-assisted deep eutectic solvent extraction of anthocyanins from raspberry using response surface methodology coupled with genetic algorithm. *Foods* **2020**, *9*, 1409. [CrossRef] [PubMed]
29. Ereminas, G.; Majiene, D.; Sidlauskas, K.; Jakstas, V.; Ivanauskas, L.; Vaitiekaitis, G.; Liobikas, J. Neuroprotective properties of anthocyanidin glycosides against H_2O_2-induced glial cell death are modulated by their different stability and antioxidant activity in vitro. *Biomed. Pharmacother.* **2017**, *94*, 188–196. [CrossRef]
30. Bendokas, V.; Stanys, V.; Mažeikienė, I.; Trumbeckaite, S.; Baniene, R.; Liobikas, J. Anthocyanins: From the field to the antioxidants in the body. *Antioxidants* **2020**, *9*, 819. [CrossRef]
31. Tena, N.; Asuero, A.G. Antioxidant capacity of anthocyanins and other vegetal pigments. *Antioxidants* **2020**, *9*, 665. [CrossRef]
32. Jokioja, J.; Linderborg, K.M.; Kortesniemi, M.; Nuora, A.; Heinonen, J.; Sainio, T.; Viitanen, M.; Kallio, H.; Yang, B. Anthocyanin-rich extract from purple potatoes decreases postprandial glycemic response and affects inflammation markers in healthy men. *Food Chem.* **2020**, *310*, 125797. [CrossRef]

33. Han, Y.; Guo, Y.; Cui, S.W.; Li, H.; Shan, Y.; Wang, H. Purple sweet potato extract extends lifespan by activating autophagy pathway in male drosophila melanogaster. *Exp. Gerontol.* **2021**, *144*, 111190. [CrossRef] [PubMed]
34. Ji, C.; Zhang, Z.; Zhang, B.; Chen, J.; Liu, R.; Song, D.; Li, W.; Lin, N.; Zou, X.; Wang, J.; et al. Purification, characterization, and in vitro antitumor activity of a novel glucan from the purple sweet potato *Ipomoea Batatas* (L.) Lam. *Carbohydr. Polym.* **2021**, *257*, 117605. [CrossRef] [PubMed]
35. Baenas, N.; Ruales, J.; Moreno, D.A.; Barrio, D.A.; Stinco, C.M.; Martínez-Cifuentes, G.; Meléndez-Martínez, A.J.; García-Ruiz, A. Characterization of Andean blueberry in bioactive compounds, evaluation of biological properties, and in vitro bioaccessibility. *Foods* **2020**, *9*, 1483. [CrossRef] [PubMed]
36. Kang, J.; Xie, C.; Li, Z.; Nagarajan, S.; Schauss, A.G.; Wu, T.; Wu, X. Flavonoids from acai (*Euterpe Oleracea Mart.*) pulp and their antioxidant and anti-inflammatory activities. *Food Chem.* **2011**, *128*, 152–157. [CrossRef]
37. Vergara, C.; Pino, M.T.; Zamora, O.; Parada, J.; Pérez, R.; Uribe, M.; Kalazich, J. Microencapsulation of anthocyanin extracted from purple flesh cultivated potatoes by spray drying and its effects on in vitro gastrointestinal digestion. *Molecules* **2020**, *25*, 722. [CrossRef]
38. Ercoli, S.; Parada, J.; Bustamante, L.; Hermosín-Gutiérrez, I.; Contreras, B.; Cornejo, P.; Ruiz, A. Noticeable quantities of functional compounds and antioxidant activities remain after cooking of colored fleshed potatoes native from Southern Chile. *Molecules* **2021**, *26*, 314. [CrossRef]
39. Gutiérrez-Quequezana, L.; Vuorinen, A.L.; Kallio, H.; Yang, B. Impact of cultivar, growth temperature and developmental stage on phenolic compounds and ascorbic acid in purple and yellow potato tubers. *Food Chem.* **2020**, *326*, 126966. [CrossRef]
40. Oertel, A.; Matros, A.; Hartmann, A.; Arapitsas, P.; Dehmer, K.J.; Martens, S.; Mock, H.P. Metabolite profiling of red and blue potatoes revealed cultivar and tissue specific patterns for anthocyanins and other polyphenols. *Planta* **2017**, *246*, 281–297. [CrossRef]
41. De Masi, L.; Bontempo, P.; Rigano, D.; Stiuso, P.; Carafa, V.; Nebbioso, A.; Piacente, S.; Montoro, P.; Aversano, R.; D'Amelia, V.; et al. Comparative phytochemical characterization, genetic profile, and antiproliferative activity of polyphenol-rich extracts from pigmented tubers of different solanum tuberosum varieties. *Molecules* **2020**, *25*, 233. [CrossRef]
42. Espada-Bellido, E.; Ferreiro-González, M.; Carrera, C.; Palma, M.; Álvarez, J.A.; Barbero, G.F.; Ayuso, J. Extraction of antioxidants from blackberry (*Rubus Ulmifolius* L.): Comparison between ultrasound- and microwave-assisted extraction techniques. *Agronomy* **2019**, *9*, 745. [CrossRef]
43. Aourach, M.; González-de-Peredo, A.V.; Vázquez-Espinosa, M.; Essalmani, H.; Palma, M.; Barbero, G.F. Optimization and comparison of ultrasound and microwave-assisted extraction of phenolic compounds from cotton-lavender (*Santolina Chamaecyparissus* L.). *Agronomy* **2021**, *11*, 84. [CrossRef]
44. Vázquez-Espinosa, M.; González de Peredo, A.V.; Ferreiro-González, M.; Barroso, C.G.; Palma, M.; Barbero, G.F.; Espada-Bellido, E. Optimizing and comparing ultrasound- and microwave-assisted extraction methods applied to the extraction of antioxidant capsinoids in peppers. *Agronomy* **2019**, *9*, 633. [CrossRef]
45. Zhu, Z.; Jiang, T.; He, J.; Barba, F.; Cravotto, G.; Koubaa, M. Ultrasound-assisted extraction, centrifugation and ultrafiltration: Multistage process for polyphenol recovery from purple sweet potatoes. *Molecules* **2016**, *21*, 1584. [CrossRef] [PubMed]
46. Pop, A.; Fizeșan, I.; Vlase, L.; Rusu, M.E.; Cherfan, J.; Babota, M.; Gheldiu, A.-M.; Tomuta, I.; Popa, D.-S. Enhanced recovery of phenolic and tocopherolic compounds from walnut (*Juglans Regia* L.) male flowers based on process optimization of ultrasonic assisted-extraction: Phytochemical profile and biological activities. *Antioxidants* **2021**, *10*, 607. [CrossRef] [PubMed]
47. Mansinhos, I.; Gonçalves, S.; Rodríguez-Solana, R.; Ordóñez-Díaz, J.L.; Moreno-Rojas, J.M.; Romano, A. Ultrasonic-assisted extraction and natural deep eutectic solvents combination: A green strategy to improve the recovery of phenolic compounds from lavandula pedunculata subsp. lusitanica (Chaytor) Franco. *Antioxidants* **2021**, *10*, 582. [CrossRef]
48. Chen, C.; Zhao, Z.; Ma, S.; Rasool, M.A.; Wang, L.; Zhang, J. Optimization of ultrasonic-assisted extraction, refinement and characterization of water-soluble polysaccharide from Dictyosphaerium sp. and evaluation of antioxidant activity in vitro. *J. Food Meas. Charact.* **2020**, *14*, 963–977. [CrossRef]
49. Pereira, D.T.V.; Zabot, G.L.; Reyes, F.G.R.; Iglesias, A.H.; Martínez, J. Integration of pressurized liquids and ultrasound in the extraction of bioactive compounds from passion fruit rinds: Impact on phenolic yield, extraction kinetics and technical-economic evaluation. *Innov. Food Sci. Emerg. Technol.* **2021**, *67*, 102549. [CrossRef]
50. Bachtler, S.; Bart, H.J. Increase the yield of bioactive compounds from elder bark and annatto seeds using ultrasound and microwave assisted extraction technologies. *Food Bioprod. Process.* **2021**, *125*, 1–13. [CrossRef]
51. Majid, H.; Silva, F.V.M. Kanuka bush leaves for Alzheimer's disease: Improved inhibition of β-secretase enzyme, antioxidant capacity and yield of extracts by ultrasound assisted extraction. *Food Bioprod. Process.* **2021**, *128*, 109–120. [CrossRef]
52. Baiano, A. Recovery of biomolecules from food wastes—A review. *Molecules* **2014**, *19*, 14821–14842. [CrossRef]
53. Celotti, E.; Stante, S.; Ferraretto, P.; Román, T.; Nicolini, G.; Natolino, A. High power ultrasound treatments of red young wines: Effect on anthocyanins and phenolic stability indices. *Foods* **2020**, *9*, 1344. [CrossRef]
54. Antoniou, C.; Kyratzis, A.; Rouphael, Y.; Stylianou, S.; Kyriacou, M.C. Heat- and ultrasound-assisted aqueous extraction of soluble carbohydrates and phenolics from carob kibbles of variable size and source material. *Foods* **2020**, *9*, 1364. [CrossRef]
55. Jiménez, N.; Bassama, J.; Bohuon, P. Estimation of the kinetic parameters of anthocyanins degradation at different water activities during treatments at high temperature (100–140 °C) using an unsteady-state 3D model. *J. Food Eng.* **2020**, *279*, 109951. [CrossRef]

56. Maran, J.P.; Manikandan, S.; Thirugnanasambandham, K.; Nivetha, C.V.; Dinesh, R. Box-behnken design based statistical modeling for ultrasound-assisted extraction of corn silk polysaccharide. *Carbohydr. Polym.* **2013**, *92*, 604–611. [CrossRef]
57. Vázquez-Espinosa, M.; González de Peredo, A.V.; Ferreiro-González, M.; Carrera, C.; Palma, M.; Barbero, G.F.; Espada-Bellido, E. Assessment of ultrasound assisted extraction as an alternative method for the extraction of anthocyanins and total phenolic compounds from maqui berries (*Aristotelia Chilensis (Mol.) Stuntz*). *Agronomy* **2019**, *9*, 148. [CrossRef]
58. González de Peredo, A.V.; Vázquez-Espinosa, M.; Piñeiro, Z.; Espada-Bellido, E.; Ferreiro-González, M.; Barbero, G.F.; Palma, M. Development of a rapid and accurate UHPLC-PDA-FL method for the quantification of phenolic compounds in grapes. *Food Chem.* **2021**, *334*, 127569. [CrossRef] [PubMed]
59. González-de-Peredo, A.V.; Vázquez-Espinosa, M.; Espada-Bellido, E.; Carrera, C.; Ferreiro-González, M.; Barbero, G.F.; Palma, M. Flavonol composition and antioxidant activity of onions (*Allium Cepa* L.) based on the development of new analytical ultrasound-assisted extraction methods. *Antioxidants* **2021**, *10*, 273. [CrossRef] [PubMed]
60. Vázquez-Espinosa, M.; González-de-Peredo, A.V.; Espada-Bellido, E.; Ferreiro-González, M.; Toledo-Domínguez, J.J.; Carrera, C.; Palma, M.; Barbero, G.F. Ultrasound-assisted extraction of two types of antioxidant compounds (TPC and TA) from black chokeberry (*Aronia Melanocarpa* L.): Optimization of the individual and simultaneous extraction methods. *Agronomy* **2019**, *9*, 456. [CrossRef]
61. Kita, A.; Bakowska-Barczak, A.; Hamouz, K.; Kułakowska, K.; Lisińska, G. The effect of frying on anthocyanin stability and antioxidant activity of crisps from red- and purple-fleshed potatoes (*Solanum Tuberosum* L.). *J. Food Compos. Anal.* **2013**, *32*, 169–175. [CrossRef]
62. Brand-Williams, W.; Cuvelier, M.E.; Berset, C. Use of a free radical method to evaluate antioxidant activity. *LWT Food Sci. Technol.* **1995**, *28*, 25–30. [CrossRef]
63. Miliauskas, G.; Venskutonis, P.R.; Van Beek, T.A. Screening of radical scavenging activity of some medicinal and aromatic plant extracts. *Food Chem.* **2004**, *85*, 231–237. [CrossRef]
64. Molyneux, P. The use of the stable free radical diphenylpicrylhydrazyl (DPPH) for estimating antioxidant activity. *Songklanakarin J. Sci. Technol.* **2004**, *26*, 211–219.
65. Carrera, C.; Aliaño-González, M.J.; Rodríguez-López, J.; Ferreiro-González, M.; Ojeda-Copete, F.; Barbero, G.F.; Palma, M. Optimization of an ultrasound-assisted extraction method for the analysis of major anthocyanin content in Erica australis flowers. *Molecules* **2021**, *26*, 2884. [CrossRef] [PubMed]
66. Mariychuk, R.; Eliasova, A.; Porubska, J.; Poracova, J.; Simko, V. Isolation and lyophilisation of anthocyanins from fruits of blackcurrant. *Acta Hortic.* **2016**, *1133*, 329–333. [CrossRef]
67. Tang, Y.; Cai, W.; Xu, B. Profiles of phenolics, carotenoids and antioxidative capacities of thermal processed white, yellow, orange and purple sweet potatoes grown in Guilin, China. *Food Sci. Hum. Wellness* **2015**, *4*, 123–132. [CrossRef]
68. Kita, A.; Bakowska-Barczak, A.; Lisińska, G.; Hamouz, K.; Kułakowska, K. Antioxidant activity and quality of red and purple flesh potato chips. *LWT Food Sci. Technol.* **2015**, *62*, 525–531. [CrossRef]
69. Lachman, J.; Hamouz, K.; Šulc, M.; Orsák, M.; Pivec, V.; Hejtmánková, A.; Dvořák, P.; Čepl, J. Cultivar differences of total anthocyanins and anthocyanidins in red and purple-fleshed potatoes and their relation to antioxidant activity. *Food Chem.* **2009**, *114*, 836–843. [CrossRef]

Article

Native Chilean Berries Preservation and In Vitro Studies of a Polyphenol Highly Antioxidant Extract from Maqui as a Potential Agent against Inflammatory Diseases

Tamara Ortiz [1], Federico Argüelles-Arias [2,3], Belén Begines [4], Josefa-María García-Montes [2], Alejandra Pereira [5], Montserrat Victoriano [6], Victoria Vázquez-Román [1], Juan Luis Pérez Bernal [7], Raquel M. Callejón [8], Manuel De-Miguel [1,*] and Ana Alcudia [4,*]

[1] Departamento de Citología e Histología Normal y Patológica, Universidad de Sevilla, Avda. Sánchez-Pizjuán s/n, 41009 Sevilla, Spain; tamara.ortiz.cerda@gmail.com (T.O.); mvazquez2@us.es (V.V.-R.)
[2] Departamento de Medicina, Universidad de Sevilla, Avda. Sánchez-Pizjuán s/n, 41009 Sevilla, Spain; farguelles1@us.es (F.A.-A.); jfgarcia@us.es (J.-M.G.-M.)
[3] Departamento de Gastroenterología, Hospital Universitario Virgen Macarena, c/Dr. Fedriani n° 3, 41009 Sevilla, Spain
[4] Departamento de Química Orgánica y Farmacéutica, Universidad de Sevilla, c/Prof García González n° 2, 41012 Sevilla, Spain; bbegines@us.es
[5] Departamento de Nutrición y Dietética, Escuela de Ciencias de la Salud, Universidad de Desarrollo Concepción Barrios Arana1735, Concepción 4070146, Chile; alejandra.pereira@udd.cl
[6] Departamento de Nutricion y Dietetica, Facultad de Farmacia, Universidad de Concepción, Concepción, Chile. Barrio Universitario s/n, Concepción 4070146, Chile; mvictoriano@udec.cl
[7] Departamento de Química Analítica, Universidad de Sevilla, c/Prof García González n° 2, 41012 Sevilla, Spain; juanluis@us.es
[8] Departamento de Nutrición y Bromatología, Toxicología y Medicina Legal, Universidad de Sevilla, c/Prof García González n° 2, 41012 Sevilla, Spain; rcallejon@us.es
* Correspondence: mmiguel@us.es (M.D.-M.); aalcudia@us.es (A.A.); Tel.: +34-955-421-025 (M.D.-M.); +34-954-556-740 (A.A.)

Abstract: The best conservation method for native Chilean berries has been investigated in combination with an implemented large-scale extract of maqui berry, rich in total polyphenols and anthocyanin to be tested in intestinal epithelial and immune cells. The methanolic extract was obtained from lyophilized and analyzed maqui berries using Folin–Ciocalteu to quantify the total polyphenol content, as well as 2,2-diphenyl-1-picrylhydrazyl (DPPH), ferric reducing antioxidant power (FRAP), and oxygen radical absorbance capacity (ORAC) to measure the antioxidant capacity. Determination of maqui's anthocyanins profile was performed by ultra-high-performance liquid chromatography (UHPLC-MS/MS). Viability, cytotoxicity, and percent oxidation in epithelial colon cells (HT-29) and macrophages cells (RAW 264.7) were evaluated. In conclusion, preservation studies confirmed that the maqui properties and composition in fresh or frozen conditions are preserved and a more efficient and convenient extraction methodology was achieved. In vitro studies of epithelial cells have shown that this extract has a powerful antioxidant strength exhibiting a dose-dependent behavior. When lipopolysaccharide (LPS)-macrophages were activated, noncytotoxic effects were observed, and a relationship between oxidative stress and inflammation response was demonstrated. The maqui extract along with 5-aminosalicylic acid (5-ASA) have a synergistic effect. All of the compiled data pointed out to the use of this extract as a potential nutraceutical agent with physiological benefits for the treatment of inflammatory bowel disease (IBD).

Keywords: maqui berry extract; preservation methods; antioxidant activity; polyphenols and anthocyanins content; oxidative stress; inflammation; RAW 264.7 cells; HT-29 cells; inflammatory bowel disease

1. Introduction

Inflammatory bowel disease (IBD) includes a group of diverse inflammatory diseases with chronic relapsing disorders and uncontrolled inflammation of the gastrointestinal tract such as ulcerative colitis (UC) or Crohn's disease (CD). Additionally, the prevalence and incidence of IBD in adults and children has increased worldwide during the last decades, mainly in Westernized countries of Europe and North America, and more recently in industrialized countries. In Europe, more than 2.2 million patients with IBD have been reported and it is increasing more and more in the last decade [1,2]. Unfortunately, no clear or definitive evidence has been found related to the pathogenesis' origin mainly due to its complexity. Urban populations, lifestyle, stress, smoking, and non-healthy diets based on high intakes of fats and poor intakes in vegetables or fruits are associated with alteration of intestinal microbiota that increase the risk of IBD in genetically susceptible individuals [3,4]. On the other hand, physiological concentrations of reactive oxygen species (ROS) are essential for cell survival and several physiological processes, including protein phosphorylation, activation of transcription factors, cell differentiation, apoptosis and cell immunity, etc. [5]. The effect of oxidative stress (OS) may have an impact on the pathophysiology of numerous chronic diseases, by reactions with the fatty acids of membranes, proteins, and DNA damage [6]. At the level of the gastrointestinal tract, OS can result in harmful effects such as stimulation of leukocytes infiltration, epithelial damage, mucus content depletion, and rupture of colonic barriers, along with inflammatory mediator release including inflammatory cytokines and arachidonic acid metabolites, as well as ROS, leading to oxidative damage and contributing to the development of IBD [7].

In this context, nutraceuticals or functional foods have gained popularity worldwide due to their high-value bioactive compounds [8]. Fruits and especially edible berries are rich sources of a wide variety of antioxidant phenols and include a high content of polyphenols and flavonoids [9]. In general, berries exhibit antioxidant and anti-inflammatory properties, decreasing OS and increasing intestinal protection, and can promote the prolongation of the remission phase and the duration/intensity of acute phases [10,11]. Among the most interesting berries for their high antioxidant capacity (AC) are the maqui and murta [12]. *Aristotelia chilensis* (Molina) Stuntz or maqui, endemic from Chile and widely consumed by its inhabitants, is an indigenous plant used as medicine. It contains an exceptional contribution both in concentration and in a variety of polyphenols (anthocyanins and non-anthocyanins) [13], with an extraordinary antioxidant power and a remarkable anti-inflammatory effect [14]. Moreover, *Ugni molinae*, known as murtilla or murta, is a wild berry native plant occurring in the lowlands of the southern mountains of Chile [15] and it is an interesting berry to be compared to maqui. A study reported by Speisky et al. [16] showed that the AC of maqui is still much higher than other cultivated berries and natural products. In general, berries are commonly consumed not only in fresh and frozen forms but also as processed tablets or pills [9]. For this reason, several studies have evaluated the effects and impact of different methods of food preservation on the maintenance of bioactive compounds and antioxidant activity. For example, the thermal effect caused by the different sterilization methods (thermal, microwave, and ultrasonic processing) on canned berries decreases the antioxidant activity and total content of anthocyanins and phenolics, whereas a low temperature storage better preserves the bioactive compounds and antioxidant activity [17,18]. Once the fruit has been preserved to retain its best characteristics as an antioxidant, an extraction to separate the desired natural products from the raw materials is needed. In this sense, and according to the highest antioxidant power of the maqui berry, various ways of production of maqui extracts have been described aiming to preserve not only the content of bioactive compounds but also the highest antioxidant activity [19,20]. Among them, the methanolic extract method has shown to preserve the highest AC, measured by oxygen radical absorbance capacity (ORAC), the radical scavenging activity using 2,2-diphenyl-1-picrylhydrazyl (DPPH) [21], ferric reducing antioxidant power (FRAP) [22], and total polyphenols content (TPC), measured by the Folin-Ciocalteu method [21]. In total, three groups of polyphenolic compounds have been described: Anthocyanins (delphinidin 3-*O*-sambubioside-5-*O*-glucoside, delphinidin 3,5-*O*-diglucoside, cyaniding 3-*O*-sambubioside-5-*O*-glucoside, cyaniding 3,5-*O*-diglucoside, delphinidin 3-*O*-sambubioside, delphinidin 3-*O*-glucoside, cyaniding 3-*O*-glucoside, and cyaniding 3-*O*-sambubioside), flavonols (quercetin, myricetin, kaemphenol, and its derivates), and phenolic acids (gallic acid). Recent studies have shown that the most abundant compounds are delphinidin and cyanidin derivatives that confer possible health benefits such as anti-inflammatory and antioxidant effects [23,24]. The characterization of the anthocyanin profile in maqui fruit has been performed using high-performance liquid chromatography-diode array detection and mass spectroscopy (HPLC-DAD-MS) [25], with delphinidin-3-sambubioside-5-glucoside as the majority delphinidin derivative, accounting for 34% of the total anthocyanidin level in maqui [26,27].

Interestingly, different authors have described a beneficial biological effect of maqui extract in cellular culture. For example, pre-incubation of fibroblast and Caco-2 cells with anthocyanin rich maqui berry extract has shown a protective effect against OS along with a high capacity to inhibit human low-density lipoprotein (LDL) oxidation [28]. Additionally, the maqui extract exhibits an important anti-inflammatory effect in RAW 264.7 macrophages cells [29] by inhibition of gene expression of *iNOS*, *COX-2*, *TNF-α*, *IL-6*, and *IL-10* [24,30].

The aim of the present study is to estimate the TPC and evaluate the AC of two native Chilean berries in different types of preservation methods: Fresh, refrigerated, and frozen. Furthermore, the correlations between the phenolic content and antioxidative activity were evaluated. To obtain a large-scale amount of maqui berry extract, richer in polyphenols content and with a more important antioxidant capacity than murta was an objective, together with exploring the curative benefits as a potential agent in the treatment of IBD. The maqui extract was evaluated in vitro, testing its viability, cytotoxicity, and antioxidant properties in colonic epithelial cell lines (HT-29) and macrophages (RAW 264.7).

2. Materials and Methods

2.1. Reagents and Apparatus

Chemical reagents and solvents for the extraction (MeOH, HCl), total polyphenols content analysis (Folin—Ciocalteu reagent, gallic acid, and sodium carbonate), DPPH assay, 2,2′-azobis(2-amidinopropane) dihydrochloride (AAPH) and reagents for FRAP analysis (sodium acetate trihydrate/glacial acetic acid, Trolox (6-hydroxy-2,5,7,8-tetramethylchroman-2-carboxylic acid), TPTZ (2,4,6-tripyridyl-s-triazine), and iron(III) sulphate ($Fe_2(SO_4)_3 \cdot 7H_2O$)) were purchased from Sigma-Aldrich (Saint Louis, MO, USA) and used without further purification. Determination of maqui's anthocyanins was performed by ultra-high-performance liquid chromatography (UHPLC-MS/MS) coupled to a hybrid quadrupole-orbitrap mass spectrometry system (Qexactive, Thermo Fisher, Waltham, MA, USA) with electrospray ionization (HESI-II). Separation was carried out using an Acquity UPLC BEH C18 (2.1 × 100 mm i.d., 1.7 um particle size) column (Waters, Milford, MA, United States) set to 40 °C. Celite® Hyflo Supercel and membrane filter (filter paper, MFS) for filtration were from Merk (Darmstadt, Germany) and Biorad (California, USA), respectively. Fetal bovine serum (FBS), trypsin/ethylenediaminetetraacetic acid (EDTA), penicillin/streptomycin (P/S), McCoy's 5a, and dulbecco's modified eagle medium (DMEM) high glucose culture medium were purchased from Biowest (Nuaillé, France). *N*-acetyl-L-cysteine 5 (NAC), 5-aminosalicylic acid (5-ASA), and lipopolysaccharide (LPS) were obtained from Sigma-Aldrich (Saint Louis, MO, USA). The Alamar Blue® cell viability reagent and 2′,7′–dichlorofluorescin diacetate (DCFDA) were purchased from Thermo Fisher Scientific (Waltham, MA, USA). The ultrasound, rotary evaporator with controlled heating, lyophilizer, optical microscope, and microplate reader were purchased from Hielscher Ultrasound Technology (UP400S, Wanaque, NJ, USA), Telstar Cryodos (Josep Tapiolas, Terrassa, Spain), Leica (Wetzlar, Germany) and Tecan (Männedorf, canton of Zürich, Switzerland) companies, respectively.

2.2. Plant Material and Preservation Methods

Wild maqui and murta berries have been collected from areas near the Andes Mountains: Bio-Bio and Araucanía Region, Chile. The fruits were separated in three groups for analytical procedures: (a) Refrigeration at 4 °C for 1 week, (b) freezing at −18 °C for 1 month, and (c) fresh fruits. The fresh samples were analyzed immediately and treated according to the procedure described by Rubilar et al. [12]. All of the berries were dried at 35 °C at an equative weight. Subsequently, grinding and sifting steps were conducted. The samples were macerated with ethanol (50% v/v in water, 20% w/v at room temperature for 2 h). The extracts were filtered through a membrane filter, centrifuged at 2600 relative centrifugal force (RCF) for 30 min at 20 °C, and lyophilized.

2.3. Preparation of Polyphenolic Maqui Extract (Ach)

For the large-scale extract, lyophilized wild maqui was used including seeds, skin, and pulp obtained from a powder wild harvested in Patagonia, marketed by "ISLA NATURA DE CHILE®" (Los Lagos, Chile). In order to quantitatively extract the raw maqui polyphenols, the extraction procedure was carried out using acidic methanol (MeOH/H$^+$) following a method published by Genskowsky et al. [20], although some modifications were implemented (Figure 1). The entire procedure was performed in darkness and at room temperature as described by Miranda Rottmann et al. [28] to avoid decomposition processes. A suspension of lyophilized maqui powder (50 g) in 0.1% HCl in MeOH (250 mL) was stirred for 15 min. Afterwards, the mixture was sonicated with an ultrasound device at maximum power for 2 min. The sample was centrifuged at 2600 RCF for 10 min and allowed to precipitate for 5 min. Finally, the supernatant was recovered in an Erlenmeyer (Thermo Fisher Scientific, Waltham, MA, USA). The previous process was repeated 5 times identically. Once all the supernatants were compiled, the solution was placed in a rotary evaporator for 2–3 h at 35 °C, in order to evaporate the organic solvent and concentrate the extract. The extract was resuspended in pure water, centrifuged at 2600 RCF for 4 min, and the supernatant recovered. Then, the filtering process was performed using a vacuum, with a short pad of celite, then passed through filter paper #1 (porosity 100–150 MM) and then #3 (porosity 40–100 MM). Finally, the lyophilization process was carried out for 24 h to prevent enzymatic and chemical changes, protein denaturation, loss of aromas, and easily oxidizable components, to get a highly hydroscopic powder that was stored at −20 °C.

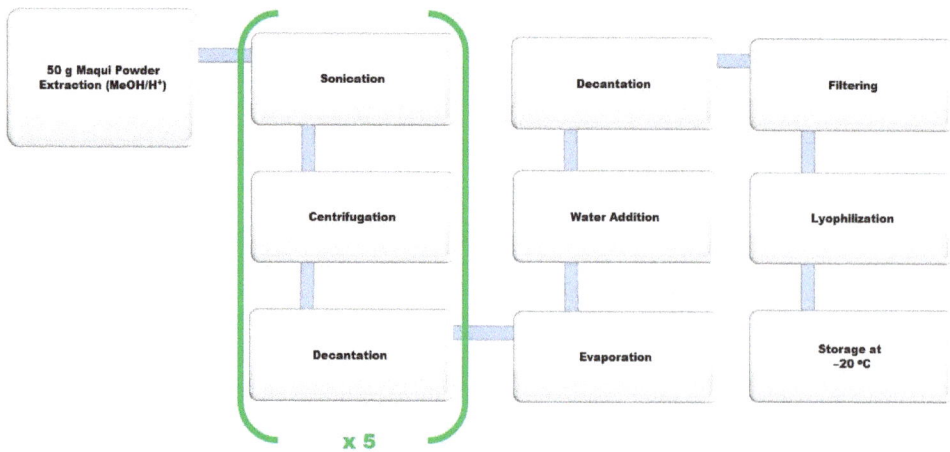

Figure 1. Conceptual scheme of maqui extraction procedure for 50 g.

2.4. The 2,2-Diphenyl-1-Picrylhydrazyl (DPPH) Free Radical Scavenging Activity Assay

This assay is based on the decoloration of DPPH free radical solution due to the free radical scavenging effect of antioxidants. The radical scavenging activity was estimated by the method described by Brand Williams [31] with minor modifications: 800 µL of extracts were mixed with 3.9 mL of methanolic solution containing 0.1 mmol of radical DPPH. After 30 min of incubation in a dark chamber, absorbance was measured at 520 nm. Results were expressed as the half maximal effective concentration EC50, defined as the concentration that causes a 50% decrease in the initial DPPH concentration.

2.5. Oxygen Radical Absorbance Capacity (ORAC) Assay

The ORAC assay was performed according to the method described by Zheng and Wang [32]. This assay estimates the ability of antioxidant components to inhibit the decline of oxidative degradation of a fluorescent molecule ((R)-phycoerythrin (R-PE)) induced by peroxyl radical generator AAPH. The assay was prepared by mixing 1.7 mL of phosphate buffer (pH 7.0), 100 µL of R-PE (3.4 mg/L), 100 µL of AAPH, and 100 µL of each sample. Samples without radicals were preincubated at 37 °C for 15 min. Fluorescence was recorded every 5 min at 540 nm excitation until the last fluorescence reading decreased 5% from the first reading. The antioxidant activity was expressed as mg of Trolox equivalents (TE) per gram of dry mass (DM) of plant material used for extraction (mg TE/g DM), according to the standard curve previously prepared.

2.6. Ferric Reducing Antioxidant Power Estimation (FRAP) Assay

AC was determined by FRAP, according to Benzie and Strain [33] with some modifications. The FRAP assay is a simple, reproducible, rapid, and inexpensive method that measures the reductive ability of antiradical species and is evaluated by the transformation of ferric ion Fe^{3+} to ferrous Fe^{2+}. This last one gives a blue complex by the reaction with TPTZ and is compared against the AC of Trolox as a measure of total antioxidant capacity. The FRAP reagent was prepared by mixing in this order, acetate buffer (0.3 M, pH 3.6), 10 mM TPTZ in 40 mM HCl, and 20 mM $Fe_2(SO_4)_3 \cdot 7H_2O$. A fresh acetate buffer was previously prepared by dissolving 3.1 g of sodium acetate trihydrate and 16 mL of acetic acid in 1 L of distilled water. The ready-to-use FRAP reagent was also freshly prepared by mixing the acetate buffer, TPTZ, and $Fe_2(SO_4)_3 \cdot 7H_2O$ in 10:1:1 proportion. Two solutions at different concentrations of extract were made from a stock solution of 1.28 mg/mL to determine its antioxidant activity. In summary, 50 µL of extracts were added to the diluted FRAP reagent in MeOH (1 mL FRAP reagent mixed with 2 mL MeOH). The measure of absorbance was recorded at 593 nm after 30 min in darkness at room temperature against a blank (FRAP diluted reagent previously prepared without the extract). The standard curve was prepared using different concentrations of Trolox and the results were expressed as µmol TE/100 g.

2.7. Determination of Total Polyphenols Content (TPC)

The Folin–Ciocalteu method is a type of assay based on an electron transfer mechanism. To provide basic conditions, Na_2CO_3 is used in the solution to dissociate phenolic protons that yield phenolate anions. Phenolate anions can reduce the Folin–Ciocalteu reagent through the reduction of Mo^{6+} to Mo^{5+} that results in a blue chromophore constituted by a phosphotungstic-phosphomolybdenum complex that can be measured spectrophotometrically [34]. The following methodology described by Slinkard and Singleton [35] was used for measurement. Then, 20 µL of the sample, 1580 µL distilled water, 100 µL Folin–Ciocalteu reagent, and 300 µL Na_2CO_3 (200 g/L) were added to a glass tube. These solutions were mixed and incubated at 40 °C for 30 min in a water bath. Absorbance was measured at 740 nm against a blank solution. Gallic acid was used as a standard. All results were expressed as milligrams of gallic acid equivalent (GAE) per gram of dry matter (DM).

2.8. Determination of Anthocyanin Profile by UHPLC-HRMS/MS

Determination of maqui's anthocyanins was performed with a Thermo Scientific Liquid Chromatography system consisting of a binary UHPLC Dionex Ultimate 3000 RS connected to a quadrupole-orbitrap Qexactive hybrid mass spectrometer (Thermo Fisher Scientific, Waltham, MA, USA) with a heated electrospray ionization probe (HESI-II). The Xcalibur software (Thermo Fisher Scientific, Waltham, MA, USA) was used for instrument control and data acquisition. Separation was carried out using an Acquity ethylene bridged hybrid (BEH) C18 (2.1 × 100 mm, 1.7 µm particle size) column (Waters, Milford, MA, USA) set to 35 °C at a flow rate of 0.4 mL/min. A binary gradient consisting of (A) water/formic acid 95:5 (v/v) and (B) acetonitrile/formic acid 95:5 (v/v) was used with the following elution profile: 0–2 min 5% B, 2–12 min from 5% to 100% B, 12–13 min 100% B, and 13–15 min 5% B. The injected volume was 5 µL. The lyophilized extract was dissolved in 1 mL of mobile phase A and microfiltered with a 0.2 µm nylon filter. A Data Dependent Acquisition method (TOP5) was used in a positive mode at a resolution of 70,000 and 17,500 at m/z 200 full-width half-maximum (FWHM) for Full Scan and Product Ion Scan, respectively. HESI source parameters were: Spray voltage, 3.5 kV; S lens level, 50; capillary temperature, 320 °C; sheath, auxiliary, and sweep gas flow, 50, 13, and 3, respectively (arbitrary units); and probe heater temperature, 425 °C. For data treatment, the TraceFinder 5.1. software (Thermo Fisher Scientific, Waltham, MA, USA) was used. The identification was made by comparing (maximum deviation of 5 ppm) the exact masses of the pseudomolecular ion and their fragment ions with the data contained in an anthocyanins database with 12 possible compounds. The retention time of standard compounds and isotopic pattern scores higher than 80% were also required. The following anthocyanin standards were used for identification purposes: Cyanidin-3-glucoside, cyanidin-3-galactoside, malvidin-3-glucoside, peonidin-3-glucoside, and delphinidin-3-glucoside. Anthocyanin compounds were quantified using the areas of the aglycone counterparts.

2.9. Cell Culture

Human colorectal adenocarcinoma, epithelial, adherent (HT-29 cells) provided by the Department of Pharmacology (Seville University, Seville, Spain), were cultured in McCoy's 5a culture medium, and murine macrophage cells, adherent (RAW 264.7 cells) were obtained from Centro Andaluz de Biología Molecular y Medicina Regenerativa (CABIMER, Seville, Spain) and cultured in DMEM high glucose. Both cell lines were supplemented with 10% FBS 1% P/S and incubated at 37 °C and 5% CO_2, with the medium changing every 2 days. Cells were maintained in 75 cm^2 flasks and, when they reached 80–90% confluence in the flask, were harvested using trypsin/EDTA (0.5% v/v).

2.10. Viability and Cytotoxic Assays

Cell viability was determined using the Alamar Blue® reagent (Invitrogen, Carlsbad, CA, USA) based on the quantitative colorimetric assay following the manufacturer's protocols. For all experiments, HT-29 and RAW 264.7 cells were seeded in 24-well plates (5 × 105 cells/well) and incubated for 24 h to allow the cells to adhere to the plate surface. Ach was prepared in distilled H_2O and diluted in a cell culture medium. The concentrations of Ach in the in vitro experiments were based on data previously published on fruit extract of other berries [36,37]. Thus, treating the cells with 100, 200, and 300 µg/mL of Ach was chosen. HT-29 cells were exposed to H_2O_2 at 0.05% and incubated with or without Ach at different concentrations during 24 or 48 h. NAC, a known antioxidant, was used as a control. Cell viability of LPS (1.0 µg/mL)-induced RAW 264.7 cells was tested to determine the potential cytotoxic effect of increasing concentrations of Ach for 12 h. The 5-aminosalicylic acid (5-ASA), an anti-inflammatory drug, was used as a control according to a previously published work [38]. The culture supernatants were collected, and the absorbance of each one was measured at 570 nm using a microplate reader. The cell viability was calculated as follows:

$$\text{Cell viability (\%)} = \frac{Abs\ (experiment) - Abs\ (blank)}{Abs\ (control) - Abs\ (blank)} \times 100 \quad (1)$$

2.11. Determination of Reactive Oxygen Species (ROS) by the DCFH-DA Assay

A fluorescent 2′,7′-dichlorofluorescein diacetate (DCFH-DA) assay was performed to determine the intracellular ROS concentrations according to the method reported by Miranda-Rottmann [28] with modifications. Cells were seeded at 1×10^5 cells/well in 96-well plates in a final volume of 100 µL of culture medium per well. When the cells reached a confluence of 80%, HT-29 and RAW 264.7 cells were incubated with H_2O_2 and LPS, respectively in the presence or absence of Ach and internal control (NAC and 5-ASA) for 1 h with 5% CO_2 at 37 °C. After removing the medium, cells were washed twice with 50 µL/well of PBS, incubated at 37 °C for 30 min with 25 µM DCFH-DA, and stored under at −20 °C. For the DCFH-DA assay, a culture medium was used without phenol red and without supplementation to avoid interference with fluorescence emission. The fluorescence was measured at Ex/Em: 485/530 nm using a fluorescence microplate reader.

$$\text{Oxidation (\%)} = \frac{Fluorescence\ (experiment) - Fluorescence\ (w/o\ cells)}{Fluorescence\ (control) - Fluorescence\ (w/o\ cells)} \times 100 \quad (2)$$

2.12. Statistical Analysis

The results were expressed as the mean ± standard error of the mean (SEM) and ± the standard deviation (SD) according to the particular characteristics of the tests. All experiments were replicated three times. Heterogeneity was tested using Levene's test and the Shapiro-Wilk test for normality. The continuous variables that are normally distributed were compared through the parametric Student t-test (two sample t-test) and proportions were tested using the two samples test calculator (prtesti). The analysis of differences between groups was evaluated by one-way analysis of variance (ANOVA) tests when the variables were normally distributed, followed by Bonferroni's post-hoc test. Descriptive statistics and tests were performed at a significance level of 0.05 using the STATA software (version 12, 2011, StataCorp, College Station, TX, USA).

3. Results

3.1. Antioxidant Capacity (AC) by DPPH and ORAC of Maqui and Murta in Different Preservation Methods

The results of the analysis of the AC of maqui and murta, as estimated by the free radical scavenging through the DPPH method and ORAC assay, are shown in Table 1. Through the DPPH assay it is observed that the AC of maqui, regardless of all different types of preservation methods, is four times higher than murta. In the same way, the ORAC values indicate that maqui has a significantly higher AC than murta in fresh, refrigerated, and frozen samples. When AC of maqui and murta was estimated by these methods, results showed that both berries have greater AC in fresh and frozen storage, in comparison with the refrigerated state.

Table 1. DPPH and ORAC assay results for maqui and murta extracts in three categories of preservation.

Sample	EC50 for DPPH (mg/mL)			ORAC Value (mg TE/g DM)		
	Fresh ($x \pm$ SD)	Refrigerated ($x \pm$ SD)	Frozen ($x \pm$ SD)	Fresh ($x \pm$ SD)	Refrigerated ($x \pm$ SD)	Frozen ($x \pm$ SD)
Maqui	372.37 ± 2.52 *	349.05 ± 1.05 *,†	364.76 ± 1.96 *	6.02 ± 0.04 *	5.93 ± 0.12 *,†	5.97 ± 0.56 *
Murta	79.68 ± 1.53	74.72 ± 0.58 †	78.02 ± 2.04	3.97 ± 0.06	3.84 ± 0.06 †	4.01 ± 0.04

Data are presented as mean ± standard deviation (SD); mg TE/g DM: Milligrams of Trolox equivalents per gram of dry mass. * Significant difference ($p < 0.05$) between berries per category of preservation. † Significant decrease ($p < 0.05$) of antioxidant capacity of maqui and murta refrigerated in comparison with fresh and frozen samples. EC50: the half maximal effective concentration; DPPH: 2,2-diphenyl-1-picrylhydrazyl; ORAC: oxygen radical absorbance capacity.

3.2. Total Polyphenols Content (TPC) of Maqui and Murta in Different Preservation Methods

The TPC values for maqui and murta estimated by the Folin-Ciocalteu method are shown in Table 2. It can be observed that maqui, with all the preservation methods, showed a significantly higher polyphenol content than murta. The TPC data from refrigerated samples presented the lowest values in comparison with fresh and frozen samples, showing statistical significance. On the other hand, when the data of maqui and murta were compared in fresh versus frozen state, the values were very similar.

Table 2. Quantification values of TPC in maqui and murta by the Folin-Ciocalteu method in three categories of preservation.

Sample	Fresh ($x \pm$ SD) mg GAE/g FW	Refrigerated ($x \pm$ SD) mg GAE/g FW	Frozen ($x \pm$ SD) mg GAE/g FW
Maqui	75.348± 1.53 *	69.652± 0.58 *,†	73.786± 3.06 *
Murta	48.041± 1.00	42.780± 0.56 †	47.309± 2.09

Data are presented as mean ± standard deviation (SD); mg GAE/g FW: Milligrams of gallic acid equivalent per gram of fresh weight; TCP: Total polyphenolic content. * Significant difference ($p < 0.05$) between berries per category of preservation. † Significant decrease ($p < 0.05$) of TCP of maqui and murta refrigerated in comparison with fresh and frozen samples.

3.3. Total Polyphenols Content (TPC) and Antioxidant Capacity (AC) of Polyphenolic Maqui Extract (Ach)

The most important factors that affect the polyphenol content in fruits are usually genotype, environment, geographical location, and altitude, as well as season, soil, maturity stage at harvest, and post-harvest conditions [20]. The Folin-Ciocalteu method is the most widely used assay for the estimation of TPC in a wide range of samples, including fruits, juices, and wine [39,40]. The results of the Folin-Ciocalteu method show that the lyophilized extracts of the maqui fruit content have an average TPC of 39.02 mg/g of polyphenols expressed as gallic acid. The results regarding the AC of the maqui extract to reduce Fe^{3+} to Fe^{2+}, indicate that the Ach has an AC of 69.48 mmol Trolox/100 g sample.

3.4. Characterization of Extracts

Characterization of the maqui's anthocyanins extract was performed by the UHPLC-MS/MS with an electrospray ionization that enables comparison of m/z signals and fragment ions of the anthocyanin pattern. Regarding the maqui composition, delphinidin-3-O-glucoside was the major component (m/z: 465.1, 34.3%), followed by delphinidin-3,5-O-diglucoside (m/z: 627.1, 21.2%), cyanidin-3-O-glucoside (m/z: 449.1, 8.3%), delphinidin-3-O-sambubioside-5-O-glucoside (m/z: 759.1, 7.7%), delphinidin-3-O-sambubioside (m/z: 597.1, 7.0%), cyanidin-3,5-O-diglucoside (m/z: 611.1, 6.2%), cyanidin-3-O-sambubioside-5-O-glucoside (m/z: 743.2, 3.4%), cyanidin-3-O-sambubioside (m/z: 581.1, 2.3%), as main components of the extraction procedure performed. Other similar structures identified were malvidin-3-O-glucoside (m/z: 493.1, 0.7%), pelargonidin-3-O-glucoside (m/z: 433.1, 0.1%), and peonidin-3-O-glucoside (m/z: 463.1, 0.05%). The anthocyanin profile is shown in Figure 2 and is compared with a previously reported one [20], without finding statistical differences.

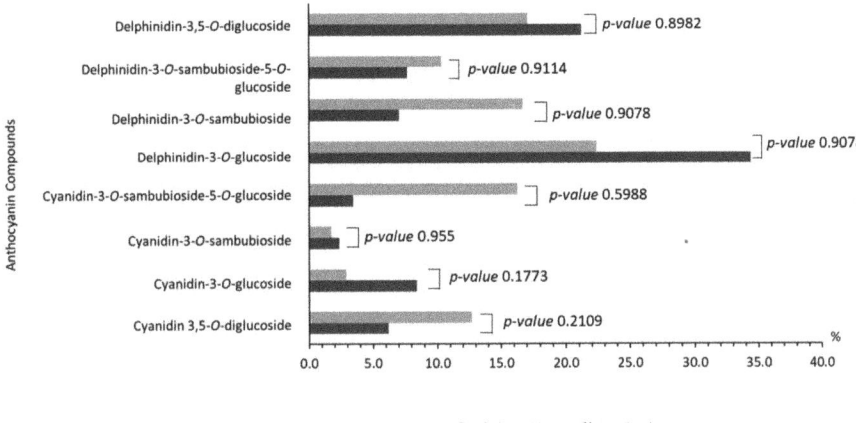

Figure 2. Determination of individual anthocyanin compounds in maqui extract by ultra-high-performance liquid chromatography (UHPLC-MS/MS) Orbitrap. Our data (new extract; dark gray) are presented according to the percentage of anthocyanin compound and compared with the anthocyanins profile previously reported (previous extract; light gray) [20].

3.5. Viability and Cytotoxicity

To test the protective effect of maqui, HT-29 cells were incubated with increasing concentrations of Ach for 24 and 48 h and compared to NAC (Figure 3). Initially, the effect of H_2O_2 on epithelial cells was determined revealing a significant reduction (p-value < 0.01) of the viability. When HT-29 cells were incubated with Ach the results showed that all concentrations (100, 200, and 300 µg/mL) tested caused a significant increase of viability at 24 and 48 h. The effect of Ach on cell viability was higher than that of 5 mM NAC. To examine the cytotoxic effects of Ach, the viability of RAW 264.7 macrophages was evaluated at concentrations of 100, 200, and 300 µg/mL of Ach for 24 h. As shown in Figure 4, no notable effects on cell viability in LPS-stimulated RAW 264.7 cells were observed with or without treatment. When comparing Ach with different concentrations of 5-ASA (0.1, 0.5, and 1 mM) in the presence of LPS, no significant cytotoxicity was found, even when mixing 5-ASA (0.1 mM) and Ach (100 µM/mL).

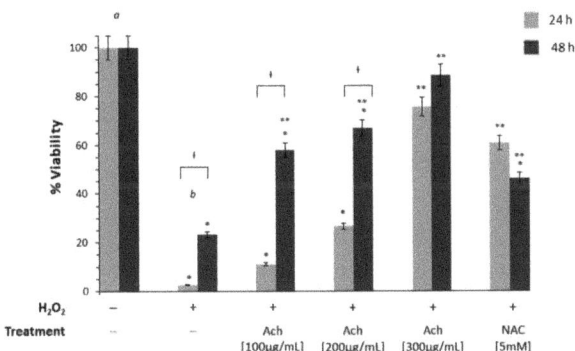

Figure 3. Ach presents a protective effect on the viability of HT-29 cells exposed to 0.05% H_2O_2 and treated during 24 and 48 h with different concentrations of Ach. The viability of cells without H_2O_2 was 100%. Data are expressed as the mean ± SEM. A p-value < 0.01 was considered statistically significant: * With respect to a, ** with respect to b, † between 24 and 48 h. Positive sign (+): presence of treatment; negative sign (−): absence of treatment.

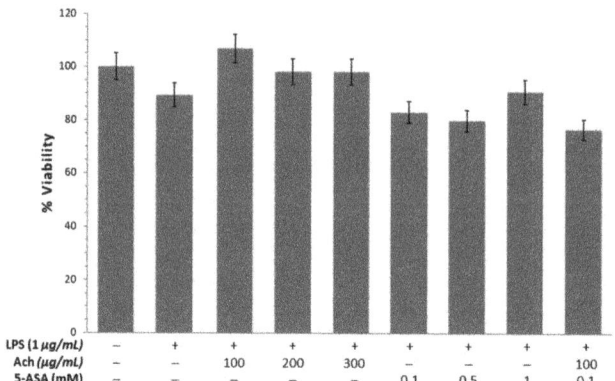

Figure 4. Ach shows no cytotoxicity effects in RAW 264.7 cells treated with lipopolysaccharide (LPS) (1.0 µg/mL). Data are expressed as the mean ± SEM. The viability of cells without LPS was 100%. Positive sign (+): presence of treatment; negative sign (−): absence of treatment.

3.6. Oxidative Stress

Figure 5 shows a significant increase of oxidation in HT-29 cells when H_2O_2 (p-value < 0.05) was added. These results were inhibited by all concentrations of Ach in a dose-dependent way, reaching significance at 300 µg/mL. In parallel, when epithelial cells exposed to H_2O_2 were treated with NAC, it was possible to observe a decrease in oxidation, although not significant. In RAW 264.6 macrophages, LPS stimulation led to the increment of oxidation, demonstrating the relationship between inflammation and OS (Figure 6). Ach effectively suppressed the LPS-induced oxidation in a dose-dependent manner, showing a significant reduction with the highest doses (200 and 300 µg/mL). At the same time, these results indicated that the suppression of oxidation could not be attributable to the direct cytotoxic effect of Ach. Of interest is the fact that the 0.1 mM of 5-ASA plus 300 µg/mL of Ach concentration had an inhibition effect comparable to the control conditions (cells in culture medium without LPS), achieving values of basal oxidative state.

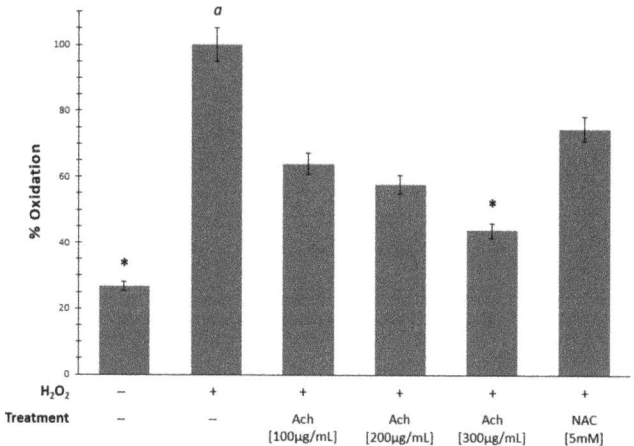

Figure 5. Supplementation with Ach to HT-29 cells stressed with 0.05% H_2O_2 decreases oxidation status in a dose-dependent manner, and more efficiently than NAC. Data are expressed as the mean ± SEM and the measured percent oxidation were expressed as a percentage from the positive control (cells induced by H_2O_2). *A p-value < 0.05 indicates significant differences with respect to a. Positive sign (+): presence of treatment; negative sign (−): absence of treatment.

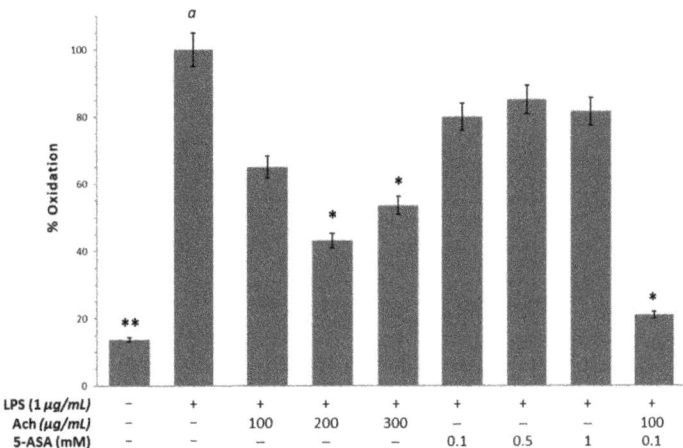

Figure 6. Ach inhibits oxidative stress in RAW 264.7 macrophages stimulated with LPS (1.0 µg/mL). The cells incubated with Ach (100 µg/mL) plus 5-aminosalicylic acid (5-ASA) (5 mM) presented the lowest percentage of oxidation. Data are expressed as the mean ± SEM and the measured percent oxidation were expressed as a percentage from the positive control (cells induced by H2O2). * A p-value < 0.05 and ** p-value < 0.01 indicate significant differences with respect to a. Positive sign (+): presence of treatment; negative sign (−): absence of treatment.

4. Discussion

The polyphenolic content in several berries has been widely proved to be the major contributor to AC [39,41] and the way of consumption (fresh, processed or stored for long-term) could be relevant to preserve polyphenol compounds and its beneficial effects, including other phytochemicals that also play a critical role. In the present work, it was demonstrated that maqui has greater AC than murta, and significant differences of TCP in fresh fruit values, with 75.3 mg GAE/g FW for maqui and 48.0 mg GAE/g FW for murta. These findings are in agreement with Salvia-Trujillo et al. [42] who reported a depletion of antioxidant "Vitamin C" during refrigerated storage, which was correlated with AC. Additionally, other studies reported that polyphenols from crude extracts of maqui were also much higher than crude extracts from murta [12], and that maqui fruit does not decrease polyphenol and anthocyanin contents during desiccation, cooling or freezing processes [43].

Maqui has been selected for further investigation based on AC and TPC results. The raw material was a marketed maqui product (native, fresh, and lyophilized maqui powder). In order to evaluate the activities of maqui berry extracts as an antioxidant and anti-inflammatory potential agent, a modified procedure was implemented to develop a multigram and more efficient methodology to provide appropriated quantities. The extraction procedure described here is based on a 50 g scale for the first time, and many laboratory operations have been reduced to simplify and minimize economic costs and increase safety, since less solvents are manipulated. The reproducibility of this procedure was confirmed via all the analyses carried out of the numerous batches obtained, preserving in all cases the TPC and AC characteristics due to low temperatures and darkness along all the processes. The multigram procedure involves an improvement of the previous method published by Viuda-Martos et al. [20], in which 3 g of lyophilized and ground maqui berry were used to test antibacterial properties. Other studies submitted 1 g to solid-liquid extraction, such as in the procedure recently described by Silvia Rossi et al. [30], that uses this extract embedded in an appropriate gel for a better delivery for inflammatory bowel disease. Moreover, 1.5 g were extracted by Fredes et al. [40] to evaluate how the maturation affects the TPC. In general, all the procedures are based on the same type of extraction and all of them used an aqueous methanol acidic mixture, choosing either HCl or formic acid to accomplish a better polyphenolic extraction. Interestingly, no high temperatures above 50 °C are used in any of them, since as previously described, it might be expected that the temperature during thermal drying technologies contributed to the loss of phenols. Temperatures above 60 °C are not favorable owing to inducing oxidative condensation/decomposition of thermolabile components. Additionally, the drying method such as lyophilization preserves TPC as described by Quispe et al. [21]. It is well known that the Folin-Ciocalteu method is the most common analysis to estimate the TPC in many fruits and foods and also shows a good linear correlation to the anthocyanin content described by HPLC [44]. Although this method often overestimates the phenol content due to the presence of L-ascorbic acid (vitamin C) fortunately, it has been well established that maqui would not present this interference since no traces of vitamin C has been detected in maqui fruit, according to Miranda-Rottmann et al. [28]. In this sense, the TPC obtained in this study was 39.02 mg/g. Genskowsky et al. [20] described the three main groups of polyphenolic compounds in maqui berries such as phenolic acids, flavonoids, and anthocyanins, identifying a TPC of 49.49 g GAE/kg, while Reyes-Farias et al. [45] published a TPC of 19.06 mg/g, much higher values when compared with other native Chilean berries and traditional berries such as blueberries [39]. Moreover, in different maqui genotypes belonging to four different geographical regions in Chile, TPC values ranged between 11.1 and 14.5 mg/g. Brauch et al. reported that data for fresh and dry maqui berries, collected in the Aysén region (Patagonia, Chile), displaying values between 19.7 and 32.0 mg/g, were always referred to as gallic acid equivalents. On the other hand, and as previously mentioned, TPC is affected by various factors related to maqui such as genotype, environment, storage, and processing or stage at harvest. Hence, fruits with different maturity stages have different types or contents of polyphenols. Our maqui berry extract comes from the sylvester variety, which interestingly has been described as the one that contains more polyphenols if compared with the other varieties named as "Luna Nueva" and "Morena" [19]. Moreover, when the phenolic content of maqui is compared with different varieties of berries, including red wine as it is a known rich source of dietary phenols, it is possible to observe that the maqui fruit is superior, even when compared with fruits of habitual consumption, showing the maqui difference values of up to 80 times [28,45].

In addition, the AC expressed as FRAP of our maqui berry extract was determined following the method previously described. The results indicate that the maqui extract has an AC of 69.48, expressed as mmol Trolox equivalent/100 g of dried sample regarding the AC of the maqui extract to reduce Fe^{3+} to Fe^{2+}. When the AC of Ach is compared with other berries, it is possible to observe significant differences, with a ferric reduction potential of 69.48 mmol Trolox/100 g compared to blueberries, which have a value of 5.9 mmol Fe^{+2}/100 g dry weight [45]. Moreover, our data is much better than the other maqui extracts previously reported, that have shown FRAP values of 10.07 mmol Trolox/100 g [20]. Interestingly, maqui has several phytochemicals, particularly anthocyanins, a set of water-soluble pigments, which represent more than 65% of the total polyphenols that are responsible for the intense red to purplish-blue color of several fruits and that are influenced by pH, light or temperature. Recent studies have demonstrated that the most abundant compounds of maqui are delphinidin derivatives and cyanidin derivatives. In this sense, in the analytical UHPLC-MS/MS semi quantitative determination, the profile obtained from the maqui composition extract concerning this study was compared to the previously described study by Genskowsky et al. [20], who used a smaller extraction procedure (Figure 2). In general terms, the concentration obtained in both cases for most of the components coincide. Specifically, among the most abundant are delphinidin-3-O-glucoside, delphinidin-3,5-O-diglucoside, cyanidin-3-O-glucoside or delphinidin-3-O-sambubioside-5-O-glucoside. Other similar structures identified for the first time as part of a maqui extract, to our knowledge, were malvidin-3-O-glucoside, pelargonidin-3-*O*-glucoside, and peonidin-3-*O*-glucoside, that represent in sum 0.85% of the total composition.

The effect of polyphenolic maqui extract, related to the different mechanisms associated with inflammation and oxidation has also been described in the literature (Table 3). Once the total chemical structural determination and characterization were demonstrated, in vitro studies were carried out to investigate the biological role of these entities as potential therapeutic agents relative to OS. In this sense, it has been suggested that OS is involved in the pathogenesis and development of IBD through many levels such as cell transformation, apoptosis, DNA damage, and pro-inflammatory response [7]. Additionally, polyphenols suppress inflammation-related gene expression, downregulate proinflammatory cytokine expression, and increase the production and effects of anti-inflammatory cytokines [46].

Table 3. Anti-inflammatory and antioxidant effect of maqui berry extract described on cell cultures.

Reference	Cell Culture	Model	Extract	Concentration	Effects
Zhou G. et al. 2019 [29]	RAW 264.7 macrophage cells	Inflammatory model with LPS stimulated for 24 h	Water fraction extract with ethyl acetate rich in phenols	2–20 µg ml^{-1}	↓COX-2 ↓ IL-6
Tenci M. et al. 2019 [30]	Fibroblasts and Caco-2	Oxidant model with H_2O_2 (1 mM) for 24 h	MBE with acid MeOH 0.1% (H_2O:MeOH/10:90 v/v) rich in anthocyanins	MBE solution (0.5% w/w) diluted at 1:2, and 1:5 v/v	No cytotoxic effect Viability under oxidative damage
Moon HD. et al. 2019 [47]	RAW 264.7 macrophage cells	Inflammatory model with LPS stimulated for 24 h (0.1 µg/mL)	Water extract of maqui rich in anthocyanins	62.5, 125, 250, 500, 1000, and 2000 µg/ml	↓NO
Céspedes, C.L., et al. 2017 [24]	RAW 264.7 macrophage cells	Inflammatory model with LPS stimulated for 24 h (1 µg/mL)	Pulp extract with acid MeOH 0.1% HCl:H_2O/6:4 v/v); Acetone/MeOH; Ethyl acetate	100 µg/ml	No cytotoxic effect ↓Oxidation ↓NO ↓iNOS ↓COX-2
Reyes-Farias, M., et al. 2015 [45]	RAW 264.7 macrophage cells	Inflammatory model with LPS stimulated for 24 h (5 µg/mL) or with CM from fully differentiated 3T3-L1 adipocytes	TCP: from Ripe fruits extract with acid MeOH:H_2O/1:1 v/v)	100 µM	**CM:** ↓NO; iNOS ↑TNF-α; IL-10 **LPS:** ↓NO; iNOS; TNF-α; IL-10
Miranda-Rottmann S. et al. 2002 [28]	Primary culture of HUVEC	Vascular OS model with 500 µM H_2O_2	Aqueous fraction juice extract with ethyl acetate at pH 2.0 rich in anthocyanins	0.1–10 µM	↓ intracellular OS
		Copper-induced LDL oxidation in vitro		1 µM GAE	↓ LDL oxidation

(LPS): lipopolysaccharide; CM: conditioned media; OS: oxidative stress; LDL: low-density lipoprotein; MBE: maqui berry extract; TCP: total polyphenolic content; COX-2: cyclooxygenase-2; IL-6: interleukin-6; NO: nitric oxide; iNOS: inducible nitric oxide synthase; TNF-α: tumor necrosis factor-α; IL-10: interleukin-10.

Studies in culture cells exposed to H_2O_2 are a good model of OS since H_2O_2 is catalyzed by Fe^{2+} (Fenton reaction) to the highly reactive hydroxyl radical (HO•). However, the loss of cell viability or lipid peroxidation in epithelial cells are achieved when cells are incubated at a concentration above 250 µM H_2O_2 [48]. Therefore, a higher concentration of 500 µM (0.05%) for inducing oxidation and death cell in HT-29 cells was used. In the present study, we found a significant improvement in cell viability and reduction of OS when the HT-29 cells were treated with Ach, finding the main effects at doses of

300 µg/mL and at 48 h. These results indicated that the effects of Ach on HT-29 cells are both concentration and time dependent, probably due to the cellular uptake sustained over time, allowing the recovery of cell damage. These results are consistent with previous ones reported by Miranda-Rottmann et al. [28], where incubation with Ach juice prior to the addition of 500 µM H_2O_2 in HUVEC cells significantly reduces oxidation in a dose-dependent manner compared to endothelial cells exposed to H_2O_2 without treatment. Furthermore, a previous study has reported that the red wine polyphenol extract did not affect the viability of HT-29 cells at several concentrations (200, 400, and 600 µg/mL^{-1}) [49], suggesting the absence of cytotoxic profile. However, higher concentrations of polyphenols such as resveratrol could induce cell death by diverse mechanisms in tumor cell lines, including colorectal cancer cells and, acting as a direct cytotoxic agent [50]. In addition, macrophages have an important role in an inflammatory response and are the central cells that initiate the production of inflammatory mediators. Macrophages stimulated with LPS, viruses, and bacterial endotoxin are able to release inflammatory factors including prostaglandin, proinflammatory cytokines, and ROS. Therefore, the regulation of the level of these factors has become vital for the treatment of inflammatory diseases [51]. In this study, we demonstrated that our extract (Ach) presented a protective activity against OS (DCF) on RAW 264.7 cells, along with the absence of cytotoxicity effect, in the same way as reported by Cespedes et al. [24]. Moreover, it has been demonstrated that other natural products decrease intracellular ROS levels in LPS-stimulated macrophages [52,53], indicating the close relationship between oxidative mechanisms and inflammation. An experimental study published by Zhou et al. [29] showed that the ethyl acetate fraction of maqui extract (MWE) contained a high content of total phenols and flavonoid and that exhibited stronger antioxidant activities than the other extracts such as n-butanol fraction. Additionally, the authors proved that MWE considerably reduced the expression of COX-2 and IL-6 in LPS-stimulated RAW 264.7 cells, demonstrating the anti-inflammatory effect. In addition, interestingly, in our work it was possible to show that when we used lower concentrations of Ach and a lower dose of 5-ASA the results on the percent oxidation was significantly lower compared to the control group and the other doses. This effect could indicate synergistic effects between Ach and 5-ASA. The finding of this synergy is very interesting since it opens the possibility of a combination therapy with lower doses of 5-ASA, which will lead to less side effects of this drug used in IBD patients. Furthermore, in agreement with other authors [49], we believe that the protective effect of our extract rich in total polyphenols may be due to the sum of all its components probably due to a synergic action of anthocyanins compounds suggesting that the use of total extracts in polyphenolic compounds, rather than pure polyphenols, could be a better alternative or complementary therapeutic option.

5. Conclusions

Our study shows that Chilean berries, specifically maqui, is a potential source of natural antioxidants and that the preservation method seems to be crucial in maintaining the AC in native berries, with fresh and frozen being the best conditions. In this study, a fresh maqui berry extract procedure has been optimized for the first time in multigram scale (50 g) via a simplified and easily reproducible laboratory procedure. The extract profile has been analyzed to quantify the TCP and AC, as well as to prove that this method fulfills the data standard previously described for smaller procedures. All these data confirm that our extraction procedure not only increases the amount of dried extract amount, but also assures an efficient process that preserves the total polyphenols, anthocyanins profile, and antioxidant properties. Interestingly, our studies in activated macrophage and epithelial colon cell cultures have shown that the maqui polyphenolic extract is a powerful antioxidant that exhibits a dose-dependent behavior and no cytotoxic effects, due to the ability to reduce ROS. In addition, our maqui extract seems to have a synergistic effect with 5-ASA, a drug of clinical use in IBD. These findings point out to a possible mechanism of action related to reduce ROS and its relationship with inflammation. However, more information concerning higher n values and better determination of endogenous AC will be reported by our group in due course.

In summary, maqui extracts could introduce bioactive candidates for further investigation as potential anti-inflammatory agents. Nevertheless, additional studies are necessary to analyze the effects on bioaccessibility or bioavailability of the bioactive compounds that are under current investigation in our group.

Author Contributions: Conceptualization, F.A.-A., M.D.-M., and A.A.; investigation, T.O., A.P., M.V., V.V.-R., J.L.P.B., R.M.C. and A.A.; writing—original draft preparation, T.O. and A.A.; writing—review and editing, T.O., B.B., J.-M.G.-M., V.V.-R. and M.D.-M.; supervision and project administration, M.D.-M. and A.A. All authors have read and agreed to the published version of the manuscript.

Funding: This work was supported by Sociedad Andaluza de Patología Digestiva 2016 (SAPD), project 2018/00000802-CTS949 (Junta de Andalucía, Spain); Ministerio de Ciencia e Innovación, project PID2019-109371GB-I00, project 2017/FQM-135 (Junta de Andalucía, Spain), and CITIUS facilities (University of Seville, Plan Propio de Investigación); Universidad del Desarrollo, project 2011090814104630378.

Institutional Review Board Statement: Not applicable.

Informed Consent Statement: Not applicable.

Data Availability Statement: Not applicable.

Conflicts of Interest: The authors declare no conflict of interest.

References

1. Hanauer, S.B. Inflammatory bowel disease: Epidemiology, pathogenesis, and therapeutic opportunities. *Inflamm. Bowel Dis.* **2006**, *12*, S3–S9. [CrossRef] [PubMed]
2. Kaplan, G.G. The global burden of IBD: From 2015 to 2025. *Nat. Rev. Gastroenterol. Hepatol.* **2015**, *12*, 720–727. [CrossRef]
3. Garber, A.; Regueiro, M. Extraintestinal Manifestations of Inflammatory Bowel Disease: Epidemiology, Etiopathogenesis, and Management. *Curr. Gastroenterol. Rep.* **2019**, *21*, 31. [CrossRef] [PubMed]
4. Cui, G.; Yuan, A. A Systematic Review of Epidemiology and Risk Factors Associated With Chinese Inflammatory Bowel Disease. *Front. Med.* **2018**, *5*, 183. [CrossRef]
5. Patlevic, P.; Vaskova, J.; Svorc, P., Jr.; Vasko, L.; Svorc, P. Reactive oxygen species and antioxidant defense in human gastrointestinal diseases. *Integr. Med. Res.* **2016**, *5*, 250–258. [CrossRef] [PubMed]
6. Masoodi, H.; Villaño, D.; Zafrilla, P. A comprehensive review on fruit Aristotelia chilensis (Maqui) for modern health: towards a better understanding. *Food Funct.* **2019**, *10*, 3057–3067. [CrossRef]
7. Tian, T.; Wang, Z.; Zhang, J. Pathomechanisms of Oxidative Stress in Inflammatory Bowel Disease and Potential Antioxidant Therapies. *Oxid. Med. Cell. Longev.* **2017**, *2017*, 4535194. [CrossRef]
8. Michalska, A.; Lysiak, G. Bioactive Compounds of Blueberries: Post-Harvest Factors Influencing the Nutritional Value of Products. *Int. J. Mol. Sci.* **2015**, *16*, 18642–18663. [CrossRef] [PubMed]
9. Nile, S.H.; Park, S.W. Edible berries: Bioactive components and their effect on human health. *Nutrition* **2014**, *30*, 134–144. [CrossRef]
10. Mijan, M.A.; Lim, B.O. Diets, functional foods, and nutraceuticals as alternative therapies for inflammatory bowel disease: Present status and future trends. *World J. Gastroenterol.* **2018**, *24*, 2673–2685. [CrossRef]
11. Hu, Y.; Chen, D.; Zheng, P.; Yu, J.; He, J.; Mao, X.; Yu, B. The Bidirectional Interactions between Resveratrol and Gut Microbiota: An Insight into Oxidative Stress and Inflammatory Bowel Disease Therapy. *BioMed Res. Int.* **2019**, *2019*, 5403761. [CrossRef] [PubMed]
12. Rubilar, M.; Jara, C.; Poo, Y.; Acevedo, F.; Gutierrez, C.; Sineiro, J.; Shene, C. Extracts of Maqui (*Aristotelia chilensis*) and Murta (Ugni molinae Turcz.): Sources of antioxidant compounds and alpha-Glucosidase/alpha-Amylase inhibitors. *J. Agric. Food Chem.* **2011**, *59*, 1630–1637. [CrossRef] [PubMed]
13. Romero-González, J.; Shun Ah-Hen, K.; Lemus-Mondaca, R.; Muñoz-Fariña, O. Total phenolics, anthocyanin profile and antioxidant activity of maqui, Aristotelia chilensis (Mol.) Stuntz, berries extract in freeze-dried polysaccharides microcapsules. *Food Chem.* **2020**, *313*, 126115. [CrossRef]
14. Ortiz, T.; Argüelles-Arias, F.; Illanes, M.; García-Montes, J.M.; Talero, E.; Macías-García, L.; Alcudia, A.; Vázquez-Román, V.; Motilva, V.; De-Miguel, M. Polyphenolic Maqui Extract as a Potential Nutraceutical to Treat TNBS-Induced Crohn's Disease by the Regulation of Antioxidant and Anti-Inflammatory Pathways. *Nutrients* **2020**, *12*, 1752. [CrossRef] [PubMed]
15. Junqueira-Goncalves, M.P.; Yanez, L.; Morales, C.; Navarro, M.; Contreras, R.A.; Zuniga, G.E. Isolation and characterization of phenolic compounds and anthocyanins from Murta (Ugni molinae Turcz.) fruits. Assessment of antioxidant and antibacterial activity. *Molecules* **2015**, *20*, 5698–5713. [CrossRef] [PubMed]

16. Speisky, H.; Lopez-Alarcon, C.; Gomez, M.; Fuentes, J.; Sandoval-Acuna, C. First web-based database on total phenolics and oxygen radical absorbance capacity (ORAC) of fruits produced and consumed within the south Andes region of South America. *J. Agric. Food Chem.* **2012**, *60*, 8851–8859. [CrossRef]
17. Syamaladevi, R.M.; Andrews, P.K.; Davies, N.M.; Walters, T.; Sablani, S.S. Storage effects on anthocyanins, phenolics and antioxidant activity of thermally processed conventional and organic blueberries. *J. Sci. Food Agric.* **2012**, *92*, 916–924. [CrossRef]
18. Jiang, B.; Mantri, N.; Hu, Y.; Lu, J.; Jiang, W.; Lu, H. Evaluation of bioactive compounds of black mulberry juice after thermal, microwave, ultrasonic processing, and storage at different temperatures. *Food Sci. Technol. Int.* **2015**, *21*, 392–399. [CrossRef]
19. Brauch, J.E.; Reuter, L.; Conrad, J.; Vogel, H.; Schweiggert, R.M.; Carle, R. Characterization of anthocyanins in novel Chilean maqui berry clones by HPLC–DAD–ESI/MSn and NMR-spectroscopy. *J. Food Compos. Anal.* **2017**, *58*, 16–22. [CrossRef]
20. Genskowsky, E.; Puente, L.A.; Perez-Alvarez, J.A.; Fernandez-Lopez, J.; Munoz, L.A.; Viuda-Martos, M. Determination of polyphenolic profile, antioxidant activity and antibacterial properties of maqui (*Aristotelia chilensis* (Molina) Stuntz) a Chilean blackberry. *J. Sci. Food Agric.* **2016**, *96*, 4235–4242. [CrossRef]
21. Quispe-Fuentes, I.; Vega-Gálvez, A.; Aranda, M. Evaluation of phenolic profiles and antioxidant capacity of maqui (*Aristotelia chilensis*) berries and their relationships to drying methods. *J. Sci. Food Agric.* **2018**, *98*, 4168–4176. [CrossRef] [PubMed]
22. Céspedes, C.; El-Hafidi, M.; Pavon, N.; Alarcon, J. Antioxidant and cardioprotective activities of phenolic extracts from fruits of Chilean blackberry *Aristotelia chilensis* (Elaeocarpaceae), Maqui. *Food Chem.* **2008**, *107*, 820–829. [CrossRef]
23. Chang, S.K.; Alasalvar, C.; Shahidi, F. Superfruits: Phytochemicals, antioxidant efficacies, and health effects—A comprehensive review. *Crit. Rev. Food Sci. Nutr.* **2019**, *59*, 1580–1604. [CrossRef] [PubMed]
24. Cespedes, C.L.; Pavon, N.; Dominguez, E.; Alarcon, J.; Balbontin, C.; Kubo, I.; El-Hafidi, M.; Avila, J.G. The chilean superfruit black-berry *Aristotelia chilensis* (Elaeocarpaceae), Maqui as mediator in inflammation-associated disorders. *Food Chem. Toxicol.* **2017**, *108*, 438–450. [CrossRef] [PubMed]
25. Escribano-Bailon, M.T.; Alcalde-Eon, C.; Munoz, O.; Rivas-Gonzalo, J.C.; Santos-Buelga, C. Anthocyanins in berries of Maqui (*Aristotelia chilensis* (Mol.) Stuntz). *Phytochem. Anal.* **2006**, *17*, 8–14. [CrossRef] [PubMed]
26. Gironés-Vilaplana, A.; Mena, P.; García-Viguera, C.; Moreno, D.A. A novel beverage rich in antioxidant phenolics: Maqui berry (*Aristotelia chilensis*) and lemon juice. *LWT* **2012**, *47*, 279–286. [CrossRef]
27. Céspedes, C.L.; Valdez-Morales, M.; Avila, J.G.; El-Hafidi, M.; Alarcón, J.; Paredes-López, O. Phytochemical profile and the antioxidant activity of Chilean wild black-berry fruits, Aristotelia chilensis (Mol) Stuntz (Elaeocarpaceae). *Food Chem.* **2010**, *119*, 886–895. [CrossRef]
28. Miranda-Rottmann, S.; Aspillaga, A.A.; Perez, D.D.; Vasquez, L.; Martinez, A.L.; Leighton, F. Juice and phenolic fractions of the berry Aristotelia chilensis inhibit LDL oxidation in vitro and protect human endothelial cells against oxidative stress. *J. Agric. Food Chem.* **2002**, *50*, 7542–7547. [CrossRef]
29. Zhou, G.; Chen, L.; Sun, Q.; Mo, Q.-G.; Sun, W.-C.; Wang, Y.-W. Maqui berry exhibited therapeutic effects against DSS-induced ulcerative colitis in C57BL/6 mice. *Food Funct.* **2019**, *10*, 6655–6665. [CrossRef]
30. Tenci, M.; Rossi, S.; Giannino, V.; Vigani, B.; Sandri, G.; Bonferoni, M.C.; Daglia, M.; Longo, L.M.; Macelloni, C.; Ferrari, F. An In Situ Gelling System for the Local Treatment of Inflammatory Bowel Disease (IBD). The Loading of Maqui (Aristotelia Chilensis) Berry Extract as an Antioxidant and Anti-Inflammatory Agent. *Pharmaceutics* **2019**, *11*, 611. [CrossRef]
31. Brand-Williams, W.; Cuvelier, M.E.; Berset, C. Use of a Free Radical Method to Evaluate Antioxidant Activity. *LWT Food Sci. Technol.* **1995**, *28*, 25–30. [CrossRef]
32. Zheng, W.; Wang, S.Y. Antioxidant activity and phenolic compounds in selected herbs. *J. Agric. Food Chem.* **2001**, *49*, 5165–5170. [CrossRef]
33. Benzie, I.F.; Strain, J.J. The ferric reducing ability of plasma (FRAP) as a measure of "antioxidant power": The FRAP assay. *Anal. Biochem.* **1996**, *239*, 70–76. [CrossRef] [PubMed]
34. Matic, P.; Sabljic, M.; Jakobek, L. Validation of Spectrophotometric Methods for the Determination of Total Polyphenol and Total Flavonoid Content. *J. AOAC Int.* **2017**, *100*, 1795–1803. [CrossRef] [PubMed]
35. Slinkard, K.; Singleton, V.L. Total Phenol Analysis: Automation and Comparison with Manual Methods. *Am. J. Enol. Vitic.* **1977**, *28*, 49.
36. Seeram, N.P.; Adams, L.S.; Zhang, Y.; Lee, R.; Sand, D.; Scheuller, H.S.; Heber, D. Blackberry, black raspberry, blueberry, cranberry, red raspberry, and strawberry extracts inhibit growth and stimulate apoptosis of human cancer cells in vitro. *J. Agric. Food Chem.* **2006**, *54*, 9329–9339. [CrossRef]
37. Kim, E.J.; Lee, Y.-J.; Shin, H.-K.; Park, J.H.Y. Induction of apoptosis by the aqueous extract of Rubus coreanum in HT-29 human colon cancer cells. *Nutrition* **2005**, *21*, 1141–1148. [CrossRef] [PubMed]
38. Qu, T.; Wang, E.; Jin, B.; Li, W.; Liu, R.; Zhao, Z.B. 5-Aminosalicylic acid inhibits inflammatory responses by suppressing JNK and p38 activity in murine macrophages. *Immunopharmacol. Immunotoxicol.* **2017**, *39*, 45–53. [CrossRef]
39. Li, C.; Feng, J.; Huang, W.Y.; An, X.T. Composition of polyphenols and antioxidant activity of rabbiteye blueberry (*Vaccinium ashei*) in Nanjing. *J. Agric. Food Chem.* **2013**, *61*, 523–531. [CrossRef]
40. Fredes, C.; Montenegro, G.; Zoffoli, J.P.; Gómez, M.; Robert, P. Polyphenol content and antioxidant activity of Maqui (*Aristotelia chilensis* [Molina] Stuntz) during fruit development and maturation in central Chile. *J. Agric. Res.* **2012**, *72*, 582–589. [CrossRef]
41. Hwang, S.J.; Yoon, W.B.; Lee, O.H.; Cha, S.J.; Kim, J.D. Radical-scavenging-linked antioxidant activities of extracts from black chokeberry and blueberry cultivated in Korea. *Food Chem.* **2014**, *146*, 71–77. [CrossRef] [PubMed]

42. Salvia-Trujillo, L.; Morales-de la Pena, M.; Rojas-Grau, A.; Martin-Belloso, O. Changes in water-soluble vitamins and antioxidant capacity of fruit juice-milk beverages as affected by high-intensity pulsed electric fields (HIPEF) or heat during chilled storage. *J. Agric. Food Chem.* **2011**, *59*, 10034–10043. [CrossRef] [PubMed]
43. González, B.; Vogel, H.; Razmilic, I.; Wolfram, E. Polyphenol, anthocyanin and antioxidant content in different parts of maqui fruits (*Aristotelia chilensis*) during ripening and conservation treatments after harvest. *Ind. Crops. Prod.* **2015**, *76*, 158–165. [CrossRef]
44. Brauch, J.E.; Buchweitz, M.; Schweiggert, R.M.; Carle, R. Detailed analyses of fresh and dried maqui (Aristotelia chilensis (Mol.) Stuntz) berries and juice. *Food Chem.* **2016**, *190*, 308–316. [CrossRef] [PubMed]
45. Reyes-Farias, M.; Vasquez, K.; Ovalle-Marin, A.; Fuentes, F.; Parra, C.; Quitral, V.; Jimenez, P.; Garcia-Diaz, D.F. Chilean native fruit extracts inhibit inflammation linked to the pathogenic interaction between adipocytes and macrophages. *J. Med. Food* **2015**, *18*, 601–608. [CrossRef] [PubMed]
46. Hossen, I.; Hua, W.; Ting, L.; Mehmood, A.; Jingyi, S.; Duoxia, X.; Yanping, C.; Hongqing, W.; Zhipeng, G.; Kaiqi, Z.; et al. Phytochemicals and inflammatory bowel disease: A review. *Crit. Rev. Food Sci. Nutr.* **2019**, 1–25. [CrossRef] [PubMed]
47. Moon, H.-D.; Kim, B.-H. Inhibitory effects of Aristotelia chilensis water extract on 2,4-Dinitrochlorobenzene induced atopic-like dermatitis in BALB/c Mice. *Asian Pac. J. Allr. Immunol.* **2019**. [CrossRef]
48. Wijeratne, S.S.; Cuppett, S.L.; Schlegel, V. Hydrogen peroxide induced oxidative stress damage and antioxidant enzyme response in Caco-2 human colon cells. *J. Agric. Food Chem.* **2005**, *53*, 8768–8774. [CrossRef]
49. Nunes, C.; Teixeira, N.; Serra, D.; Freitas, V.; Almeida, L.; Laranjinha, J. Red wine polyphenol extract efficiently protects intestinal epithelial cells from inflammation via opposite modulation of JAK/STAT and Nrf2 pathways. *Toxicol. Res.* **2016**, *5*, 53–65. [CrossRef]
50. San Hipólito-Luengo, Á.; Alcaide, A.; Ramos-González, M.; Cercas, E.; Vallejo, S.; Romero, A.; Talero, E.; Sánchez-Ferrer, C.F.; Motilva, V.; Peiró, C. Dual Effects of Resveratrol on Cell Death and Proliferation of Colon Cancer Cells. *Nutr. Cancer* **2017**, *69*, 1019–1027. [CrossRef]
51. Zengin, G.; Locatelli, M.; Ferrante, C.; Menghini, L.; Orlando, G.; Brunetti, L.; Recinella, L.; Chiavaroli, A.; Leone, S.; Leporini, L.; et al. New pharmacological targets of three Asphodeline species using in vitro and ex vivo models of inflammation and oxidative stress. *Int. J. Environ. Health Res.* **2019**, *29*, 520–530. [CrossRef] [PubMed]
52. He, J.; Han, S.; Li, X.-X.; Wang, Q.-Q.; Cui, Y.; Chen, Y.; Gao, H.; Huang, L.; Yang, S. Diethyl Blechnic Exhibits Anti-Inflammatory and Antioxidative Activity via the TLR4/MyD88 Signaling Pathway in LPS-Stimulated RAW264.7 Cells. *Molecules* **2019**, *24*, 4502. [CrossRef] [PubMed]
53. Ren, J.; Su, D.; Li, L.; Cai, H.; Zhang, M.; Zhai, J.; Li, M.; Wu, X.; Hu, K. Anti-inflammatory effects of Aureusidin in LPS-stimulated RAW264.7 macrophages via suppressing NF-κB and activating ROS- and MAPKs-dependent Nrf2/HO-1 signaling pathways. *Toxicol. Appl. Pharm.* **2019**, *387*, 114846. [CrossRef]

Review

Up-To-Date Analysis of the Extraction Methods for Anthocyanins: Principles of the Techniques, Optimization, Technical Progress, and Industrial Application

Noelia Tena *[ ] and Agustin G. Asuero [ ]

Departamento de Química Analítica, Facultad de Farmacia, Universidad de Sevilla, Prof. García González 2, 41012 Sevilla, Spain; asuero@us.es
* Correspondence: ntena@us.es

Abstract: Nowadays, food industries are concerned about satisfying legal requirements related to waste policy and environmental protection. In addition, they take steps to ensure food safety and quality products that have high nutritional properties. Anthocyanins are considered high added-value compounds due to their sensory qualities, colors, and nutritional properties; they are considered bioactive ingredients. They are found in high concentrations in many by-products across the food industry. Thus, the non-conventional extraction techniques presented here are useful in satisfying the current food industry requirements. However, selecting more convenient extraction techniques is not easy. Multiple factors are implicated in the decision. In this review, we compile the most recent applications (since 2015) used to extract anthocyanins from different natural matrices, via conventional and non-conventional extraction techniques. We analyze the main advantages and disadvantages of anthocyanin extraction techniques from different natural matrices and discuss the selection criteria for sustainability of the processes. We present an up-to-date analysis of the principles of the techniques and an optimization of the extraction conditions, technical progress, and industrial applications. Finally, we provide a critical comparison between these techniques and some recommendations, to select and optimize the techniques for industrial applications.

Keywords: anthocyanins; non-conventional extraction techniques; anthocyanin yields; industrial application; optimization

1. Introduction

In recent years, interest in food quality has increased. Consumers are increasingly becoming aware about the correlation between nutrition and health. Thus, many consumers are opting for foods that contain bioactive ingredients, which have health-promoting properties [1]. Vegetables and fruits have high amounts of antioxidant compounds that are considered bioactive ingredients. Different chemical compounds present in vegetables, fruits, and flowers have antioxidant properties. Some of these compounds belong to the group of anthocyanins, which are not only antioxidant compounds, but are also natural colorants (with aglycones of anthocyanins being the most common plant pigments). The properties of anthocyanins have been extensively studied in the literature [2–10]. These properties are directly related to their chemical structures. A flavylium cation acts as an acid, and it gives anthocyanins a high chemical reactivity. The structures and properties of anthocyanins depend on different factors, such as temperature, pH, and solvent. Figure 1 shows the base structure of anthocyanins. These factors must be considered in the extraction processes to minimize changes in the quality and activity of the resulting extract [2,3].

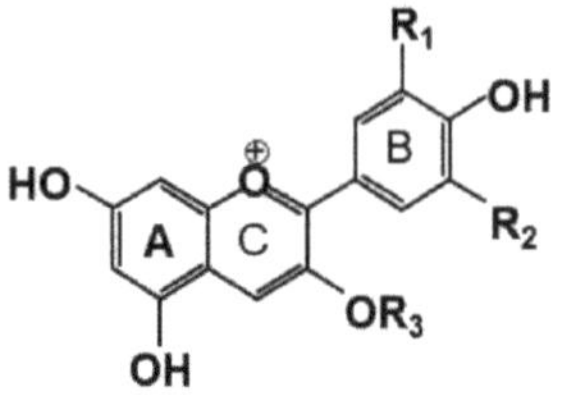

Name	R_1	R_2	R_3
Delphinidin	OH	OH	H
Petunidin	OH	OCH_3	H
Malvidin	OCH_3	OCH_3	H
Cyanidin	OH	H	H
Peonidin	OCH_3	H	H
Pelargonidin	H	H	H

Figure 1. Structure of anthocyanins R_3 = sugar and anthocyanidins R_3 = H (taken from Tena et al. [2]).

The interest in anthocyanins can be attributed to the fact that they can be used as natural coloring agents rather than as synthetic coloring agents [11–14]. Anthocyanins have attractive colors, ranging from red to purple, and their use as food coloring agents has been authorized under code E163 by the European Food Safety Authority (EFSA) [11,15]. This has expanded the use of these natural coloring agents at the industrial level; they can be included in the food industry because they are innocuous and safe molecules. Moreover, anthocyanins have gained attention due to their potential health benefits, e.g., (i) anti-inflammatory, antioxidant, anti-diabetic, and anti-cancer properties; (ii) preventing cardiovascular diseases, neurological disorders, obesity, and for eye health; (iii) improving the gut microbiome; and (iv) decreasing H_2O_2-induced cell apoptosis of the human normal liver cell (LO2 cell) line [2,16–18]. Thus, anthocyanins provide excellent added value to foods due to their dual nature as coloring agents and antioxidants, with beneficial effects on health. Therefore, they are determining factors in the quality and value of fruits and vegetables, and in processed food derived from those. In addition, they are widely used in the food and cosmetics industries.

However, their supply currently depends, to a large extent, on the complex extractions of these compounds from the matrix. At present, expensive raw materials and production technologies make the extraction of natural anthocyanins relatively expensive [1,19]. For this reason, new sources of anthocyanins are being studied, in order to extract these compounds at lower costs. Recent studies have explored the extraction performances of anthocyanins in different matrices, such as microalgae [20], potatoes [21,22], and rice [19,23]. In order to comply with current regulations regarding environmental sustainability, to improve economic performance and reduce waste in the food industry, there is growing interest in investigating the possibility of extracting anthocyanins from by-products and waste generated at different industrial food productions [17,24–29]. Anthocyanins recovered from food waste could have high potential in being used in different food and biotechnological applications, e.g., as food supplements, nutraceuticals, and/or food additives.

The food industry is currently looking for new sources of bioactive compounds, such as anthocyanins, to meet consumer demands. By-products and waste from some food industries represent low-cost sources of anthocyanins that are of interest to the industry. The extraction methods applied to anthocyanin extraction in different natural matrices have been extensively studied [6,17,30–35]. Conventional techniques, such as maceration and heat-assisted extraction (HAE), do not require sophisticated instrumentation and are easy to apply at the industrial level. However, they have a number of limitations, such as the following: the toxicity of the solvents used, possible solvent residues in the extracts, safety risks associated with the use of large volumes of solvents, deterioration of the extracts due to heating, and low yields in the extraction of anthocyanins. The latter limitation may be due to the fact that anthocyanins are found in the vacuoles of plant cells. In order to extract them with cost-effective yields, it is necessary to apply extraction methods that reduce the mass transfer resistance of the plant cell wall. To avoid the resistance of the cell wall, emerging extraction technologies have been proposed, such as ultrasound-assisted extraction (UAE), microwave-assisted extraction (MAE), supercritical fluid extraction (SFE), high-pressure liquid extraction (HPLE), pulsed electric fields (PEFE), high voltage electrical discharge (HVED), and enzyme assisted extraction (EAE). These techniques require more sophisticated instrumentation than conventional techniques, but have some advantages: in general, they do not use heat for extraction, reducing energy costs, or improving the stability of the extracts. In addition, the volume of solvent used is lower (or zero) compared to conventional techniques, reducing greenhouse gas emissions and complying with the legal requirements of green chemistry. Thus, selecting the most appropriated extraction techniques is the primary goal for many industries, so that they can increase profitability by decreasing energy costs, and be responsible with the "green chemistry".

This review provides a critical comparison on the extraction methods used for anthocyanin extraction in recent years (2015 onwards). We classified the extraction methods into conventional and non-conventional techniques. For each one, we describe the principles of the techniques, the optimal parameters for anthocyanin extraction (with concrete examples), the most recent innovations in the techniques, and we comment on the industrial applications.

2. Extraction Methods Used to Extract Anthocyanins

The first step is extraction. In this way, the desired natural products can be separated from the raw materials. Solvent extraction, distillation method, pressing, and sublimation are some extraction methods, according to the extraction principles. The selection of the extraction process, to extract anthocyanins, is based on the preservation of the stability and shelf life of these compounds, which is directly related to the beneficial properties that these compounds provide. Sometimes, prior to applying the selected extraction technique, it is necessary to eliminate other compounds present in the sample matrix that may hinder the extraction of the anthocyanins, such as lipids, proteins, or contaminants. In this section, the conventional extraction techniques applied to extract anthocyanins are described and classified, with their advantages and disadvantages highlighted.

2.1. Conventional Extraction Techniques

Depending on the procedure used to mix the powdered solid sample with the solvent, the conventional techniques are commonly classify as: (i) maceration, when the powdered crude sample is mixed with solvent; (ii) infusion, when a maceration is carried out with water; (iii) digestion, when a maceration is carried out with mild heating, also known as heat-assisted extraction (HAE); (iv) decoction, when an infusion is made with boiling water; (v) percolation and filtration, when the powdered sample is mixed with a continuously renewed solvent in a percolator, with a filtration process applied afterwards. More recently, the Soxhlet extraction technique emerged, which consists of mixing the powdered solid with the solvent inside a "Soxhlet apparatus", allowing continuous cyclical repetitions of the extractions during a controlled period of time.

These techniques are based on the use of different types of solvents and/or heat. Considering the law that "like dissolves like", the solvents commonly used to extract anthocyanins are: methanol, ethanol, water, acetone, or mixtures thereof. Acid solutions are often added to these solvents to help stabilize the flavylium cation, which is stable in highly acidic conditions (pH ~ 3). To achieve this, the use of weak acids (e.g., formic acid, citric acid, or acetic acid) is recommended, since the use of strong concentrated acids may lead to destabilizing the anthocyanin molecule. In view of the polar structure of the anthocyanins, the addition of water to the solvent mixture can improve the extractive yield.

In addition to the solvents, the powder size of the solid, the solvent-to-solid ratio, and the time and temperature of the extraction are other analytical parameters that should be optimized to ensure the maximum yields. Different authors have optimized some of these analytical parameters to extract the maximum yield of total anthocyanin content (TAC) in different natural matrices. Paludo et al. [36] concluded that the optimal conditions for extracting anthocyanins and phenolic compounds from the skin and seed of Jabuticaba (*Plinia cauliflora*) fruits, respectively, was a solvent mixture of methanol/water/acetic acid (80:20:0.5 $v/v/v$), with a solid–liquid ratio of 0.01 g/mL, with two hours of constant agitation. Other authors consider the use of temperature necessary to optimize the extraction yield, as is the case of Albuquerque et al. [11], who claim that the optimal conditions to extract anthocyanins in the skins of Jabuticaba fruits involved a solvent mixture of ethanol (9.1% v/v) acidified with citric acid at pH 3, mixed and centrifuged with the powdered sample (~20 mesh), in a solid–liquid ratio of 50 g/L during 21.8 min at 47.1 °C. They conclude that the total amount of anthocyanins extracted increase with mild temperatures and decrease with high ethanol concentrations and with long times of extraction. The latter could be explained by the fact that long times could lead to the breakdown of the structures of sensitive compounds, such as cyanidin-3-O-glucoside [11]. Other studies carried out with other natural matrices found similar extraction conditions in the optimization of anthocyanin extraction. Thus, Demirdöven et al. [37] determined that a solvent mixture of ethanol (42.39% v/v) acidified with formic acid in a solid–liquid ratio of (1:3 w/v) heated at 40 °C for 75 min were optimal conditions to extract anthocyanins in red cabbage. Backes et al. [38] affirmed that the optimal analytical parameters to improve the yield of TAC in fig skin was a solvent mixture of ethanol (100% v/v) acidified with citric acid (pH = 3), mixed and centrifuged with the powdered sample (~20 mesh) in a solid–liquid ratio of 50 g/L for 13.74 min at 35.64 °C. These authors revealed that the high content of ethanol and the temperature increase the yield of the extraction. In the particular case of cyaniding-3-rutinoside, the maximum extraction was obtained when 100% of ethanol was used, in contrast to what was stated by Albuquerque et al. [11]. It is well-known that high values of ethanol in the solvents increase the extraction of bioactive compounds from plant materials. However, these studies pointed out the importance of the amount of ethanol in the solvent mixture in the selective extractions of individual anthocyanins [11,12,37]

These studies show that, at the laboratory level, the analytical parameter, solid–liquid S/L ratio, has little effect on TCA extraction [12]. However, Backes et al. [38] conducted a study to optimize the solid-to-liquid (S/L) ratio in order to apply HAE at the industrial scale. They determined that a S/L ratio higher than 200 g/L does not allow a homogenous mixture. Furthermore, they established that, depending on the TAC in the natural material, the optimal S/L ratio changes, since highly concentrated samples will saturate the solvent earlier and need lower S/L than less concentrated samples. Their results applied to fig skin concluded that a S/L ratio higher that 100 g/L provoked saturation of the solvent, leading to a decrease in cyanidin-3-rutinoside levels. Fernandes et al. [12] affirmed that, in basil leaves, when the S/L ratio increased from 15 to 30 g/L, any significant proportional decrease was not observed. Thus, they concluded that ratios lower than 30 g/L do not provoke saturation in the extraction of anthocyanins from basil leaves.

These types of extractions are currently the most widely used in the industry, particularly in natural dye industries, likely because these extraction methods have low instrumentation costs. Despite their wide use, these methods also have some disadvantage: (i) high energy consumption; (ii) the use of environmentally unfriendly organic solvents; (iii) the need for expensive and high purity solvents; (iv) the use of a large volume of solvents; (v) the application of a long extraction time to extract compounds with lower yields; (vi) the need for moderate–high temperatures in some cases, which could cause deterioration of antioxidants; (vii) the need for evaporation of a huge amount of solvents; and (viii) the low selectivity of extraction [24,29,39]. Thus, conventional extraction techniques are currently applied in the industry. However, they do not provide high yields in the extraction of anthocyanins, although the yield can improve by increasing the temperature of the extraction. It can also provoke change in the color or in the properties of the extract. The most important factors to optimize are the S/L ratios, which depend on the TAC in the matrix and the solvent used for the extraction. The optimization of the solvent amount is essential to minimize the use of contaminants and work, in agreement with the legal requirements, and in order to reduce the evaporation costs.

2.2. Non-Conventional Extraction Techniques

To overcome the above-mentioned disadvantages of conventional extraction methods, new and promising extraction techniques have been introduced over the year. These techniques are more environmentally-friendly and have important industrial focuses, as they aim to improve the extraction efficiency and yield. However, they have not been employed on a massive scale yet. Among these extraction methods, the most applied techniques to extract anthocyanins are: ultrasound-assisted extraction (UAE), microwave-assisted extraction (MAE), supercritical fluid extraction (SFE), high-pressure liquid extraction (HPLE), pulsed electric fields (PEFE), high voltage electrical discharge (HVED), and enzyme assisted extraction (EAE):

2.2.1. Ultrasound-Assisted Extraction (UAE)

The UAE technique is based on an ultrasound force that is able to break the cell wall due to that cavitation phenomenon that occurs in the tissue of the sample [10,40]. Figure 2 shows a schematic representation of ultrasound-assisted extraction (UAE) equipment, with an explanation about the cavitation process that occurs as a consequence of the ultrasound force. The ultrasound moves through the liquid as waves formed by a process of compression and rarefaction, generating cavitation bubbles with the liquid. The sizes of the bubbles increase with the repetition of a few cycles of compression and rarefaction, up to a critical size (Figure 2B). Then, the bursting of the bubbles, producing a large amount of energy in the form of pressure and temperature, destroys the cell walls, improving the mass transport of the anthocyanins from the plant cell walls to the solvent [41]. UAE could be performed using an ultrasonic bath (BUE) or ultrasonic probe (PUE). Bath ultrasonic extraction (BUE) is more economical and easier to handle. However, the energy produced is not homogeneously distributed in the bath, reducing the efficiency of the extraction. In addition, it has lower reproducibility than PUE. Probe ultrasonic extraction (PUE) consists of a probe connected to a transducer. The probe is immersed in the extraction vessel and disperses the ultrasound in the media with a minimum energy loss. PUE provides higher ultrasonic intensity than the bath system. In PUE, the ultrasound energy is concentrated in a specific zone of the sample, making the extraction more efficient. PUE is commonly preferred to extract bioactive compounds when compared with the bath system [17]. Sabino et al. [42] carried out a comparative study between BUE, PUE, and pressurized liquid extraction (PLE) to determine which technique was most efficient at extracting anthocyanins from jambolan fruit. They concluded that PUE promoted the largest recovery of anthocyanins compared to PLE and BUE. In addition, they pointed out that the three studied techniques extracted the same type of anthocyanin [42].

UAE is widely used in the extraction of bioactive compounds in different natural matrices [10,21,29,41–45]. However, to obtain the maximum yield of anthocyanins by UAE, it is necessary to optimize the extraction conditions. The parameters to be optimized are temperature, time, solvent composition, the liquid/solid (L/S) ratio, moisture content in the sample, particle size, and ultrasound power [21]. Table 1 shows some current examples selected from the literature. It shows the optimal extraction conditions in different natural matrices and the yields obtained by UAE. Demirdöven et al. [37] optimized the extraction conditions, time, temperature and solvent composition (percentage of ethanol) to maximize the efficiency of anthocyanin extraction from red cabbage by UAE (Table 1).

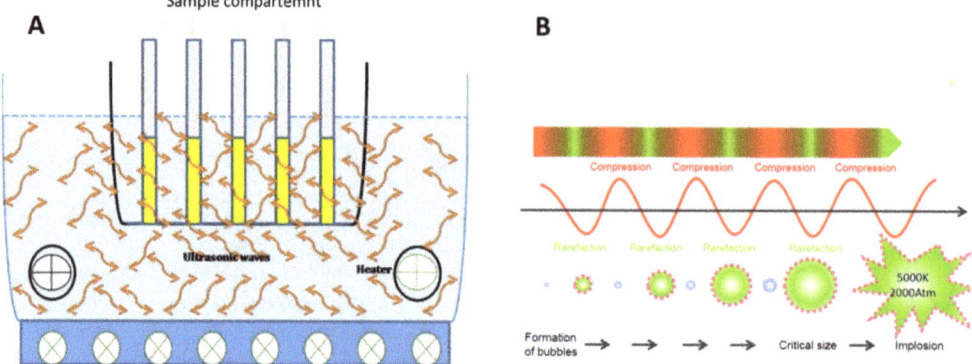

Figure 2. (**A**) Schematic representation of ultrasound-assisted extraction (UAE) equipment, in which a fixed amount of crushed natural material is placed inside the sample compartment. (**B**) Explanation of the formation of the bubbles by alternative compression–rarefaction effects generated by ultrasound waves. (**A**) adapted from Belwal et al. [29], with permission from Elsevier, and (**B**) taken from Morata et al. [10]).

Espada-Bellido et al. [45] showed that temperature and solvent concentrations were the most influential extraction conditions for the extraction of anthocyanins from black mulberry pulps by UAE. However, some studies have demonstrated that the ultrasonic power or force and the pulse cycle applied to the extraction play important roles in the yields of the anthocyanins extractions. Thus, Ravanfar et al. [43], in a study involving red cabbage, reported that time, temperature, and the ultrasonic power were the parameters that most affected the extraction yield, while the pulsation during sonication had no significant effect [43]. Agcam et al. [44] demonstrated the importance of the ultrasonic power in the increment of the anthocyanin extracted from black carrot pomace. Their study also demonstrated the synergy of the combination of ultrasonication and temperature to increase the yield of anthocyanin extraction in black carrot pomace [44]. Different studies have revealed that the ultrasound power, defined by its frequency and intensity, together with time and temperature, are parameters that have a direct effect on the efficiency and yield of anthocyanins extraction [11,41].

Studies have demonstrated that UAE increases the yield of the extraction of antioxidant compounds versus conventional techniques (by at least 20%) and increases the antioxidant activities of the extracts, improving the extraction efficiency of antioxidant compounds [41]. Thus, Backes et al. [37] and Demirdöven et al. [38] showed that they obtained better yields by UAE than by HAE when extracting anthocyanins from red cabbage and fig peel, respectively (Table 1). However, the study carried out by Albuquerque et al. [11] revealed that HAE gave better yields (76 mg/g) than UAE (32 mg/g), when anthocyanins were extracted from the epicarp of Jabuticaba. They attribute these results to the less agitation (and, therefore, less homogenization) between the solvent and the sample in the UAE extraction than in the HAE. Other authors have obtained similar results. These results indicate that the nature of the anthocyanin (e.g., its anthocyanidin and sugar molecule) may also be a factor influencing UAE extraction [11].

Table 1. Current examples of the application of UAE for the extraction of anthocyanins from different natural matrices. Optimal extraction conditions to achieve maximum yield.

Natural Matrices	T^a (°C)	Solvent (%)	Time (min)	L/S Ratio (mL/g)	Power (W)/ Frequency (kHz)/Solid Amount (g)	Steps after Separation. Comments on the Extract.	Recovery	Ref.
Jabuticaba epicarp.	30–35	Ethanol 34.47%	24.44	100:5	500/20/2.5	• Centrifugation. • No selectivity in extraction was verified.	32 mg of D3G + C3G/g of extract.	[11]
Purple sweet potato.	60	Ethanol 90% (0.1% HCl)	60	100:5	200/-/10	• Centrifugation and evaporation. • Content of nonacyl and monoacyl anthocyanins higher than diacyl anthocyanins in the UAE extract.	214.92 mg of C3GE/100 g of potato DW.	[22]
Red cabbage.	40	Ethanol 42.39%	75	3:1	-/37/-	• Filtration, vacuum filtration, and evaporation. • No selectivity studies. The UAE extract has 12% more anthocyanins than the CE extract.	58.67 mg of C3G/L of extract.	[37]
Fig (*Ficus carica* L.) peel.	30–35	Ethanol 100%	21	100:15	310/-/2.5	• Centrifugation filtration, and evaporation. • Purity of the extract: 9.1 mg of C3R/g of extracted residue. This results were better than the obtained by MAE.	4.32 mg C3R/g of fig peel DW.	[38]
Jambolan (*Syzygium cumini* L.) fruit.	30	Ethanol 79.6%	7.5	15:1	Power density: 112.5 W/L/40/4	• Filtration and evaporation. • Ethanolic extract obtained by BUE was rich in D3,5DG, Pt3, 5DG, and M3,5DG.	(BUE) 54.2 mg C3GE/g of fruit DW.	[42]
Jambolan (*Syzygium cumini* L.) fruit.	30	Ethanol 79.6%	7.5	15:1	Power density: 5000 W/L/20/4	• Filtration and evaporation. • Ethanolic extract obtained by PUE was rich in diglucosides of delphinidin, petunidin, and malvidin. This extract was richer in these anthocyanins than the extract obtained by BUE by 15–25%.	(PUE) 60.5 mg C3GE/g of fruit DW.	[42]
Red cabbage.	30	Water	15	100:2	100/30/2	• Filtration and centrifugation. • No selectivity studies.	20.9 mg of P3G/L of extract.	[43]

Table 1. Cont.

Natural Matrices	T^a (°C)	Solvent (%)	Time (min)	L/S Ratio (mL/g)	Power (W)/ Frequency (kHz)/Solid Amount (g)	Steps after Separation. Comments on the Extract.	Recovery	Ref.
Black carrot pomace.	50	Water	20	3:1	102/24/75	• Centrifugation. • Study of the extraction yield of individual anthocyanin, two non-acylated anthocyanins, and three monoacylated anthocyanins in black carrot.	12.4 mg of C3XGG/L of extract. 69.7 mg of C3XG/L of extract. 16.0 mg of C3XGGS/L of extract. 73.4 mg of C3XGGF/L of extract. 34.2 mg of C3XGGC/L of extract.	[44]
Mulberry (*Morus nigra*) pulps.	48	Methanol 76% pH = 3	10	12:1.5	200/24/1.5	• Filtration and dilution with the same solvent. • No selectivity studies.	149.95 µg of C3G + C3R + C3MG + C3DG/g of mulberry FW.	[45]
Haskap (*Lonicera caerulea* L.) berries.	35	Ethanol 80%, (0.5% formic acid)	20	25:1	100/40/-	• Centrifugation and filtration. • C3,5DG, C3G, C3R, P3G, PE3G, were identify in the extract. The anthocyanin/phenolic ratio in the extract was from 62.5 to 92.19%.	22.45 mg C3GE/g of berries DW.	[46]
Blueberries (*V. Angustifolium* Aiton).	65	Ethanol 60% acidified	11.5	50:1	100/40/-	• Centrifugation and filtration. • No selectivity studies.	13.22 mg C3GE/g of blueberries DW.	[47]
Blackthorn (*Prunus spinosa* L.) Fruit Epicarp.	Room T^a	Ethanol 47.98% acidified (citric acid, pH = 3).	5	100:5	400/40/2.5	• Centrifugation, filtration, and drying. Purity of the extract: 18.17 mg of C3R + P3R/g of extracted residue.	11.76 mg of C3R+P3R/g of fruit epicarp DW.	[48]
Purple Majesty potato.	33	Ethanol 70%	5	200:5	35/20/5	• Filtration. • Better yields were obtained from raw freeze dried potato than from microwaved or raw sliced potato.	364.3 mg C3G/kg of potato FW.	[49]

Note: CE, conventional extraction; C3DG, cyanidin-3-O-(6″-dioxalyl-glucoside); C3,5DG, cyanidin 3,5-diglucoside; C3G, cyanidin-3-O-glucoside; C3GE, cyaniding 3-glucoside equivalents; C3MG, cyanidin-3-O-(6″-malonyl-glucoside) C3R, cyaniding-3-O-rutinoside; C3XG, cyanidin-3-xyloside-galactoside; C3XGG, cyanidin-3-xylosyl-glucosyl-galactoside; C3XGGC, cyanidin-3-xylosyl-glucosyl-galactoside-coumaric acid; C3XGGF, cyanidin-3-xyloside-galactoside-glucoside-ferulic acid; C3XGGS, cyaniding-3-xylosyl-glucosyl-galactosidesinapic acid; D3,5DG, delphinidin-3,5-diglucoside; D3G, Delphinidin-3-O-glucoside; DW, dried weight; FW, fresh weight; M3,5DG malvidin-3,5-diglucoside; PE3G, peonidin 3-glucoside; P3G, pelargonidin-3-glucoside; P3R, peonidin 3-rutinoside; Pt3, 5DG, petunidin-3,5-diglucoside; Ref. reference; T^a, temperature.

Backes et al. [38] not only obtained the highest yield for fig peel extraction by EAU than by HAE and MAE, but they also showed that the purity of fig peel extract increased as the L/S ratio decreased [38]. Mane et al. [49], as well as other authors, confirmed that, in the extraction of anthocyanins from the purple potato by UAE, the extraction conditions interfering with the extraction yield were the potato shape, time, solvent composition, solvent ratio, and ultrasound power. They affirmed that a higher organic solvent:water ratio increases the TAC extracted.

UAE is proven to be a useful technology for the extraction of anthocyanins. However, overly-long extractions can destroy these bioactive compounds. More recently, some modifications to the ultrasound-assisted technique have been introduced to increase extraction performance and to reduce time and energy consumption. Mane et al. [49] explored the effect of the combination of UAE with a previous microwave treatment of the sample—in this case, potato. They observed that this combination decreased the volume of the solvent and the time needed to obtain the same extraction yield [49]. Another proposal is the case of the pulsed ultrasound assisted technique (PUAE), which, instead of applying the ultrasound continuously, it is applied intermittently, producing less heat and saving costs [50]. Another alternative has been the combination of Ultra-Turrax with ultrasound-assisted extraction—Ultra-Turrax-based ultrasound-assisted extraction (UT-UAE). Ultra-Turrax allows producing a narrow and uniform particle size distribution that can increase the extraction speed [51]. Another innovation is the combination of an ionic liquid with UAE—ionic liquid-based ultrasound-assisted extraction (IL-UAE). This combination produces a synergy effect that increases the extraction efficiency due to the characteristic of the ionic liquid, such as high polarity, high ionic conductivity, and chemical stability [52].

From the results shown in this section, it can be concluded that UAE, under optimal conditions, can be considered an easy and economical tool for the extraction of anthocyanins, considering that the anthocyanins are labile at a basic pH, at high temperatures, and when exposed to light [2]. The extraction of anthocyanins should be carried out by avoiding the exposition of the sample to these factors; thus, faster extraction techniques are recommended when compared with conventional techniques, in order to avoid deterioration of the anthocyanins. The UAE, using a cavitation process, allows the solvent to easily penetrate into the solid matrix by increasing the mass transfer between the solid matrix and the solvent. This allows higher yields to be obtained in short periods of time and at low temperatures. Consequently, this technique is not only ideal because it is faster, but it also consumes less solvent, is more environmentally friendly, and has a lower cost with higher yields compared to conventional extraction techniques. In addition, this technique is simple (Figure 2) and does not require complex maintenance [10,21]. At the industrial scale, UAS is a promising technique to displace conventional techniques, due to less extraction times, higher extraction yields, and lower operating temperatures.

2.2.2. Microwave-Assisted Extraction (MAE)

The MAE technique uses electromagnetic radiation energy in the microwave region. This energy is absorbed by the polar molecules present in the solvent and the food, producing a dipolar rotation in the molecules and the migration of ions. Microwaves act selectively on plant cells, vaporizing the water in the matrix and, thereby, generating high pressure in the cell wall. This effect produces heat, causes changes in the physical properties of the cell wall, and eventually leads to cell wall rupture [29,41,53]. Thus, the solvent penetrates more easily into the plant cell, favoring the mass transport from the cell to the solvent. The heat produced during this process is transferred from the inside of the plant cell to the outside, which occurs in the opposite direction in UAE. Figure 3 shows a schematic representation of microwave-assisted extraction (MAE) equipment, along with an explanation about the heating produced by ionic conduction and dipole rotation.

The extraction conditions that should be optimized to increase the yield of the bioactive compounds by MAE are: the composition of solvent, which must be sufficiently polar to absorb the microwaves radiation, L/S ratio, temperature, time, microwaves power, and particle size of the matrix. MAE has been used to extract anthocyanins in recent studies. Table 2 shows some recent examples selected from the literature on the application of MAE, to extract anthocyanins from different natural matrices.

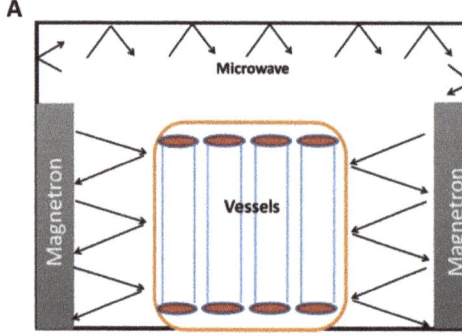
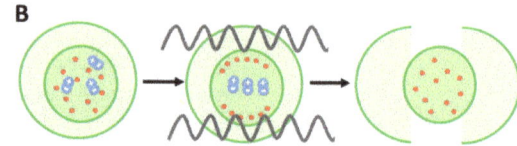

Effect of microwaves on the plant cell during MAE

1. **Drying of the cell** due to heating of the internal water.
2. **Stretching of the cell** due to the internal temperature and pressure reached.
3. **Cell wall rupture** releasing bioactive compounds.

Figure 3. (**A**) Schematic representation of microwave-assisted extraction (MAE) equipment, in which a fixed amount of crushed natural matrix is placed inside the vessels, together with the solvent. (**B**) Explanation of how microwave heating of the cell by ionic conduction and dipolar rotation causes cell wall rupture. ((**A**) adapted from Belwal et al. [29], with permission from Elsevier).

Xue et al. [54] applied MAE to extract anthocyanins from blueberries. They studied the effects of microwave powers on the yield of extraction and observed that rising microwave powers had little effect on the distribution of microwave energy, but increased the temperature in the center location in the vessels. They concluded that the highest anthocyanin yield was obtained at a critical temperature of 50.75 °C and suggested that the control of microwave power contributes to the improvement of anthocyanin yield and efficiency of microwave energy [54]. Farzaneh and Carvalho [55] applied MAE to extract anthocyanin from *Lavandula pedunculata* L. fresh plants. They studied the effect of microwave irradiation power, irradiation time, and L/S ratio on the yield. They observed that the highest antioxidant activity was obtained with a lower power (300 W) and time of irradiation (107.3 s) and a higher L/S ratio (34.807 mL/g) than those applied to obtain the highest TAC (Table 2) [55]. These results could be explained by the fact that a higher irradiation power and time increases the yield of the extraction, as well as increases the temperature of the sample. It could also damage the thermally labile antioxidants, which might result in the reduction of their antioxidant activities. Thus, Sun et al. [56] applied MAE to extract anthocyanins from powdered blueberry. They demonstrated that the acquirement and degradation of anthocyanins occur simultaneously with exponential trends. They determined that the lowest degradation of anthocyanins were products at a temperature of 53.3 °C. Lower temperatures favored the self-aggregation of the anthocyanins and higher temperatures provoked the degradation of them. They proved that the acquired anthocyanins were higher and the degradation was lower by MAE than by hot reflux extraction (HRE), a conventional technique [56]. Thus, microwave power and the time of irradiation are complementary variables that influence the extraction process. High power can increase the heating effect decreasing the microwave irradiation time and increasing the yield, but it also might provoke a degradation of thermally labile compounds [41]. Thus, Xue et al. [57] optimized the irradiation time and the resulting temperature to achieve the highest yield of anthocyanins from cranberry. They concluded that an irradiation time of 8 s and an extraction temperature of 50 °C provided the highest yield (Table 1). Their results showed that increasing the temperature was beneficial for anthocyanin extraction when the extraction temperature was below 323 K (50 °C) [57]. Despite the need to control and optimize the thermal conditions under which MAE is carried out to avoid a negative effect on the antioxidant properties and the color of the extract obtained, several studies have been carried out to explore the possible application of this technique in the extraction of anthocyanins [Table 2].

Table 2. Current examples of the application of MAE for the extraction of anthocyanins from different natural matrices. Optimal extraction conditions to achieve maximum yield.

Natural Matrices	T^a (°C)	Solvent (%)	Time (s)	L/S Ratio (mL/g)	Irradiation Power (W)/Solid Amount (g)	Steps after Separation. Comments on the Extract.	Recovery	Ref.
Fig (*Ficus carica* L.) peel.	62.4	Ethanol 100% (pH 3)	300	100:5	400/0.5	• Centrifugation, filtration, and evaporation. • Purity of the extract: 7.43 mg of C3R/g of extracted residue. It was lower than the obtained by UAE.	411 mg of C3R/100 g of fig peel DW.	[38]
Lavender (*Lavandula pedunculata* L.) fresh plants.	-	Water	114.3	30.32:1	464.9/1	• Cooling, centrifugation, filtration, and evaporation. • No selectivity studies.	273.3 mg of C3G/L.	[55]
Cranberry.	50	Ethanol 52% (pH = 3)	8	28:1	-/2	• Centrifugation and filtration. • No selectivity studies.	306 mg of C3G/100 g of cranberry.	[57]
Red cabbage.	100	Water (pH = 3–3.3)	300	30:1	200/5	• Cooling and filtration. • Extraction efficiency comparable to that of the EC.	110.0 mg of C3G/L.	[58]
Red cabbage.	90	Ethanol 50% (pH = 3–3.3)	600	20:1	600/5	• Cooling and filtration. • Extraction efficiency lower than with EC.	220.2 mg of C3G/L.	[58]
Purple sweet potato.	-	Ethanol 30% (citric acid pH = 2)	500	3:1	320/10	• Filtration and centrifugation. • Anthocyanin composition comparable to that of the extract obtained with aqueous solvent containing citric acid.	31 mg of C3GE/100 g of potato.	[59]
Black raspberry Korean.	-	Ethanol 74% (pH = 2)	66	30:1	148/5	• Filtration. • No selectivity studies.	372 mg of C3G/100 g of fruit.	[60]
Eggplant Peel.	-	Ethanol 80%	40	50:1	480/-	• Centrifugation and filtration. • No selectivity studies.	881 mg of C3G/100 g of peel.	[61]

Table 2. Cont.

Natural Matrices	Tª (°C)	Solvent (%)	Time (s)	L/S Ratio (mL/g)	Irradiation Power (W)/Solid Amount (g)	Steps after Separation. Comments on the Extract.	Recovery	Ref.
Grape juice waste.	55	Double distilled water	138.6	19.2:1	435/1	• Cooling to room temperature and filtration. • No selectivity studies.	132 mg of M3G/100 g of grape juice waste DW.	[28]
Blackcurrant.	-	Ethanol 60% (pH = 2.5)	984	28.3:1	551/-	• Cooling and centrifugation. • No selectivity studies.	47.37 mg of C3G + D3R + C3R + D3DG/100 g of blackcurrant.	[62]
Red rice.	-	Ethanol 85% acidified	100	22:1	400/-	• Cooling and centrifugation. • No selectivity studies.	3.82 mg of C3G/100 g of rice.	[23]
Rosa pimpinellifolia L. fruits.	60	Ethanol 26.85% $(NH_4)_2SO_4$ 19.15%	1037.4	40:1	400/20	• Cooling and centrifugation. • The purity of the extract obtained by MA-ATPE was 1.65-fold greater than MAE using 80% ethanol.	1373.04 mg C3GE/g of fruit DW.	[63]

Note: C3G, cyanidin-3-glycoside; C3GE, cyanidin-3-glycoside equivalent; C3R, cyanidin 3-rutinoside; D3DG, delphinidin 3-O-β-d-glucoside; D3R, delphinidin 3-O-rutinoside; DW, dried weight; MA-ATPE, microwave-assisted aqueous two-phase extraction; M3G, malvidin-3-glucoside. Ref. reference; Tª, temperature.

Several studies have shown that MAE has some advantages compared to conventional extraction [58–62]. These studies claim that MAE lasts for less time and consumes less solvent [58], and allows higher extraction yields than a conventional extraction [41,56]. In addition, it is an energy-saving method due to its short processing time. However, the main limitation of this technology is the negative effect of heating, caused by microwave energy, on the properties of anthocyanins.

Nowadays, in order to overcome the possible limitations of the MAE, some modifications have been introduced, such as working at low pressures or working at atmospheric pressure. Changes have also been made to the type of solvent used in MAE. Applications have been developed in which oxygen flow or different aqueous phases are used for extraction. Some of the most current proposals could be classified as the following [30]:

MAE for thermolabile substances, which include: nitrogen-protected microwave-assisted extraction (NPMAE) and vacuum microwave-assisted extraction (VMAE). This is the most convenient alternative to extract anthocyanins. These techniques are based on using low oxygen content inside the extraction tanks, combined with moderate or low temperatures. In this way, they can avoid degradation of thermolabile, and oxygen-sensitive plant phytochemicals compared to conventional MAE [30,41].

Microwave heating extraction, which includes: focused microwave-assisted Soxhlet extraction (FMASE), ultrasonic microwave-assisted extraction (UMAE), microwave hydro-distillation (MWHD or MAHD), and microwave steam distillation (MSD).

Green extraction without solvent, which includes: solvent-free microwave extraction (SFME), vacuum microwave hydro-distillation (VMHD), microwave hydro-diffusion and gravity (MHG).

Microwave extraction with no solvent pressure, which includes: pressurized solvent-free microwave extraction (PSFME).

Odabas and Koca [63] proposed microwave assisted aqueous two-phase extraction (MA-ATPE) to extract anthocyanins from *Rosa pimpinellifolia* L. fruits. They applied MA-ATPE to simultaneously extract and purify the anthocyanins with an ethanol/ammonium sulfate aqueous two-phase system. The optimal anthocyanin yields (1373.04 mg C3G/g DW) were achieved in conditions of 0% HCl (w:w), 26.85% ethanol (w/w), 19.15% ammonium sulfate (w/w), L/S ratio 40, 17.29 min, 60 °C and 400 W. They showed that the purity of the anthocyanin extract was 1.65 times higher with MA-ATPE than with MAE, using 80% ethanol [63]. SFME is also an applied alternative to extract flavonoids. This alternative is a green technique that uses the in-situ water in plant cells to absorb the microwave energy and break the cell wall. The extraction could be carried out at atmospheric pressure. Some applications of this technique can be found in the literature for the extraction of flavonoids from different matrices, such as onion [64] or mandarin leaves [65]. Another possibility consists of applied SFME under pressure PSFME. Thus, Michel et al. [66] applied this technique to extract antioxidants from sea buckthorn (*Hippophae rhamnoides* L.) berries. They demonstrated that PSFME, compared to other conventional extraction techniques, such as pressing, maceration, and pressurized liquid extractions lead to the most active and richest extract in phenolic content. Molecules, such as quercetin and isorhamnetin, were extracted in this way, in contrast to other previously applied techniques [66]. Solvent-free hydrodiffusion and gravity MHG have also been applied to extract antioxidants from food by-products from sea buckthorn obtaining similar results [25].

This technique and its variants work well in terms of solvent consumption, extraction time, and extraction yield, and they are considered as possible substitutes for conventional methods. The yield of natural components from vegetable matrices, for implementation as food ingredients, must be carried out under the best extraction conditions, to promote application at an industrial scale, competing against the low economic costs of producing artificial compounds. The characteristics of the MAE and its effectiveness cannot be applied in a generalized way to all matrices, demanding specific optimization for each particular case. In addition, some factors might affect anthocyanin stability and, so, their deterioration rate, reinforcing the importance of determining the conditions that maximize the extraction yield of these compounds. Thus, this technique could be well applied at industrial scale and some studies demonstrate that MAE has been applied to extract anthocyanins from food waste, such as by-products of the wine industry and grape juice [28,29]. However, there are few studies on the application of this technique at an industrial scale. This could be explained by critical points in industrial microwave design. Therefore, to promote the use of MAE in the food and drug industries, it is necessary that research points to the technical issues related to the design of microwave extractors and their suitability for the isolation of bioactive components from the vegetal matrix [30].

2.2.3. Pressure Fluids Extraction Techniques

Extraction under pressure is another alternative used among the non-conventional methods to facilitate mass transfer from the inside to the outside of the plant cell, in addition to the use of electromagnetic radiation in the ultrasound or microwave region. In the case of high-pressure fluid extractions, a reduction in extraction times and in the amount of solvent used has been observed. In addition, the pressure applied influences the selectivity of the extraction. Currently, different techniques based on pressurized fluid extractions have been applied for anthocyanin extraction. They can be classified as follows: supercritical fluid extraction (SFE), pressurized liquid extraction (PLE), and high pressure liquid extraction (HPLE).

Supercritical Fluid Extraction (SFE)

SFE has been widely applied in the last decades, as it is considered one of the most sustainable green technologies. This technique commonly uses supercritical CO_2 (critical temperature (CT) = 31.3 °C; critical pressure (CP) = 72.9 atm) as solvent, since it is nontoxic, not very expensive, preserves the extracts from atmospheric oxidation, and has a moderate CT. The disadvantage of this solvent is that it is non-polar. This is why a co-solvent can be added to promote the extraction of polar compounds, as in the case of anthocyanins [31]. The most common co-solvents used are ethanol, methanol, and aqueous solutions of these alcohols, in concentration within 1% to 15% [67]. In general, SFE and PLE are carried out under medium-to-high pressures. However, SFE operates using solvents at temperatures and pressures above their critical points, and PLE is based on the use of liquids at temperatures above their normal boiling points. SFE uses the properties of the supercritical fluid to extract the compound of interest. Basically, the process consists of two fundamental stages: firstly, the compound is extracted by the supercritical fluid and then the fluid is rapidly removed by a change in pressure and/or temperature. Figure 4 shows a schematic representation of SFE equipment with CO_2 as solvent at a specific pressure and temperature. The extraction process is commonly carried out, combining a static period where the solvent is permanently in contact with the solid, and a dynamic period where the solvent is continually passing through the solid [68].

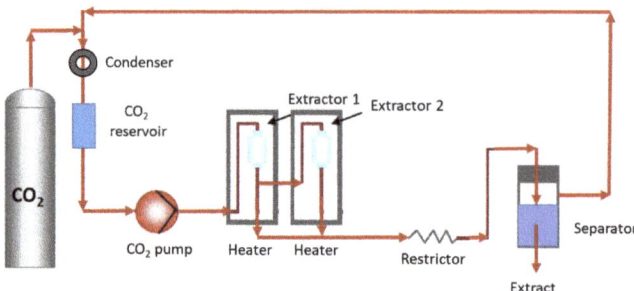

Figure 4. Schematic representation of a supercritical fluid extraction (SFE) equipment, in which a fixed amount of crushed sample is placed inside the extractor 1 and 2.

The extraction conditions that should be optimized to guaranty the maximum efficiency in the extraction by SFE are: temperature, pressure, particle size, amount of co-solvent, moisture content of natural material, time of extraction, flow rate of CO_2, and L/S ratio. Different studies have been carried out applying SFE to extract flavonoids and anthocyanins in different matrices [68–75]. Table 3 shows some recent examples selected from the literature on the application of SFE in the extraction of anthocyanins from different natural matrices, together with the optimal extraction conditions to achieve maximum yields. Maran et al. [69] observed that at pressures higher than 200 bar, the yields slightly decrease in the extraction of total monomeric anthocyanins in Indian blackberry. They pointed out that temperatures above 50 °C led to a slight decrease in the yield, with the optimum temperature being 50 °C. Furthermore, they demonstrated that the optimum solvent flow rate was set at 2 g/min, showing that higher flow rates had no significant effect on the extraction yield. Other studies, shown in Table 3, have been carried out to optimize these parameters, in order to obtain the maximum yield in the extraction of anthocyanins in different natural matrices. These studies conclude that the recovery of antioxidant compounds improved when the pressure was moderately high (higher than 100 bar), the temperature was moderated (not higher than 100 °C), and the addition of a co-solvent was considered. The amount of co-solvent depends on the time of the extraction and the amount of the powdered sample. The most commonly used co-solvent is ethanol [70]. Jiao et al. [68] compared conventional extraction with SFE with CO_2 and water as a co-solvent to extract anthocyanins from Haskap berry paste. After optimizing the extraction conditions for SFE (Table 3), they concluded that SFE extraction offers higher anthocyanin extraction efficiency (52.7% vs. 38.3%) and improves antioxidant activity (89.8% vs. 72.2%) compared to conventional extraction [68].

Table 3. Current examples of the application of SFE with CO_2 for the extraction of anthocyanins from different natural matrices. Optimal extraction conditions to achieve maximum yield.

Natural Matrices	Tª (°C)	P. (bar)	Co-Solvent	Flow Rate	Steps after Separation. Comments on the Extract.	Recovery	Ref.
Haskap (*Lonicera caerulea* L.) berry pulp paste.	65	450	Water L/S ratio 5.4/3.2 (w/w)	15 min static time 20 min dynamic time at 10 mL/min.	• The extract was collected in a vial free of CO_2. • 527 mg of C3G in the extract by g of C3G in starting material.	25 mg of C3G/g of paste DW.	[68]
Indian blackberry (*Syzygium Cumini*) fruit pulp.	50	162	Ethanol (10 g of sample)	2 g/min.	• The extract was separated and collected at an ambient temperature and atmospheric pressure. • A total of 7 different anthocyanins were extracted together with 8 different bioactive phenols.	231.28 mg C3G/100 g of fruit.	[69]
Colombian blueberry (*Vaccinium meridionale*) fresh and mature fruit.	40	300	None (160 g of sample and 800 g of sample)	32 g/min.	• The extract was separated and collected at ambient temperature and atmospheric pressure. • No electivity studies. Better yields were obtained from freeze-dried samples using water and ethanol as co-solvent.	26.7 mg of extract/g of sample.	[72]
Bilberry (*Vaccinium myrtillus* L.) dried fruits.	45	250	(1) 6% of 30% distilled water, 70% ethanol. (2) 6% of 50% distilled water, 50% ethanol at 6 mL/min. (3) 9% of 90% distilled water, 10% ethanol. (430 g of sample)	Multistage supercritical/subcritical extraction: (1) SC-CO_2 8 kg/h (2) SubC-CO_2 6 kg/h (3) SubC-CO_2 6 kg/h.	• The extract was separated and collected at ambient temperature and atmospheric pressure by a cyclonic separator. • SubC-CO_2 selectively extracted C3G and C3A.	60 mg of C3G/100 g of fruit DW.	[73]
Roselle (*Hibiscus Sabdariffa* L.) dry calyces.	70	89	Ethanol 75% (1.5 g of sample)	6 mL/min (modifier flow rates 9.5%)	• Dried at 40 °C to maintain the compounds structure. • No selectivity studies. Study the percentage of the red color extracted from Roselle calyces.	26.7 g of dried extract/100 g of sample.	[74]
Merlot red grape (*Vitis vinifera*) pomace.	95	100	Ethanol 10 mL/min (30 g of sample)	32 g/min	• The extracts were collected by a cyclone separator system. • The extraction efficiency was around 36%. This can increase to 63% if the extraction time is increased from 80 to 180 min.	700 mg of M3G/kg of grape DW.	[75]

Note: C3A, cyanidin-3-O-arabinoside; C3G, cyanidin-3-glucoside; DW, Dried weight; M3G, Malvidin-3-O-glucoside; P., pressure; Ref. reference; SC-CO_2, supercritical CO_2 extraction; SubC-CO_2, subcritical CO_2 extraction; Tª, temperature.

Thus, SFE might be considered an effective green technology to extract anthocyanins. One of the main advantages of this extraction method compared with others carried out under low pressure is that the amount of solvent used is greatly smaller [41]. In addition, this technique allows obtaining extract with higher antioxidant activity, and has shown higher efficiency, selectivity, and a lower extraction time [68,71]. Pazir et al., [75] compared the yield of the conventional extraction with the yield of SFE with CO_2 and ethanol to extract anthocyanins from Merlot red grape pomace. They conclude that, in a short period of time, 80 min, 36% of the anthocyanins could be extracted by $SC-CO_2$. However, they failed to obtain recovery percentages similar to those obtained by conventional extraction (63%) when they increased the extraction time to 180 min. However, the increase in the amount of extractable anthocyanins with increasing extraction time corresponded strongly with the increase in total antioxidant activity [75].

In order to improve the efficiency and selectivity of the SFE, different alternatives have been proposed. An alternative is to increase the capacity and selectivity of the supercritical extraction with CO_2 in the extraction of anthocyanins, based on the application of subcritical CO_2 extraction. The latter consists of using supercritical CO_2 followed by supercritical CO_2 with a certain percentage of ethanol as co-solvent. During the SubC-CO2, the solvent flow through the extraction vessel changes direction, e.g., it may change from an upward to a downward direction. The first and second extraction steps were carried out for 1 h, and the last step for 3 h. In a study carried out by Babova et al. [73] with bilberry, they applied supercritical extraction with CO_2 followed by subcritical CO_2 with 10% v/v ethanol as co-solvent (Table 3). They observed that the subcritical CO_2 selectively extracted cyanidin-3-O-glucoside and cyanidin-3-O-arabinoside, delphinidin-3-O-glucoside, ellagic acid pentoside, feruloyl hexoside, and several quercetin glycosides. Furthermore, the extract obtained with subcritical CO_2 extraction showed high antioxidant activity [73]. Other proposals are based on the synergistic effects of combining several extraction techniques. For example, enzyme-assisted supercritical fluid extraction (EASCFE) has been proposed by Mushtaq et al. [76] for extracting antioxidants from pomegranate peel. This approach increases the extraction yield and selectivity, since the enzymatic treatment facilitates the breakdown of the cell wall, allowing the fast penetration of the supercritical fluid and increasing the mass transfer [67]. Another proposal consists of combining ultrasound with SFE. Pasquel Reategui et al. [77] applied this combination to extract the antioxidant form blackberry bagasse. They obtained a higher yield with this combination. This may be because the ultrasound treatment breaks down the cell walls, facilitating the rapid penetration of the supercritical fluid and the extraction of the antioxidants. These proposals increase the efficiency and the selectivity of the extraction, but also decrease the extraction time and, consequently, the extraction costs. These proposals are presented as promising alternatives to conventional techniques for anthocyanin extraction, from the point of view of environmentally responsible extraction technologies.

Although, a priori, the application of supercritical fluid extraction can be considered too costly to be implemented at a large scale, studies on its viability show that a good optimization of the working parameters (mainly pressure and temperature) allows for obtaining a competitive extraction compared to conventional extraction processes [78]. However, more studies should be performed to optimize other parameters that also have an effect in the efficiency of the extraction. Future studies should focus on the optimization of the extraction time for different geometries of plant materials because the mass transfer could be affected by the nature of the food matrix. Furthermore, the type of co-solvent and the flow rate should be optimized in combination with the extraction time to improve the cost efficiency of the extraction. Finally, further studies, including the economic aspect of extraction are needed to estimate the operating costs of an industrial-scale extraction, accordingly.

Pressurized Liquid Extraction (PLE)

PLE, also known as accelerated solvent extraction (ASE), is based on the use of solvents at a temperature between room temperature and 200 °C, and pressure between 35 and 200 bar. These conditions allow bioactive compounds to be extracted relatively easily from various natural matrices. This is due to the fact that, at high pressure, the solvent remains liquid at temperatures above its boiling point, favoring the solubility of the analytes. This extraction technique receives a different name when the solvent used is water. In this case, the technique is known as sub-critical water extraction (SWE). These types of extractions

(PLE or SWE) are carried out in an accelerated solvent extractor whose scheme is similar to the one shown in Figure 4.

The SWE method has been used as an economical, green, and sustainable extraction process to extract anthocyanins in different natural matrices. The experimental conditions that should be optimized to obtain the maximum yield of the extraction are temperature, pressure, static time, and number of cycles. Table 4 shows some recent examples selected from the literature on the application of SWE in the extraction of anthocyanins from different natural matrices, together with the optimal extraction conditions to achieve maximum yields. Thus, Wang et al. [79] optimized the pressure, temperature, and time to extract the maximum anthocyanin content from the raspberry by SWE. They concluded that the maximum amount of total anthocyanins (8.15 mg/g) was obtained when SWE was carried out at 70 bar and 130 °C during 90 min. Their results demonstrated that the extraction efficient and the antioxidant activity of anthocyanin obtained from the raspberry by SWE were significantly higher than those obtained by conventional extraction with hot water or methanol [79]. Kang et al. [80] also determined the optimal extraction conditions for extracting anthocyanins from blueberries and chokeberries by SWE. They optimized the temperature (110 °C, 130 °C, 150 °C, 170 °C, 190 °C, and 200 °C), the extraction time (1, 3, 5 and 10 min) and the pH of the solvent (water and 1% citric acid). They concluded that the optimal conditions for blueberries were 130 °C for 3 min and for chokeberries 190 °C for 1 min at a pressure of 100 bar, for both natural matrices. They also concluded that the use of SWE with 1% of citric acid increased the total content of anthocyanins and the content of malvidin-3-galactoside in the extract obtained from blueberries. The total content of anthocyanins and the content of cyanidin-3-galactoside were three times higher in the extract obtained from chokeberries, when 1% of citric acid was used in the solvent. They also concluded that the solubility of anthocyanins depended on their structures; thus, the presence of more methoxy and hydroxyl functional groups in the basic skeleton of anthocyanin will result in a lower solubility [80].

Table 4. Current examples of the application of SWE and PLE for the extraction of anthocyanins from different natural matrices. Optimal extraction conditions to achieve maximum yield.

Natural Matrices	Tª (°C)	P. (bar)	Time (min)	Solvent/Flow Rate (mL/min) (Amount of Sample)	Steps after Separation. Comments on the Extract.	Recovery	Ref.
SWE							
Raspberry.	130	70	90	Double distilled water/3 (20 g FW).	• The extract was collected and analyzed immediately. • The content of individual anthocyanins, C3S, C3G, and C3CS, in the extract was higher for SWE than for hot water extraction and methanol extraction.	815 mg of C3GE/100 g of Raspberry FW.	[79]
Blueberries.	130	100	3	Water 1% citric acid/- (1 g DW).	• The extract was filtered using nitrogen gas. • The amount of anthocyanins in the extract obtained by SWE was about 4.5 times higher than that obtained by pressed juice and 1.5 times higher than that obtained by hot water.	50 mg of anthocyanin pigment/100 g of blueberries FW. 18 mg of M3G/100 g of blueberries FW.	[80]

Table 4. Cont.

Natural Matrices	T^a (°C)	P. (bar)	Time (min)	Solvent/Flow Rate (mL/min) (Amount of Sample)	Steps after Separation. Comments on the Extract.	Recovery	Ref.
Chokeberries.	190	100	1	Water 1% citric acid)/- (1 g DW).	• The extract was filtered using nitrogen gas. • The amount of anthocyanins in the extract obtained by SWE was about 9.5 times higher than that obtained by pressed juice and 1.7 times higher than that obtained by hot water.	66 mg of anthocyanin pigment/100 g of chokeberries FW. 134 mg of C3Ga/100 g of chokeberries FW.	[80]
Barberry (*Berberis vulgaris*) Fruit.	157.5	29.64	170	Water.	• The extract was filtrated under vacuum condition. • No selectivity studies.	9.84 mg C3G/mL of sample.	[81]
PLE							
Purple sweet potatoes.	90	-	15 (2 cycles)	Ethanol 80% (acidified 0.1% HCl)/- (10 g).	• The extract was collected and adjusted the volume by evaporation. • In the extract anthocyanin yield followed the order PLE > UAE > CE, which was opposite to the total phenolic and flavonoid yield CE > UAE > ASE. The extract obtained by PLE contained more diacyl anthocyanins and less nonacyl and monoacyl anthocyanins than CE or UEA extracts.	252.34 mg of C3GE/100 g potatoes DW.	[22]
Jambolan (*Syzygium cumini* L.) fruit.	90	117.2	5 rinsing time; 10 extraction time/cycle (2 cycles)	Ethanol 80% (acidified 0.1% TFA)/- (4 g).	• The extract was concentrated in rotavapor. • The extract obtained by PLE presented the lowest proportional abundance of diglucosides of delphinidin, petunidin, and malvidin compared to CE and UAE.	47.05 mg C3GE/g of fruit DW.	[42]
Broken black bean (*Phaseolus vulgaris* L.) hulls.	60	100	26	Ethanol: citric acid 30:70 (pH = 3.4)/5 (5 g DW).	• The extract was lyophilized. • The extract obtained by PLE has the highest value of total monomeric anthocyanins compared to UAE and maceration extraction.	3.96 mg C3GE/g of sample DW.	[82]

Note: C3CS, cyanidin-3-(6′-citryl)–sophoroside; C3G, cyanidin-3-O-glucoside; C3Ga, cyanidin-3-galactoside; C3GE, cyanidin-3- glucoside equivalent; C3S, cyanidin 3-sophoroside; DW, dried weight; FW, fresh weight; M3G, malvidin-3-galactoside; P., pressure; PLE, pressurized liquid extraction; Ref., reference; SWE, sub-critical water extraction; T^a, temperature; TFA trifluoracetic acid.

In conclusion, SWE is a green, faster, and more-efficient method to extract anthocyanin than conventional extraction methods. Thus, SWE is presented as a feasible application for the extraction of anthocyanins, and it can be easily implemented on an industrial scale.

PLE has also been applied to increase the extraction yield of different antioxidant compounds, between them, the anthocyanins [22,42,81,82]. The parameters that should be optimized for this technique are the same as for SWE, but in this case, the selection of the solvent is also important. In the case of anthocyanins extraction, acidified aqueous ethanol or methanol are commonly used (Table 4). In relation to temperature and pressure, high pressure and temperature facilitated the penetration of the solvent in the natural matrix.

The maximum temperature is limited by the thermal instability of anthocyanins. The static time and the number of cycles should be enough to guaranty the total contact between the solvent and the bioactive compounds in the matrix. Thus, Cai et al. [22] concluded that the maximum amount of anthocyanins from purple sweet potatoes was obtained by PLE using the conditions showed in Table 4, compared with conventional extraction CE and UAE. The extract obtained by PLE contained more diacyl anthocyanins and less nonacyl and monoacyl anthocyanins than CE and UAE extracts. The PLE extract also had higher antioxidant activity than the others measured by FRAP, but not by ORAC [22]. In relation to the number of cycles, Cai et al. [22] and Sabino et al. [42] determined that increasing the number of cycles was negative in anthocyanin recovery. On the other hand, they showed that more than two cycles led to a decrease in anthocyanin content. They justified this effect by the relationship between the number of cycles and temperature. The anthocyanins presented a slight degradation with the increasing of the temperature. However, previous research on PLE has indicated a positive effect of temperature on the extraction of anthocyanins from some vegetal matrices [22,42]. They concluded that the temperature effect on anthocyanin extraction is influenced by the product matrix and the solvent used for extraction. In addition, the high pressure applied in PLE increases the covalent bond stability within molecules, preventing the thermal degradation of these compounds [42].

The results have revealed that PLE allows obtaining antioxidant-rich extracts with high potential application in food, supplements, and pharmaceutical industries [82]. However, it is not free of limitation and some studies have showed lower efficiency than UAE in the extraction of anthocyanins or extract with lower antioxidant activity [22,42]. As a consequence, some alternatives have been studied. Moirangthem et al. [19] proposed the combination of microwave-assisted sub-critical water extraction (MA-SWE), at 90 °C for 5 min, from Manipur black rice. They concluded that this combination allowed obtaining an extraction efficiency of 85.8% and extracts with higher antioxidant activity than an equivalent conventional extraction with methanol. They consider that a validation at a small-scale level is needed to assess its economic feasibility of MA-SWE [19]. Andrade et al. [83] proposed another alternative consisting of combining ultrasound-assisted with pressurized liquid extraction (US-PLE) to improve the yield of anthocyanin extraction from black chokeberry pomace. They concluded that the effects of pressure and sonication were more pronounced at low temperatures and using slightly acidified solvent, lower than 1.5%. Thus, US-PLE showed clear advantages reaching a high total extraction yield. However, an economic assessment is necessary in order to evaluate the influence of the capital investment on the process profitability at an industrial level [83].

Thus, the PLE method is an emerging technique that carries out fast and efficient extractions under high temperature and pressure conditions. The association of green technologies with green solvents makes PLE an eco-friendly technology. The PLE allowed obtaining antioxidant-rich extracts with high application potential in food, supplements, and pharmaceutical industries. In comparison to conventional methods, it affords better results for the recovery of these compounds, since the viscosity of the solvent decreased, favoring the solubilization of the compounds of interest. This allowed high penetration of the liquid solvent in the solid matrix (raw material) and increased the diffusion and mass transfer coefficient, improving the extraction performance. However, the negative effect of temperature on the extract of some natural matrices may lead the industry to opt for other techniques, such as UAE rather than PLE or SWE.

High-Pressure Liquid Extraction (HPLE)

HPLE includes techniques known as high hydrostatic pressure extraction (HHPE), ultra-high pressure extraction (UHPE), and high-pressure processing extraction (HPPE). They are characterized as using higher pressure than PLE in order to keep the solvent beyond its boiling point. This further facilitates the extraction and decreases the amount of solvent used and the extraction time [32]. The latter advantage is directly related to the lower exposure of the sample to high temperatures, which favors the quality of the extract. These techniques are some of the most recent extraction techniques developed to increase extraction yields, by applying non-thermal green technology. These techniques use pressures from 100 to 800 MPa or even more than 1000 MPa [33,41], allowing for better penetration of the solvent into the cell membrane, improving the bioaccessibility. Studies have demonstrated that the higher the hydrostatic pressure, the more components can be released and higher yield of extraction can be obtained [41,84]. The main disadvantage of these extraction methods is the cost of the energy needed to obtain the higher pressures. Currently, the limit of the high pressure at the industrial level is 600 MPa [41]. Figure 5 shows a schematic representation of high-pressure processing equipment. This equipment basically consists of a high-pressure vessel, a high-pressure pump, and a cooling system. This figure also shows a scheme on the different steps that should be followed to perform UHPE.

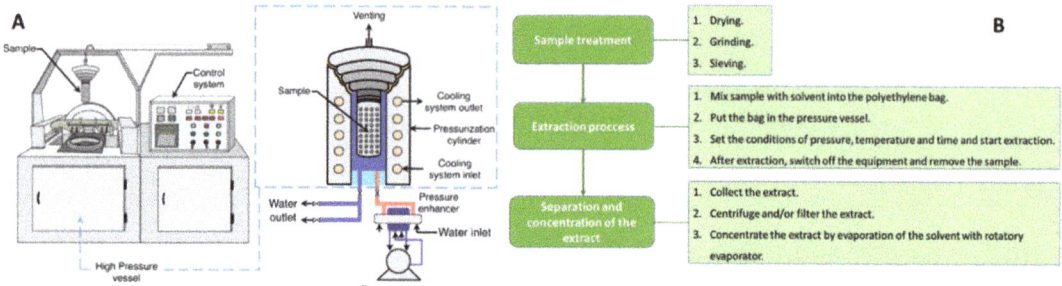

Figure 5. (A) Schematic representation of high-pressure processing equipment. (B) Schematic procedures of UHPE processing. ((A) taken from Barba et al. [85], with permission from Elsevier).

In order to guaranty the maximum efficiency of the extraction, the factors that should be optimized are the composition of the solvent, pressure, temperature, particle size, moisture content of natural material, time of extraction, and solvent-to-solid ratio. The most important of these factors are solvent, pressure, temperature, and L/S ratio. Martin and Asuero [33] recently published a table with selected applications of HHPE for the recovery and purification of anthocyanins in fruits, vegetables, and juices [33].

Table 5 shows some current examples of the optimal experimental conditions applied to extract anthocyanins from different natural matrices by HHPE. Fernandes et al. [86] carried out a study to optimize the pressure (0–500 MPa), time (5–15 min), and ethanol concentration (0–100%) to extract flavonoids, tannins, and anthocyanins from dried pansies by HPLE. They concluded that the optimum conditions were to apply a pressure of 384 MPa for 15 min with 35% (v/v) ethanol as solvent. They determined that a mixture of water and ethanol (20–70% v/v) as solvent increased the values of monomeric anthocyanins compared with only the water of ethanol as solvent. They also observed that high pressure increases the values of monomeric anthocyanins. It could be explained by the fact that high pressure has the ability to reduce the pH value of the solvent during extraction, derived from the enhanced protonation of molecules present in the extract [86]. This might increase the extraction effectiveness of anthocyanins, stable at acid pH. In addition, they determined that anthocyanins are selectively extracted depending on pressure intensity. A pressure at 200 MPa favors the extraction of anthocyanin monoglucosides, whereas a pressure at 600 MPa increases the extraction of acylglucosides [86]. In relation to the extraction time, they concluded that it is not a relevant factor in the amount of anthocyanins extracted, since the high pressures used make the permeabilization of the cell almost instantaneous [86].

Table 5. Some current examples of the application of HHPE for the extraction of anthocyanins from different natural matrices. Optimal extraction conditions to achieve maximum yield.

Natural Matrices	T^a (°C)	P. (Mpa)	Time (min)	Solvent (Amount of Sample)	Solid/Liquid Ratio	• Steps after Separation. • Comments on the Extract.	Recovery	Ref.
Pansies (Viola x wittrockiana).	Room T^a	384	15	Ethanol 35% (0.8 g DW)	1:30	• The extract was filtered. • Anthocyanins are selectively extracted, depending on pressure intensity. The extraction of anthocyanin monoglucosides were optimized at pressures of 200 MPa and the acylglucosides were optimized at 600 MPa.	6.09 mg of C3G/g of flower DW	[86]
Blueberries (O'Neal variety).	20	500	15	Acetone/water/acetic acid 70:29.5:0.5 (2g)	-	• The extract was centrifuged, filtrated, and evaporated. • The extract obtained by HHPE has the highest yields of bioactive compounds and the strongest antioxidant capacity compared to the CE extract.	117.1 mg C3GE/100 g of blueberry extract.	[87]
Haskap (Lonicera caerulea) berry.	18–22	200	10	Ethanol 60% (acidified HCl 0.1%) (1 g)	1:20	• The extract was centrifuged. • High-pressure affected the monomer composition and anthocyanin content in the extract. C3S5G was not detected at 200 MPa/5 and 10 min. P3DG was not detected at 200 MPa/10 min and 500 MPa/15 min. C3HE was not detected at 400 MPa/20 min.	336 mg C3G/100 g of sample	[88]

Note: C3G, cyanidin-3-glucoside; C3GE, cyanidin-3-glucoside equivalent; C3HE, cyanidin-3-hexoside-ethyl-catechin; C3S5G, cyanidin-3-sophoroside-5-glucoside; DW, dried weight; P., pressure; Ref., reference; T^a, temperature.

Briones-Labarca et al. [87] demonstrated that the application of HHPE to extract the bioactive compound from discarded blueberries increased the content of anthocyanins, polyphenols, and flavonoids, as well as the antioxidant capacity by DPPH and FRAP compared with conventional extraction. However, although the bio-accessibility of the polyphenols and flavonoids also increased with HHPE vs CE, the anthocyanins showed a decrease in the bio-accessibility after HHPE. In addition, they proved that the anthocyanins extracted increased from 15.1 to 39.4% with the time of the extraction from 5 to 15 min when compared to CE [87].

Compared to conventional techniques, HHPE uses less solvent volume, shorter times, and increases extractability, antioxidant capacity, and bio-accessibility of the bioactive compounds present in the natural matrices. In general, bioactive compounds remain unaffected by the pressure, while the structure of large molecules can be altered by high pressure [33]. Under UHP conditions, the differential pressure between the inside and outside of the cell is very large. This is why the solvent permeates very quickly through the ruptured membranes in the cells, increasing the mass transfer rate of solute. This could result in a very short extracting time of UHPE. The holding time should also be carefully considered because too long of a pressure holding time can damage the biological activity of extracts [41]. Pressures above 400 Mpa present the advantage that the activity of some oxidative enzymes, such as polyphenol oxidase (PPO) and peroxidase (POD), decreases, and the activity of the superoxide dismutase remains unaltered [88]. Temperature is the other important factor to control during HPLE because it directly affects the efficiency and selectivity during extraction. In general, HPLE does not use heat as an energy source. HPLE is usually carried out under refrigeration or at room temperature (Figure 5, Table 5) helping to protect thermosensitive biocompounds, such as anthocyanins. However, high temperatures (50–100 °C) improve the efficiency of the extraction because the diffusion rate is incremented, allowing for faster extractions. Thus, the choice of extraction temperature depends on the stability of the compounds and the extraction yields required. In the extraction of thermolabile compounds, high temperatures may cause the degradation of extracts [34]. However, it has been shown that, depending on the pressure and time conditions applied, some microorganisms could not be inactivated. In order to avoid this problem, some authors suggest combining HHPE with temperature, which increases the efficiency of the extraction and decreases the enzymatic action and the antimicrobial effect, preserving the antioxidant activity. Thus, the combination of HHPE and temperature for a short time of processing produce high nutritional and sensory qualities in the food [33,89–92]. Additionally, the high-pressure effect on anthocyanin content cannot be generalized, because of the composition of the matrix, the activity of the oxidative enzymes, the pressure and holding time could compromise the efficiency of the extraction. For that reason, it is recommended to not only optimize the extraction conditions to increase the amount of anthocyanins extracted, but to also carry out a bioaccessibility evaluation from bioactive compounds extracted by HHPE. The high pressure applied during HHPE could affect the conformation and the structure of macromolecules and also provoke the formation of metabolites that may exert biological action on the extract. Therefore, a bioaccessibility study of the extract is recommended.

The application of HPLE on an industrial scale is limited by the cost of the equipment. On the other hand, the technology is currently only available for batch processes, which also implies higher costs.

2.2.4. Pulsed Electric Field Extraction (PEFE)

PEFE is another novel, environmentally friendly cell membrane permeabilization technique that breaks down the cell wall of plants using electrical pulses. Different studies have showed that the increment of the temperature during the application of high-voltage electrical pulse is not higher than 10 °C [41,93]. Thus, this technique can also be classified as a non-thermal extraction method, which makes it an ideal technique to extract thermolabile bioactive compounds, such as anthocyanins. The pulsed electric field technique is based on the application of short duration pulses of moderate to high electric field strengths ranging from 0.1 to 0.3 kV/cm in batch mode and 20–80 kV/cm in continuous mode extraction at room temperature [35]. The basic principles of PEFE processing are based on innovative concepts within electrical engineering, fluid mechanics, and biology. The application of PEFE to biological cells is based on the principle of electropermeabilization, due to an induced transmembrane potential. The electroporation or electropermeabilization process enhances the permeability of cell membranes by an external electrical force [13,14,94]. The biological sample is placed between the electrodes and a high-voltage electrical pulse is applied from a few to several hundred microseconds. The electric field generated is able to induce the formation of hydrophilic pores in the cell membrane, which opens protein channels (Figure 6). This process is known as "electroporation" [35,93]. The strong electric field applied on the cell sample originates accumulation of oppositely charged ions on both sides of the membrane. When transmembrane potential exceeds a critical value of about 1 V, the repulsion between charge-carrying molecules leads to membrane thickness reduction and permeabilization to small molecules [95]. The sample experiences a force per unit of charge called "the electric field" when high-voltage electrical pulses are applied through the electrodes. Based on this process, the membrane cell loses its structural functionality, and the subsequent extraction of the bioactive compounds can be easily performed [35]. Depending on electric field strength and treatment intensity, the permeabilization might be reversible or irreversible. To achieve good extraction results in soft plant tissues, electric field strengths should be applied between 0.1 and 10 kV/cm. However, electric field strengths up to 20 kV/cm should be applied in order to achieve good extraction results from seeds and stalks where lignification can occur [95]. Figure 6 shows a schematic diagram of a high-intensity pulsed electric field (PEF) continuous extraction system together with the electroporation mechanism for the extraction.

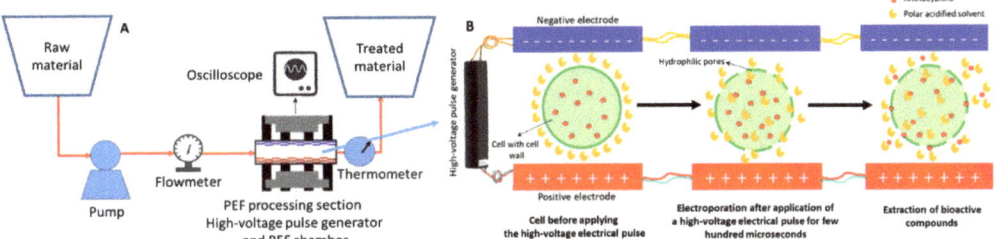

Figure 6. (**A**) Schematic diagram of a high-intensity pulsed electric field (PEF) continuous extraction. (**B**) Electroporation mechanism for bioactive compound extraction.

Different parameters should be optimized to improve the extraction efficiency of bioactive compounds applying PEFE. The most important parameters are the exposition time, the electric field strength, and the total specific energy input of the pulse. Some authors have demonstrated that the size of the chamber also affect the yield of the extraction. Thus, a chamber with a larger diameter allows the application of a higher number of pulses and increases the residence time. In addition, a large diameter chamber would allow to work in a continuous mode [96]. Another factor that affects the yield of the extraction is the initial temperature of the process. Thus, Gagneten et al. [97] evaluated the effect of the initial temperature at 10 or 22 °C. They observed that the efficiency of the extraction was higher at 22 °C. The temperature at 22 °C was associated with a higher electric current than the temperature at 10 °C. In addition, they observed that during the extraction the temperature increased 5 °C in both experiments and the correlation with the electric current was linear. Thus, cold temperatures can provoke a decrease in the electroporation efficiency [97]. This technique has been applied to improve the extraction efficiency of anthocyanins in different natural matrices. Table 6 shows some current examples of the optimal extraction condition used for different natural matrices and the yield obtained by PEFE. Lamanauskas et al. [98] concluded that the best conditions to optimize the extraction of anthocyanins from European blueberries were to expose the sample for 20 µs to monopolar square wave pulses with an electric field strength of 5 kV/cm and a total specific energy input of 10 kJ/kg. They found that an increase in the pulse electric field intensity (strength and total energy) resulted in a significant increase in the rate of cell disintegration. Under these conditions, they managed to increase the anthocyanin extraction yield by 8.3% compared to the extraction without applying PEFE [97]. Taiebirad et al. [99] optimized the extraction condition to extract the maximum yield of anthocyanins, to improve the antioxidant capacity by DPPH and to increase the extraction of the bioactive compounds for barberry fruit by PEFE. They applied an intermittent electric field at three levels of electric field intensity (0.5, 1.75, and 3 kV/cm) and three levels of number of pulses (15, 30, and 45). They observed that the total amount of anthocyanins extracted increased when the number of pulses and the intensity of the pulse increased, in agreement with the results obtained by Lamanauskas et al. [98]. However, they noted that the increase of the electric field strength and the number of pulses initially increased the flavonoid content and the DPPH capacity, but with the increase of these variables, these two properties decreased. A study carried out by Lončarić et al. [95] corroborated these results. They compared high voltage electrical discharges (HVED) PEFE and UAE, in terms of extraction yield of total and individual phenolic acids, anthocyanins, and flavanols of blueberry pomace extracts. The highest total content of anthocyanin (1757.32 µg/g of DW) was obtained in the methanol-based solvent by PEF-assisted extraction after 100 pulses and 20 kV/cm and at energy input of 41.03 kJ/kg. However, the other antioxidant compounds were extracted better in ethanol-based solvent. Other studies observed no significant effects in the anthocyanin yield when they increased the electric field intensity. Aadil et al. [100] investigated the effect of PEFE with different electric field strengths: 0, 5, 10, 15, 20, and 25 kV/cm in grapefruit juice. They observed no significant change in pH, Brix, titratable acidity, sugars, total anthocyanins, and color attributes with the increase in pulsed electric strength as compared to control treatment. However, a significant decrease in non-enzymatic browning (NEB), viscosity, and in the activity of microorganisms was observed. In addition, they realized an increase in cloud value, DPPH, total antioxidant activity, total phenolic compounds, and total carotenoids with the increase of pulsed electric strength compared with the control treatment. Finally, the authors suggest that applying PEFE at 25 kV cm^{-1} could improve the quality of grapefruit juice, although the anthocyanin content would not be significantly increased. These results are in agreement with those obtained by other authors [27,98,99,101].

Table 6. Current examples of the application of PEF for the extraction of anthocyanins from different natural matrices. Optimal extraction conditions to achieve maximum yield.

Natural Matrix	Pulses/Pulses Width/Frequency (Hz)	Electric Field Intensity (kV/cm)	T^a (°C)	Steps after Separation. Comments on the Extract.	Recovery	Ref.
By-products of Blueberry.	10/2 µs/-	20	Room T^a	• Centrifugation and the supernatant were analyzed. • No selectivity studies. The PEF extract has 0.9% more anthocyanins than the untreated extract.	223 mg of C3GE/L of sample.	[27]
Blueberry pomace.	100/2 µs/-	20	Room T^a	• Centrifugation and the supernatant were analyzed. • D3G, D3A, Pt3G, C3A, P3G, P3A, and M3G were identified. The PEF extract has 61% and 84% more anthocyanins than the HVED and UAE extracts, respectively.	175 mg of/100 g of sample DW.	[95]
Grape peels.	25/6 µs/10	25	25	• Centrifugation and the supernatant were analyzed. • The PEF extract was 34% and 420% richer in anthocyanins than the UAE at 50 °C and water extraction at 70 °C of the extracts, respectively.	78 mg of C3G/mL of sample.	[96]

Table 6. Cont.

Natural Matrix	Pulses/Pulses Width/Frequency (Hz)	Electric Field Intensity (kV/cm)	T^a (°C)	Steps after Separation. Comments on the Extract.	Recovery	Ref.
Blackcurrant.	315/100 ms/-	1.32	22	• Centrifugation and the supernatant were analyzed. • Under the best operating conditions, total monomeric anthocyanins increased by 6% in the extract.	1.38 mg of C3G/g of the extract.	[97]
Frozen/thawed European blueberry (*Vaccinium myrtillus* L.).	-/20 µs/20	1	20–25	• After the treatment, the sample was pressed, and the juice was obtained as extract. • The PEF extract has 8.3% more anthocyanins than the untreated extract.	1750 mg of C3G/L of juice.	[98]
Pinot Noir (PN) and Merlot (M) grapes.	-/300 s/344	8	Room T^a	• After the PEF treatment, no further treatment was conducted. • The PEF treatment applied to the PN sample increased the amount of anthocyanins in the extract by 46%. PEF in the M sample decreased the amount of anthocyanin in the extract by 2.7%, despite prior centrifugation.	81.5 mg of M3GE/L of pinot noir wine or must. 76.92 mg of M3GE/of merlot wine or must.	[101]
Grapefruit juice.	-/600 µs/1000	20	40	• After PEF treatment, the juice was sonicated. • The PEF treatment increased the amount of anthocyanins in the extract by 15%. The PEF+UAE treatment increased the amount of anthocyanins in the extract by 23%.	1.58 mg of C3GE/L of juice obtained after PEF treatment. 1.68 mg of C3GE/L of juice obtained after PEF treatment followed by UAE.	[102]
Strawberry juice (SJ).	13/2 µs/155	35	22–46	• After the treatment, the juice was cooled at 4 °C. • The PEF treatment increased the amount of anthocyanins in the extract by 16.9% compared to untreated extract. This increment was 7.8% and 1.7 compared to the extract obtained by UAE and HPE, respectively.	179.21 mg of Pl3G/L of juice.	[103]
Spinach juice.	4/80 µs/1000	9	30	• After the treatment, the sample was passed through a sterilized double layer muslin cloth. • The PEF treatment increased the amount of anthocyanins in the extract by 8.2% compared to the untreated extract. This increase was increased to 18% when the sample was subjected to a UAE before the PEF treatment.	38.12 mg of M3G/L of juice. 41.31 mg of M3G/L of juice when UAE is applied before PEF treatment.	[104]

Table 6. Cont.

Natural Matrix	Pulses/Pulses Width/Frequency (Hz)	Electric Field Intensity (kV/cm)	Tª (°C)	• Steps after Separation. • Comments on the Extract.	Recovery	Ref.
Valuable compounds from blackberries.	100/10 µs/-	13.3	20–35	• Supplementary extraction with hot water at 20 °C (W20) and 50 °C (W50) and with 30% ethanol at 20 °C (EE) was performed in a closed diffusion cell in the dark. • The amount of anthocyanins obtained by PEF combined with W50 increases by 53% and 185% compared to the extract obtained by HVED+W50 and UAE+W50, respectively.	100 mg of C3G/100 g of sample when W50 was applied. 90 mg of C3G/100 g of sample when EE was applied. 40 mg of C3G/100 g of sample when W20 was applied.	[105]
Date palm fruit.	30/30 µs/10	3	Room Tª	• No information. • The extract obtained by PEF had 177% more anthocyanins than the extract of untreated sample.	2.08 mg of C3GE/L of sample.	[106]
Red cherry samples.	-/20 µs/100	2.5	20	• No information. • The extract obtained after PEF treatment had 20% more C3G than the extract obtained without treatment.	0.23 mg of C3G/100 g of sample FW. 0.20 mg of C3R/100 g of sample FW. 0.02 mg of Pl3G/100 g of sample FW. 0.10 mg of P3G/100 g of sample FW.	[107]
Red grapes (Pinot Noir (PN) and Merlot (M).	-/150 s/178	7	20	• After PEF treatment, the must was seeded with selected yeasts (Lallemand); after 11 days of fermentation, the yeast was separated by open decanting. Finally, the wine was analyzed. • Relative difference to the untreated sample was 186% for the PN sample and 138% for the M sample.	Maximum absorbance = 0.67 u.a. for PN wine. Maximum absorbance = 1.45 u.a. for M wine.	[108]
Fresh blueberries.	24,000/1 µs/-	2	26	• After the PEF treatment, samples were blotted with paper towels to remove excess water, were weighed, and subject to anthocyanin extraction by CE with ethanol and further centrifugation. • The content of anthocyanins in the blueberries treated by PEF was 10% higher than the content in blueberries in sanitizing solutions.	110 mg of C3GE/g of sample FW.	[109]

Table 6. *Cont.*

Natural Matrix	Pulses/Pulses Width/Frequency (Hz)	Electric Field Intensity (kV/cm)	T^a (°C)	Steps after Separation. Comments on the Extract.	Recovery	Ref.
Merlot grapes (*Vitis vinifera*)	1033/20 µs/50	1.4	Room T^a	• Extraction by CE with acidified methanol followed by centrifugation. Four types of extract were obtained: juice without treatment (J); treated juice (PEFJ); juice after 48 h of in contact with untreated berries (J48); and juice after 48 h in contact with treated berries (PEFJ48). • M3G, D3G, Pt3G, C3G, MAG, DAG, PtAG, MCG, and PtCG were determined in the juices. In J, anthocyanins were not detected, and in PEFJ, only malvidin derivatives and C3G were detected. The J48 and PEFJ48 contained higher amounts of malvidin derivates than PEFJ, with the amount in PEFJ48 340% higher than in J48.	2.07 mg of M3G+MAG+MCG/100 mL of J48. 0.87 mg of M3G+MAG+MCG+CEG/100 mL of PEFJ. 9.88 mg of M3G+D3G+Pt3G+C3G+MA+DAG+PtAG+MCG+PtCG/100 mL of PEFJ48.	[110]

Note: C3A, cyanidin-3-arabinoside; C3GE, cyanidin-3-glucoside equivalent; C3R, cyanidin-3-rutinoside; D3A, delphinidin-3-arabinoside; DAG, delphinidin-acetyl-glucoside; D3G, delphinidin 3-glucoside; MAG, malvidin-acetyl-glucoside; MCG, malvidin-p-coumaroyl-glucoside; M3G, malvidin 3-glucoside; M3GE, malvidin-3-glucoside equivalents; P3A, peonidin 3-arabinoside; P3G, peonidin 3-glucoside; Pl3G, pelargonidin-3-glucoside; PtAG, petunidin-acetyl-glucoside; PtCG, petunidin-p-coumaroyl-glucoside; Pt3G, petunidin 3-glucoside; pulses, number of pulses applied; pulse widths, duration of each pulse; Ref., reference; T^a, temperature.

The combination of PEFE and UAE was explored in order to obtain higher anthocyanin extraction yields and to increase the quality and the antioxidant capacity of the extracts (Table 6) [102–104]. A study carried out by Barba et al. [105] on the extraction of anthocyanins from grape pomace showed that PEF treatment allowed a selective recovery of extracts with an amount of anthocyanins—22 and 55% higher than with UAE and HVED, respectively. Thus, they suggested that a combination of different extraction techniques, such as PEF + HVED or PEF + supplementary extraction + HVED, is a good alternative to use in the food industry, since it allows, in the first step, an extraction of sensitive compounds, such as anthocyanins, and in the second step, more resistant compounds, such as phenols and flavones [105]. In conclusion, it can be noted that the electric field strength plays an important role in the selectivity of the extraction. This conclusion reveals the complexity of the optimization of the different parameters, and the usefulness of statistical treatments to determine the optimal conditions depending on the extraction objective [103].

One of the main advantages of PEFE, EUS, and HPLE, is the ability to inactivate the microbiota present in food, ensuring food safety, while increasing food quality [111]. The application of pulse electric field (PEF) in the industry, for the purpose of pasteurization, has been extensively investigated, with PEF-assisted sterilization achieved under laboratory conditions [94,112,113]. However, some studies have shown that PEF processing requires more energy and is consequently more expensive than thermal processing [94,111]. This disadvantage could be reduced by improving the processing chamber to reduce inactivation areas, promoting a more homogeneous treatment, and by improving experimental planning [94]. In addition, the advantages of PEF could motivate the industry to implement this technique, despite this disadvantage. The main advantages of PEF are: it is a sustainable technique due to the non-thermal treatment, its beneficial effects on food quality (by increasing the content of bioactive compounds), and the ability to selectively inactivate microorganisms.

Yildiz et al. [102] compared different extraction techniques with microbiota inhibition capacity (HPLE, USE, and PEFE), with a conventional thermal pasteurization treatment. Their objective was to study the effect of these treatments in the quality and antioxidant properties of strawberry juice. All techniques significantly reduced the initial natural microbiota. However, they pointed out that HPLE and PEFE obtained juice with significantly higher amounts of total anthocyanin content 15–17% in comparison with untreated strawberry juice. Furthermore, they observed higher radical scavenging activity in the juice extracted by HPLE and PEFE than in the juice treated with thermal pasteurization. A statistical analysis indicated that HPLE and PEFE extracted juices with similar antioxidants content and antioxidant activity. Both techniques obtained better extracts than the extract obtained by sonication, thermal pasteurization, or untreated. The study concluded that HPLE and PEFE produce enhanced quality fruit juices [102]. Even better results were obtained when more than one extraction technique was combined. Thus, Faisal Manzoor et al. [104] investigated the combined effects of the ultrasound (US) and pulsed electric field (PEF) treatment on spinach juice. They demonstrated that both techniques provide spinach juice with an acceptable higher quality than the juice untreated. However, their results revealed that the combined treatment based on US-PEF increased the concentration of flavonoids, phenolic, flavonols, anthocyanins (Table 6), carotenoids, total chlorophyll, vitamin C, DPPH activity, and the total antioxidant capacity in the spinach juice than single treatments or the untreated sample. Furthermore, this combination of techniques that preserves antioxidants and inactivates spoilage enzymes was better than the individual treatments or the untreated sample. This study confirms that US-PEF can enhance the quality of spinach juice at an industrial scale [104]. The results indicate that the combined treatments of non-thermal technologies are useful for the food processing and preservation industries, but they are slightly costlier than using individual treatments. However, the combination of non-thermal techniques allows for high quality, healthy, and safe food to be produced, covering the demands of the consumers, although it is likely to be marketed at higher prices to cover production costs.

Additionally, to the combination of PEFE with UAE, or another extraction technique described above, other electrical techniques have recently been proposed to improve the extraction of bioactive compounds. These techniques are known as high-voltage electrostatic fields (HVEFs) and high-voltage electric discharges (HVEDs).

HVEF is a non-pulsed electrostatic field treatment. This technique can also modify the structure of animal and vegetable tissues favoring the extraction of valuable components from different materials (e.g., anthocyanins). As in the case of PEF, this technique is used for the extraction of bioactive compounds and to inactivate microorganisms and enzymes in liquid matrices, by means of a non-thermal treatment, while preserving the nutritional quality of the food [21].

HVED is based on both chemical reactions and physical processes. When a HVED is produced directly in water, it injects energy directly into an aqueous solution through a plasma channel produced by a high-current/high-voltage electrical discharge (>40 kV; >10 kA) between two submerged electrodes. The action of HVED is based on two phases: the pre-breakdown phase and the breakdown phase. During the first phase, relative weak shock waves are produced and, as a result, it the formation of a few little bubbles can be observed. Furthermore, strong UV radiations and active radicals are generated. These phenomena cause cell structure damage and accelerate extraction of intracellular compounds. Thus, it is important to control the electric field intensity to avoid the damage of the bioactive compounds and only motivate the wall cell destruction. The enhanced electrohydraulic phase occurs during the transition of pre-breakdown to breakdown phase, having several effects: strong shock waves, strong UV radiations (200–400 nm), production of highly concentrated free radicals, bubbles with plasma inside and strong liquid turbulence. They provoke mechanical destruction of cell tissues and oxidation, which could affect the antioxidant activity of bioactive compounds [95]. In general, this extraction provides a more powerful mechanical disintegration of the cell walls and, consequently, it allows achieving a more effective extraction. Thus, this technique has been applied in removing organic impurities of water [20], improving the shelf life and the antioxidant capacity of fruits [114], exploring its utility in microbial and enzymatic inactivation, winemaking, and thawing, drying, and freezing in different foods, among others [115]. This technique has also been used to increase the extraction yield of bioactive compounds from different raw materials. Thus, Lončarić et al. [26] carried out a study to evaluate the capacity of HVED to extract anthocyanins, including malvidin, delphinidin, peonidin-3-O-glucoside, and cyanidin-3-O-glucoside from indigenous fungus-resistant grape by-products compared to UAS. They concluded that the extraction carried out by HVED showed high yields of all analyzed compounds compared to UAS [26].

Studies have demonstrated the efficacy of HVED treatment in the aqueous extraction of bioactive compounds, in particular polyphenols from different matrices. [116,117]. Thus, Rajha et al. [117] studied the mechanical, electrical, and chemical effects of HVED and their roles in polyphenol extraction from vine shoots. This study demonstrates the efficacy of electrical discharges for the intensification of the extraction of bioactive molecules (polyphenols) from a specific byproduct (vine shoots). It also shows HVED as an energy-saving extraction process, relevant for industrial application, since no organic solvents were used, and the process duration and temperature were reduced. However, HVED also presents some disadvantages. The intense cell wall rupturing power of HVED extraction could cause the formation of small particles, which may hinder the subsequent separation stage [116].

2.2.5. Enzyme-Assisted Aqueous Extraction (EAE)

Another strategy to access the interior of cells and extract bioactive compounds is based on a biological technology, which involves the use of enzymes. Enzyme-assisted aqueous extraction (EAE) has been used for years and has improved over time. This technique has the following advantages: (i) the enzymes are highly selective and efficient; (ii) they work under moderate conditions of pressure and temperature; (iii) some have the properties to destroy or degrade the cell wall; and (iv) it can be qualified as a green technique [41,118–121]. Today, advances in enzyme catalysis, the availability and diversity of enzymes, as well as increasing environmental constraints, make this technology a potential tool for industrial application.

Enzymes can be used as their own extraction methods by EAE or as additional tools to other extraction processes to increase yields of extraction by weakening the vegetal cell walls. The principle of the enzymatic extraction is based on the action of the enzyme on the wall cells. In order to reach the bioactive compounds stocked in vegetal cells, several barriers have to be crossed: extracellular cell walls, cell walls, and oleosomes. Each of these barriers is composed of its own constituents. They are naturally synthesized and hydrolyzed by specific enzymes. The most common enzymes used to hydrolyze part of the constituents of vegetal cell walls could be group in four different families: cellulases, hemicelluloses, pectinases, and proteases. The three first families can be used alone or in combination, each one having a different effect in the cell wall. Thus, a selective extraction of different bioactive compounds can be achieved according to the enzyme applied [118]. Cellulases, hemicelluloses, pectinases, and proteases have been employed to enhance and accelerate pigment extraction of various plant materials, being able to extract pigments quicker and with higher efficiencies than conventional ethanol extraction [119]. However, proteases have to be used separately from the other families of enzymes because they are able to hydrolyze the enzymatic proteins, decreasing or eliminating the specific activities of the enzyme mixture. The enzymes applied in EAE can be obtained from different natural sources, such as bacteria, fungi, vegetable, and fruit extracts, or animal organs [41]. Depending on the source of the enzyme, they have different properties and act under different conditions. Thus, in general, an animal enzyme has a denaturing temperature of 40–45 °C, whereas this temperature is more than 60–65 °C for enzymes produced by microorganisms. Nevertheless, some enzymes are thermo-resistant and could tolerate more than 100 °C during several minutes. Currently, different commercial enzymes can be acquired in the market.

EAE has been applied to extract pigment, phenols, and anthocyanins in different natural matrices, improving the yield of the extraction in comparison with the conventional techniques [118–121]. Table 7 shows some current examples of the optimal extraction conditions used for different natural matrices and the yields obtained by EAE. The main extraction conditions that should be optimized to increase the yields of the extraction by EAE are: the composition of the enzymatic mixture, pH, temperature, L/S ratio, enzymes/solid ratio, and hydrolysis time.

Table 7. Current examples of the application of EAE for the extraction of anthocyanins from different natural matrices. Optimal extraction conditions to achieve maximum yield.

Natural Matrix	Enzymatic Mixture	pH/T^a (°C)	L:S Ratio/Enzymes: Mixture Ratio/ Hydrolysis Time (min).	Steps after Separation. Comments on the Extract.	Recovery	Ref.
Saffron tepals	Pectinex (containing cellulase, hemicellulase, and pectinase).	3.5/45	10:1/5:100/120	• Centrifugation and supernatant was analyzed. • Pectinex extracted 50% more anthocyanins than Cellubrix. The EAE with water extracted 33% more anthocyanins than CE with acidified ethanol.	675 mg of C3G/100 g of saffron tepals extracted.	[119]
Saffron (*Crocus sativus* L.) tepals.	Cellulolytic preparation Celluclast BG and hemicellulolytic preparation Xylanase AN (1:1).	4/50	10:1/10:100/145–185	• Cooling and centrifugation and filtration of the supernatant. • No selectivity study	2.0 g of C3GE/kg of saffron tepals DW.	[121]
Skin of the Băbească neagră grapes.	Zymorouge pectolytic enzyme EG from *Aspergillus niger*.	5.0/40	28:1/2:100/60	• Centrifugation. • UAE with 96% ethanol acidified at 50 °C extracted 68% more anthocyanins than EAE.	2.54 mg of C3G/g of sample DW.	[122]
Mulberry wine residue.	Pectinase.	5.9/45	20:1/-/58	• Filtration. • No selectivity study.	6.04 mg of C3G/g of sample.	[123]
Leaf of monguba.	α-Amylase and protease.	6.0/50	10:1/-/160	• Centrifugation and filtration. • No selectivity study.	30.59 mg of TA/100 g of sample DW.	[124]

Table 7. Cont.

Natural Matrix	Enzymatic Mixture	pH/T[a] (°C)	L:S Ratio/Enzymes: Mixture Ratio/ Hydrolysis Time (min).	Steps after Separation. Comments on the Extract.	Recovery	Ref.
Seeds (*Adenanthera pavonina* L.)	Protease and cellulose.	7.0/50	45:5/-/160	• Cooling and freeze-drying. • EAE improved the recovery of TFC than increased 47%.	14.71 μg of total phenols/g of sample DW.	[125]
Blueberry.	Pectinase.	4.5/45	8:1/-/60	• Centrifugation. • No selectivity study.	2.346 mg of TA/mL of extract.	[126]
Raspberry (*Rubus idaeus* L.) pomace.	Ultrazym AFP-L.	-/45	100:15/1:100/60	• Cooling and centrifugation; the pomace was extracted by CE and fractionated by SFE. • 17% more anthocyanins were quantified in pomace without enzymatic treatment. The Ultrazym EAE increased the C3G content of the extract by 13%.	0.32 mg of C3S+C3G+C3R/g of sample FW.	[127]
Roselle samples.	Cellulase solution with exo- and endo-β-1,4-D-glucanases.	4.8/40	40:1/16:100/60	• EAE+UAE followed by centrifugation. • EAE+UAE increased the anthocyanin content by 1.2% compared to the untreated sample.	676.03 mg of C3G/100 g of sample DW.	[128]
Raspberry wine residues.	Pectinase.	3/40	30:1/0.16:100/30	• UAE+EAE followed by centrifugation, evaporation, and freeze-drying. • UAE+EAE obtained 49% and 77% more anthocyanins than EC with acidified ethanol and hot water, respectively.	0.853 mg of C3G/g of sample.	[129]
Raspberry wine residues.	Pectic enzyme	-/52	100:1/0.2:100/66	• Centrifugation, evaporation, and freeze drying. • UAE+EAE obtained 32% and 73% more anthocyanins than EC with acidified ethanol and hot water, respectively.	0.75 mg of C3G/g of sample.	[129]

Note: C3G, cyanidin-3-glucoside; C3GE, cyaniding 3-glucoside equivalents; C3S, cyanidin-3-O-sophoroside; C3R, cyanidin-3-O-rutinoside; DW, dried weight; FW, fresh weight; Ref., Reference; SFE, solid phase extraction; T[a], temperature; TA, total anthocyanins; TFC, total flavonoid content.

The composition of the enzymatic mixture is a key factor. With a high amount of different enzymes, the yield of extraction will be high and rapidly achieved. However, the optimal enzymatic mixture must be adapted to each kind of matrix and each bioactive compound [118,122]. In some cases, when more than one compound is to be extracted, a multistep enzymatic extraction is applied. Thus, it is necessary to optimize each extraction condition, such as pH or temperature, for the hydrolysis of the walls, according to each enzyme.

The pH is another important factor to be optimized for several reasons: (i) an acid pH is important to keep the structure of the anthocyanins stable; (ii) an acid pH provides an increase in wall plasticity and intermolecular sliding; (iii) the enzyme activity (ionization) depends on the acid pH, thus, an inappropriate pH might strongly decrease the enzymatic activity. In addition, pH-related denaturation may be irreversible to some enzymes. In consequence, when an enzymatic mixture is used, in order to promote the optimum enzymatic hydrolysis for each enzyme, it is necessary to define an average pH that covers the optimum pH for each enzyme in the mixture.

As with pH, temperature is an important, because it has an effect in the viscosity of the matrix: a high temperature reduces the viscosity and facilitates the extraction. However, the stability of the extract could also be affected by temperature, decreasing if the temperature is too high. In addition, the enzymatic activity depends on the temperature. Catalytic activity of the enzyme is higher with an increasing temperature. However, since they are also proteins, thermal denaturing is possible with a heating step. Thus, the temperature reaction is an important factor that should be kept in mind for enzymatic extraction.

As in other extraction techniques, the optimal liquid/solid ratio is a parameter often discussed in the literature [119,120]. It is different, depending on the studied matrix; often, it is even different for the same matrix. Sometimes the L/S ratio is defined according to the equipment and the system for stirring available.

Likely, to the L/S ratio, the enzyme/solid ratio and hydrolysis time are discussed. In theory, hydrolysis time and enzyme concentration are well correlated. For example, the hydrolysis time can be halved by doubling the enzyme concentration, without saturating the mixture, giving similar results in the extract. In practice, the effectiveness of the enzymes decreases more-or-less quickly when the hydrolysis time is prolonged. Different inhibition factors may occur: from the matrix, from products of reactions, from the process factors, etc. Thus, some studies have proven that an increase in the enzyme dose raises the recovery rate of total anthocyanins until the middle of the experiment, and afterwards starts to decline. The negative effects of the higher enzyme dose could indicate that the preparations might possess secondary enzyme activities that catalyze the degradation of anthocyanins [119,121].

More recently, some improvements of this technique have been introduced in order to enhance anthocyanin extraction yield. Thus, Xue et al. [129] suggested combining the power of enzymatic treatment with the power of ultrasonic extraction by a new extraction technique known as ultrasound-assisted enzymatic extraction UAEE. They applied this combination to extract anthocyanins from raspberry wine residues. In this study, under the optimal conditions (Table 7), the anthocyanin yield and the extraction efficiency using UAEE were higher, 0.853 mg/g and 84.75%, respectively than those obtained by the other three conventional extraction methods (hot water extraction, acidified ethanol extraction, and EAE). In addition, they determined that the extract obtained by UAEE presented high activities of DPPH 417.15 Trolox equivalents/g extract and ABTS 520.07 Trolox equivalents/g extract, reducing power (412.79 Trolox equivalents/g extract). Thus, UAEE allows a higher yield with less energy consumption and better selectivity than the conventional extraction methods. They concluded that the UAEE of the desired anthocyanin components from natural plant resources is a rapid, efficient, and environmentally friendly extraction method [129]. Similar conclusions were reported by Oancea and Perju [128], who obtained high amounts of total anthocyanins (676.03 mg/100 g DW) and strong antioxidant activity in the crude acidified hydroethanolic extracts obtained by UAEE with cellulases.

EAE and its different combinations, such as UAEE, improve the yields of anthocyanin extraction and have a number of advantages mentioned above. However, it has a number of drawbacks that limits, in part, its use on an industrial scale. The use of enzymes has some commercial and technical limitations that should be considered. Enzyme cost and process energy are high, and enzyme activity is highly dependent on pH, temperature, and nutrient availability, as different combinations of enzyme preparations need to be tested and optimized. These factors influence the application of the EAE; thus, further research will be required to improve its application at an industrial scale [41,120].

3. Advantages and Disadvantages of the Promising Green Extraction Techniques

According to the results presented above, the most promising green techniques to improve the anthocyanin yield and the extraction efficiency from natural matrices have been presented. These techniques are considered as 'green non-conventional techniques'. All of them have the following in common: they reduce processing time, temperature, energy consumption, and the use of organic solvents, in comparison with conventional techniques. In addition, they have previously been shown to be effective in the extraction of anthocyanins; thus, which of these techniques is the most suitable for extracting anthocyanins at an industrial scale? It is not possible to give a clear answer to this question, as each technique has a number of advantages and disadvantages. In general, the selection of extraction methods depends mainly on many factors, such as the physicochemical properties of the compound and solvent, the economic value of the compound, environmental concerns, the cost of the process, the required instrumentation, among others. Table 8 shows a comparison between these techniques based on their strengths, weaknesses, and suitability to extract anthocyanins.

Table 8. Strengths, weaknesses, and suitability to extract anthocyanins of non-conventional extraction techniques.

Technique	Strengths	Weaknesses	Suitability
UAE	Versatile, flexible, low cost, and very easy to use; fast energy transfers; low solvent usage; extraction time (5–60 min); can be combined with heating to improve the yield or with enzymatic treatment to improve the anthocyanin yield and the bioactivity of the extract; available on a large scale.	Lack of homogeneity in the process improved by probe system (PUE); the large-scale application could be limited by the higher cost and nonlinearity of process; after the extraction, a filtration and clean-up step is required; the process can lead to operator fatigue.	😊😊😊
MAE	Quick and homogeneous heating; low solvent usage; extraction time (1–40 min); currently, vacuum microwave extraction has been developed to provide a MAE method with a lower reactor temperature; possible application on a large scale.	The solvent must absorb microwaves; the heating could damage the structure and the activity of some compounds; after the extraction, a filtration and clean-up step is required.	😊😊
SFE	CO_2 as a solvent; easy to remove after extraction; reduced the thermal degradation. Extraction time (up to 1 h); it does not require an alternative energy source; it is available on a large scale.	Needs a co-solvent to extract polar compounds. The amount and type of co-solvent need to be optimize together with other parameters. SWE present the limitation of need high temperature to reach the subcritical condition, ethanol could be used instead of water.	😊😊
PLE	Low solvent consumption; protection for oxygen and light sensitive compounds; it needs temperature; possible application on a large scale.	Expensive equipment required; after the extraction, a clean-up step is required; extraction time (1–2 h).	😊😊
HHPE	Short extraction time (~5 min); performed at room temperature; higher repeatability; smaller amount of solvents; possible application at large scale.	High investment cost and cost maintenance and service; high pressure could affect the structure or activity of some compounds. The parameter should be optimized to avoid it.	😊😊😊

Table 8. Cont.

Technique	Strengths	Weaknesses	Suitability
PEFE	Short extraction time (less than 1 s); performed at room temperature; low energy and monetary costs; possible application on a large scale.	Some compounds could be affected by high electric fields; it is desirable to reduce the electrical conductivity of the matrix before the extraction. For industrial application there are some problems related to: non-uniform distribution of the electric pulses, the suitable solvents are very limited and cooling system is necessary to control the temperature when extracting thermolabile compounds if high electrical pulses are applied.	☺☺☺
HVED	Low temperature; short extraction time and energy input; possible application on a large scale.	High cost maintenance and service; high voltage electrical discharges may generate chemical products and free reactive radicals, which can react with antioxidant compounds decreasing their bioactive activity.	☺☺
EAE	Moderate extraction conditions; eco-friendly; selectivity due to the specificity of enzymes; can be combined with ultrasonic extraction to improve the yield and the bioactivity of the extract.	Expensive cost of enzymes; activity of enzymes varying with the pH, temperature and nutrients of the matrix; after the extraction, a filtration and clean-up step is required. Difficulties to be applied on a large scale; extraction time (1–12 h); low availability of commercial enzyme types; sometimes they have low selectivity and variability.	☺☺

These techniques have been used not only in isolation to extract anthocyanins, but their combined use has also been explored to improve the extraction of anthocyanins. In general, this alternative produces a more effective extraction than the use of a single one, as is the case of UAE combined with microwave UMAE [49], enzyme-assisted supercritical fluid extraction (EASCFE) [76,77], microwave-assisted sub-critical water extraction (MA-SWE) [19], ultrasound-assisted with pressurized liquid extraction (US-PLE) [83], ultrasound and pulsed electric field US-PEF [104], or ultrasound-assisted enzymatic extraction (UAEE) [128,129]. In some cases, the combination of these techniques could provoke deterioration of the extract. In order to avoid the deterioration of the extract during the process, a specific optimization of the extraction parameters should be done according to the matrix and the extracted compounds. In addition, innovative approaches are needed for overcome these shortcomings. Furthermore, the combination of various techniques implies even greater complexity for large-scale implementation. The scaling of the extraction methods to industrial scale presents some disadvantages, such as insufficient recovery, degradation due to excessive heating and extraction time, which ultimately results in high-energy consumption [29]. To choose a suitable extraction process for anthocyanins it is necessary to consider the extraction efficiency, economic feasibility, and environment aspects. Extraction efficiency of non-conventional extraction techniques is clearly advantageous compared with the conventional extraction methods in regard to time, energy, and extraction yield. Nevertheless, the extraction efficiencies among non-conventional extraction techniques are different (Table 8). For these reasons, nowadays, conventional techniques are still used at the industrial scale [39].

4. Conclusions

Extraction techniques trends have evolved, as more advanced techniques making use of green extraction concepts have emerged. The advantages of such techniques include producing good quality yield, lesser solvents and energy consumption, and shorter extraction times. Scientific literature shows clear evidence that extraction procedures of target compounds from natural matrices must be assessed individually. In order to take full advantage of the technological advances in the extraction techniques, the extraction conditions need to be optimized. Mathematical solutions could increase the efficiency and profitability of the process and help to change conventional extraction approaches.

The selection of an extraction technique is not an easy task as documented above. It is necessary to take multiple factors into consideration: (i) the cost of the instrumentation, maintenance, material, and process. It is necessary to find a compromise between the amount of energy or solvent consumed and the time of the extraction; (ii) yield is another economic factor, but not the only one. The reaction must be sufficiently effective to give good yields and quality of molecules of interest in this case anthocyanins; (iii) microbiology: a mixture of water and organic matter is a very favorable environment for the development of microorganism. The time of the extraction and the conditions applied must consider this factor; (iv) product quality: some products may be sensitive to oxidation or temperature. Extraction for too long, or with too much energy, may affect the quality of the extract or the concentration of antioxidant to be extracted.

In order to consider all these factors, it is necessary to design an extraction process to optimize the extraction conditions. The steps to be taken in such a design are as follows: (i) adjusting extraction conditions (solvent choice, temperature, particle size of sample, pressure, extraction time, among others) according to the characteristics of anthocyanins and natural matrices. The selection is important for achieving high extraction efficiency with high quality extracts; (ii) it is also important to study the influence of the extraction conditions in the bioactivity of the extract. It should be investigated independently; (iii) once that extraction conditions with significant impact in the extraction are selected, the optimization of them with the minimum experimental trials should be done. Commonly, screening experimental designs and optimization experimental designs are used. Screening designs can be used to analyze the most important conditions and their interactions from all potential conditions. The most common design used in the literature for the screening purpose are: two-level full factorial, two-level factorial and Plackett–Burman design. For optimization designs—central composite design (CCD), Box–Behnken design (BBD), Taguchi design, and Doehlert design are used.

Although design processes are applied, advanced extraction methods have limitations in scaling up for pilot or industrial purposes. Value changes and experimental conditions in the laboratory are mostly not optimized for industrial use. Therefore, the development of economically viable industrial extraction methods and tools for mass extraction are needed. In addition, they are limited for industrial applications due to the high equipment costs and complicated installation procedures. Thus, establishing the balance between "energetic" and cost will be a key focus of research in the future. To take advantage of the different extraction methods and to limit their drawbacks, the combinative applications of multiple extraction technologies and the automated potential of these non-conventional extraction technologies would be development tendencies in the near future.

Author Contributions: Investigation, N.T. and A.G.A.; project administration, N.T. and A.G.A.; supervision, A.G.A.; writing—original draft, N.T.; writing—review and editing, N.T. All authors have read and agreed to the published version of the manuscript.

Funding: This research received no external funding.

Institutional Review Board Statement: Not applicable.

Informed Consent Statement: Not applicable.

Data Availability Statement: The data presented in this study are available in review.

Conflicts of Interest: The authors declare no conflict of interest.

References

1. Jezek, M.; Zorb, C.; Merkt, N.; Geilfus, C.-M. Anthocyanin Management in Fruits by Fertilization. *J. Agric. Food Chem.* **2018**, *66*, 753–764. [CrossRef]

2. Tena, N.; Martín, J.; Asuero, A.G. State of the Art of Anthocyanins: Antioxidant Activity, Sources, Bioavailability, and Therapeutic Effect in Human Health. *Antioxidants* **2020**, *9*, 451. [CrossRef]
3. Dangles, O.; Fenger, J.A. The Chemical Reactivity of Anthocyanins and Its Consequences in Food Science and Nutrition. *Molecules* **2018**, *23*, 1970. [CrossRef]
4. Martín, J.; Kuskoski, E.M.; Navas, M.J.; Asuero, A.G. Antioxidant Capacity of Anthocyanin Pigments. In *Flavonoids—From Biosynthesis to Human Health*; Justino, J., Ed.; Science, Technology and Medicine Open Access Publisher: Rijeka, Croatia, 2017; Chapter 11; pp. 205–255.
5. Martín Bueno, J.; Sáez-Plaza, P.; Ramos-Escudero, F.; Jímenez, A.M.; Fett, R.; Asuero, A.G. Analysis and antioxidant capacity of anthocyanin pigments. Part II: Chemical structure, color, and intake of anthocyanins. *Crit. Rev. Anal. Chem.* **2012**, *42*, 126–151. [CrossRef]
6. Navas, M.J.; Jiménez-Moreno, A.M.; Martín Bueno, J.; Sáez-Plaza, P.; Asuero, A.G. Analysis and antioxidant capacity of anthocyanin pigments. Part IV: Extraction of anthocyanins. *Crit. Rev. Anal. Chem.* **2012**, *42*, 313–342. [CrossRef]
7. Yang, W.; Guo, Y.; Liu, M.; Chen, X.; Xiao, X.; Wang, S.; Gong, P.; Ma, Y.; Chen, F. Structure and function of blueberry anthocyanins: A review of recent advances. *J. Funct. Foods* **2022**, *88*, 104864. [CrossRef]
8. Enaru, B.; Dretcanu, G.; Pop, T.D.; Stanila, A.; Diaconeasa, Z. Anthocyanins: Factors Affecting Their Stability and Degradation. *Antioxidants* **2021**, *10*, 1967. [CrossRef]
9. Kay, C.D.; Pereira-Caro, G.; Ludwig, I.A.; Clifford, M.N.; Crozier, A. Anthocyanins and flavanones are more bioavailable than previously perceived: A review of recent evidence. *Annu. Rev. Food Sci. Technol.* **2017**, *8*, 155–180. [CrossRef]
10. Morata, A.; Escott, C.; Loira, I.; López, C.; Palomero, F.; González, C. Emerging Non-Thermal Technologies for the Extraction of Grape Anthocyanins. *Antioxidants* **2021**, *10*, 1863. [CrossRef]
11. Albuquerque, B.R.; Pinela, J.; Barros, L.; Oliveira, M.B.P.P.; Ferreira, I.C.F.R. Anthocyanin-rich extract of jabuticaba epicarp as a natural colorant: Optimization of heat- and ultrasound-assisted extractions and application in a bakery product. *Food Chem.* **2020**, *316*, 126364. [CrossRef]
12. Fernandes, F.; Pereira, E.; Prieto, M.A.; Calhelha, R.C.; Ciric, A.; Sokovic, M.; Simal-Gandara, J.; Barros, L.; Ferreira, I.C.F.R. Optimization of the Extraction Process to Obtain a Colorant Ingredient from Leaves of Ocimum basilicum var. Purpurascens. *Molecules* **2019**, *24*, 686. [CrossRef]
13. Ngamwonglumlert, L.; Devahastin, S.; Chiewchan, N. Natural colorants: Pigment stability and extraction yield enhancement via utilization of appropriate pretreatment and extraction methods. *Crit. Rev. Food Sci. Nutr.* **2017**, *57*, 3243–3259. [CrossRef]
14. Fincan, M. Potential Application of Pulsed Electric Fields for Improving Extraction of Plant Pigments. In *Handbook of Electroporation*; Miklavčič, D., Ed.; Springer: Cham, Germany, 2017; Chapter 9; pp. 2171–2192.
15. EFSA. Scientific Opinion on the re-evaluation of anthocyanins (E 163) as a food additive. *EFSA J.* **2013**, *11*, 3145.
16. Hair, R.; Sakaki, J.R.; Chun, O.K. Anthocyanins, Microbiome and Health Benefits in Aging. *Molecules* **2021**, *26*, 537. [CrossRef]
17. Kumara, K.; Srivastava, S.; Sharanagatb, V.S. Ultrasound assisted extraction (UAE) of bioactive compounds from fruit and vegetable processing by-products: A review. *Ultrason. Sonochemistry* **2021**, *70*, 105325. [CrossRef]
18. Gao, Y.; Ji, Y.; Wang, F.; Li, W.; Zhang, X.; Niu, Z.; Wang, Z. Optimization the extraction of anthocyanins from blueberry residue by dual-aqueous phase method and cell damage protection study. *Food Sci. Biotechnol.* **2021**, *30*, 1709–1719. [CrossRef]
19. Moirangthem, K.; Ramakrishna, P.; Amer, M.H.; Tucker, G.A. Bioactivity and anthocyanin content of microwave-assisted subcritical water extracts of Manipur black rice (Chakhao) bran and straw. *Future Foods* **2021**, *3*, 100030. [CrossRef]
20. Barba, F.J.; Grimi, N.; Vorobiev, E. New Approaches for the Use of Non-conventional Cell Disruption Technologies to Extract Potential Food Additives and Nutraceuticals from Microalgae. *Food Eng. Rev.* **2015**, *7*, 45–62. [CrossRef]
21. Carrera, C.; Aliaño-González, M.J.; Valaityte, M.; Ferreiro-González, M.; Barbero, G.F.; Palma, M. A Novel Ultrasound-Assisted Extraction Method for the Analysis of Anthocyanins in Potatoes (*Solanum tuberosum* L.). *Antioxidants* **2021**, *10*, 1375. [CrossRef]
22. Cai, Z.; Qu, Z.; Lan, Y.; Zhao, S.; Ma, X.; Wan, Q.; Jing, P.; Li, P. Conventional, ultrasound-assisted, and accelerated-solvent extractions of anthocyanins from purple sweet potatoes. *Food Chem.* **2016**, *197*, 266–272. [CrossRef]
23. Jiang, H.; Yang, T.; Li, J.; Ye, H.; Liu, R.; Li, H.; Deng, Z. Response surface methodology for optimization of microwave-assisted extraction and antioxidant activity of anthocyanins from red rice. *J. Chin. Inst. Food Sci. Technol.* **2015**, *15*, 74–81.
24. Selvamuthukumaran, M.; Shi, J. Recent advances in extraction of antioxidants from plant by-products processing industries. *Food Qual. Saf.* **2017**, *1*, 61–81. [CrossRef]
25. Perino-Issartier, S.; Zill-e-Huma; Abert-Vian, M.; Chemat, F. Solvent free microwave-assisted extraction of antioxidants from sea buckthorn (*Hippophae rhamnoides*) food by-products. *Food Bioprocess Technol.* **2011**, *4*, 1020–1028. [CrossRef]
26. Lončarić, A.; Jozinović, A.; Kovač, T.; Kojić, N.; Babić, J.; Šubarić, D. High Voltage Electrical Discharges and Ultrasound-Assisted Extraction of Phenolics from Indigenous Fungus-Resistant Grape By-Product. *Pol. J. Food Nutr. Sci.* **2020**, *70*, 101–111. [CrossRef]
27. Zhou, Y.; Zhao, X.; Huang, H. Effects of Pulsed Electric Fields on Anthocyanin Extraction Yield of Blueberry Processing By-Products. *J. Food Process. Preserv.* **2015**, *39*, 1898–1904. [CrossRef]
28. Varadharajan, V.; Shanmugam, S.; Ramaswamy, A. Model generation and process optimization of microwave assisted aqueous extraction of anthocyanins from grape juice waste. *J. Food Process. Eng.* **2017**, *40*, e12486. [CrossRef]
29. Belwal, T.; Ezzat, S.M.; Rastrelli, L.I.; Bhatt, D.; Daglia, M.; Baldi, A.; Prasad Devkota, H.; Orhan, I.E.; Patra, J.K.; Das, G.; et al. A critical analysis of extraction techniques used for botanicals: Trends, priorities, industrial uses and optimization strategies. *TrAC Trends Anal. Chem.* **2018**, *100*, 82–102. [CrossRef]

30. Bagade, S.B.; Patil, M. Recent Advances in Microwave Assisted Extraction of Bioactive Compounds from Complex Herbal Samples: A Review. *Crit. Rev. Anal. Chem.* **2021**, *51*, 138–149. [CrossRef]
31. Zhang, Q.-W.; Lin, L.-G.; Ye, W.-C. Techniques for extraction and isolation of natural products: A comprehensive review. *Chin. Med.* **2018**, *13*, 20. [CrossRef]
32. Azmir, J.; Zaidul, I.S.M.; Rahman, M.M.; Sharif, K.M.; Mohamed, A.; Sahena, F.; Jahurul, M.H.A.; Ghafoor, K.; Norulaini, N.A.N.; Omar, A.K.M. Techniques for extraction of bioactive compounds from plant materials: A review. *J. Food Eng.* **2013**, *117*, 426–436. [CrossRef]
33. Martín, J.; Asuero, A.G. High hydrostatic pressure for recovery of anthocyanins: Effects, performance, and applications. *Sep. Purif. Rev.* **2021**, *50*, 159–176. [CrossRef]
34. Xi, J. Ultrahigh pressure extraction of bioactive compounds from plants—A review. *Crit. Rev. Food Sci. Nutr.* **2017**, *57*, 1097–1106. [CrossRef]
35. Ranjha, M.M.A.N.; Kanwal, R.; Shafique, B.; Arshad, R.N.; Irfan, S.; Kieliszek, M.; Kowalczewski, P.Ł.; Irfan, M.; Khalid, M.Z.; Roobab, U.; et al. A Critical Review on Pulsed Electric Field: A Novel Technology for the Extraction of Phytoconstituents. *Molecules* **2021**, *26*, 4893. [CrossRef]
36. Paludo, M.; Colombo, R.; Teixeira, J.; Hermosín-Gutiérrez, I.; Ballus, C.; Godoy, H. Optimizing the Extraction of Anthocyanins from the Skin and Phenolic Compounds from the Seed of Jabuticaba Fruits (Myrciaria jabuticaba (Vell.) O. Berg) with Ternary Mixture Experimental Designs. *J. Braz. Chem. Soc.* **2019**, *30*, 1506–1514. [CrossRef]
37. Demirdöven, A.; Özdoğan, K.; Erdoğan-Tokatli, K. Extraction of Anthocyanins from Red Cabbage by Ultrasonic and Conventional Methods: Optimization and Evaluation. *J. Food Biochem.* **2015**, *39*, 491–500. [CrossRef]
38. Backes, E.; Pereira, C.; Barros, L.; Prieto, M.A.; Genena, A.K.; Barreiro, M.F.; Ferreira, I.C.F.R. Recovery of bioactive anthocyanin pigments from *Ficus carica* L. peel by heat, microwave, and ultrasound based extraction techniques. *Food Res. Int.* **2018**, *113*, 197–209. [CrossRef]
39. Khazaei, K.M.; Jafari, S.M.; Ghorbani, M.; Kakhki, A.H.; Sarfarazi, M. Optimization of Anthocyanin Extraction from Saffron Petals with Response Surface Methodology. *Food Anal. Methods* **2016**, *9*, 1993–2001. [CrossRef]
40. Sang, J.; Ma, Q.; Li, B.; Li, C.-Q. An approach for extraction, purification, characterization and quantitation of acylated-anthocyanins from Nitraria tangutorun Bobr. fruit. *J. Food Meas. Charact.* **2018**, *12*, 45–55. [CrossRef]
41. Xu, D.P.; Li, Y.; Meng, X.; Zhou, T.; Zhou, Y.; Zheng, J.; Zhang, J.J.; Li, H.B. Natural Antioxidants in Foods and Medicinal Plants: Extraction, Assessment and Resources. *Int. J. Mol. Sci.* **2017**, *18*, 96. [CrossRef]
42. de Sousa Sabino, L.B.; Alves Filho, E.G.; Fernandes, F.A.N.; de Brito, E.S.; da Silva Júnior, I.J. Optimization of pressurized liquid extraction and ultrasound methods for recovery of anthocyanins present in jambolan fruit (*Syzygium cumini* L.). *Food Bioprod. Process.* **2021**, *127*, 77–89. [CrossRef]
43. Ravanfar, R.; Tamadon, A.M.; Niakousari, M. Optimization of ultrasound assisted extraction of anthocyanins from red cabbage using Taguchi design method. *J. Food Sci. Technol.* **2015**, *52*, 8140–8147. [CrossRef]
44. Agcam, E.; Akyıldız, A.; Balasubramaniam, V.M. Optimization of anthocyanins extraction from black carrot pomace with thermosonication. *Food Chem.* **2017**, *237*, 461–470. [CrossRef]
45. Espada-Bellido, E.; Ferreiro-González, M.; Carrera, C.; Palma, M.; Barroso, C.G.; Barbero, G.F. Optimization of the ultrasound-assisted extraction of anthocyanins and total phenolic compounds in mulberry (*Morus nigra*) pulp. *Food Chem.* **2017**, *219*, 23–32. [CrossRef]
46. Celli, G.B.; Ghanem, A.; Brooks, M.S.-L. Optimization of ultrasound-assisted extraction of anthocyanins from haskap berries (*Lonicera caerulea* L.) using Response Surface Methodology. *Ultrason. Sonochemistry* **2015**, *27*, 449–455. [CrossRef]
47. Dibazar, R.; Celli, G.B.; Brooks, M.S.-L.; Ghanem, A. Optimization of ultrasound-assisted extraction of anthocyanins from lowbush blueberries (Vaccinium Angustifolium Aiton). *J. Berry Res.* **2015**, *5*, 173–181. [CrossRef]
48. Leichtweis, M.G.; Pereira, C.; Prieto, M.A.; Barreiro, M.F.; Baraldi, I.J.; Barros, L.; Ferreira, I.C.F.R. Ultrasound as a Rapid and Low-Cost Extraction Procedure to Obtain Anthocyanin-Based Colorants from Prunus spinosa L. Fruit epicarp: Comparative study with conventional heat-based extraction. *Molecules* **2019**, *24*, 573. [CrossRef]
49. Mane, S.; Bremner, D.H.; Tziboula-Clarke, A.; Lemos, M.A. Effect of ultrasound on the extraction of total anthocyanins from Purple Majesty potato. *Ultrason. Sonochemistry* **2015**, *27*, 509–514. [CrossRef]
50. Pan, Z.; Qu, W.; Ma, H.; Atungulu, G.G.; McHugh, T.H. Continuous and pulsed ultrasound-assisted extractions of antioxidants from pomegranate peel. *Ultrason. Sonochemistry* **2011**, *18*, 1249–1257. [CrossRef]
51. Xu, W.J.; Zhai, J.W.; Cui, Q.; Liu, J.Z.; Luo, M.; Fu, Y.J.; Zu, Y.G. Ultra-turrax based ultrasound-assisted extraction of five organic acids from honeysuckle (Lonicera japonica Thunb.) and optimization of extraction process. *Sep. Purif. Technol.* **2016**, *166*, 73–82. [CrossRef]
52. Tan, Z.J.; Yi, Y.J.; Wang, H.Y.; Zhou, W.L.; Wang, C.Y. Extraction, preconcentration and isolation of flavonoids from Apocynum venetum L. Leaves using ionic liquid-based ultrasonic-assisted extraction coupled with an aqueous biphasic system. *Molecules* **2016**, *21*, 262. [CrossRef]
53. Ardestani, S.B.; Sahari, M.A.; Barzegar, M. Effect of Extraction and Processing Conditions on Organic Acids of Barberry Fruits. *J. Food Biochem.* **2015**, *39*, 554–565. [CrossRef]
54. Xue, H.; Xu, H.; Wang, X.; Shen, L.; Liu, H.; Liu, C.; Qin, Q.; Zheng, X.; Li, Q. Effects of Microwave Power on Extraction Kinetic of Anthocyanin from Blueberry Powder considering Absorption of Microwave Energy. *J. Food Qual.* **2018**, *2018*, 9680184. [CrossRef]

55. Farzaneh, V.; Carvalho, I.S. Modelling of microwave assisted extraction (MAE) of anthocyanins (TMA). *J. Appl. Res. Med. Aromat. Plants* **2017**, *6*, 92–100. [CrossRef]
56. Sun, Y.; Xue, H.K.; Liu, C.H.; Liu, C.; Su, X.L.; Zheng, X.Z. Comparison of microwave assisted extraction with hot reflux extraction in acquirement and degradation of anthocyanin from powdered blueberry. *Int. J. Agric. Biol. Eng.* **2016**, *9*, 186–199.
57. Xue, H.; Tan, J.; Fan, L.; Li, Q.; Cai, X.J. Optimization microwave-assisted extraction of anthocyanins from cranberry using response surface methodology coupled with genetic algorithm and kinetics model análisis. *Food Process Eng.* **2021**, *44*, e13688.
58. Yiğit, Ü.; Yolaçaner, E.T.; Hamzalıoğlu, A.; Gökmen, V. Optimization of microwave-assisted extraction of anthocyanins in red cabbage by response surface methodology. *J. Food Process. Preserv.* **2021**, *46*, e16120. [CrossRef]
59. Liu, W.; Yang, C.; Zhou, C.; Wen, Z.; Dong, X. An improved microwave-assisted extraction of anthocyanins from purple sweet potato in favor of subsequent comprehensive utilization of pomace. *Food Bioprod. Process.* **2019**, *115*, 1–9. [CrossRef]
60. Jiang, H.; Wang, X.; Yang, D. Comparison of extraction methods for anthocyanins from fruit of rubus coreanus maq. and optimization of microwave-assisted extraction process. *J. Food Sci. Technol. China* **2019**, *37*, 91–97.
61. Zheng, S.; Deng, Z.; Jiang, H.; Li, H. Comparison of Different Extraction Methods and Antioxidant Activity of Anthocyanins from Eggplant Peel. *J. Chin. Inst. Food Sci. Technol.* **2017**, *17*, 92–99.
62. Li, X.; Chen, F.; Li, S.; Jia, J.; Gu, H.; Yang, L. An efficient homogenate-microwave-assisted extraction of flavonolsand anthocyanins from blackcurrant marc: Optimization usingcombination of Plackett-Burman design and Box-Behnken design. *Ind. Crops Prod.* **2016**, *94*, 834–847. [CrossRef]
63. Odabas, H.I.; Koca, I. Simultaneous separation and preliminary purification of anthocyanins from *Rosa pimpinellifolia* L. fruits by microwave assisted aqueous two-phase extraction. *Food Bioprod. Process.* **2021**, *125*, 170–180. [CrossRef]
64. Zill-e-Huma, M.A.V.; Maingonnat, J.F.; Chemat, F. Clean recovery of antioxidant flavonoids from onions: Optimising solvent free microwave extraction method. *J. Chromatogr. A* **2009**, *1216*, 7700–7707. [CrossRef] [PubMed]
65. Sahin, S. A novel technology for extraction of phenolic antioxidants from mandarin (*Citrus deliciosa* Tenore) leaves: Solvent-free microwave extraction. *Korean J. Chem. Eng.* **2015**, *32*, 950–957. [CrossRef]
66. Michel, T.; Destandau, E.; Elfakir, C. Evaluation of a simple and promising method for extraction of antioxidants from sea buckthorn (*Hippophae rhamnoides* L.) berries: Pressurised solvent-free microwave assisted extraction. *Food Chem.* **2011**, *126*, 1380–1386. [CrossRef]
67. Da Silva, R.P.F.F.; Rocha-Santos, T.A.P.; Duarte, A.C. Supercritical fluid extraction of bioactive compounds. *TrAC Trends Anal. Chem.* **2016**, *76*, 40–51. [CrossRef]
68. Jiao, G.; Kermanshahi pour, A. Extraction of anthocyanins from haskap berry pulp using supercritical carbon dioxide: Influence of co-solvent composition and pretreatment. *LWT* **2018**, *98*, 237–244. [CrossRef]
69. Maran, J.P.; Priya, B.; Manikandan, S. Modeling and optimization of supercritical fluid extraction of anthocyanin and phenolic compounds from *Syzygium cumini* fruit pulp. *J. Food Sci. Technol.* **2014**, *51*, 1938–1946. [CrossRef]
70. Woz´niak, Ł.; Marszałek, K.; Skapska, S.; Jedrzejczak, R. The Application of Supercritical Carbon Dioxide and Ethanol for the Extraction of Phenolic Compounds from Chokeberry Pomace. *Appl. Sci.* **2017**, *7*, 322. [CrossRef]
71. Talmaciu, A.I.; Volf, I.; Popa, V.I. Supercritical fluids and ultrasound assisted extractions applied to spruce bark conversion. *Environ. Eng. Manag. J.* **2015**, *14*, 615–623.
72. López-Padilla, A.; Ruiz-Rodriguez, A.; Flórez, C.E.R.; Barrios, D.M.R.; Reglero, G.; Fornari, T. Vaccinium meridionale Swartz supercritical CO_2 extraction: Effect of process conditions and scaling up. *Materials* **2016**, *9*, 519. [CrossRef]
73. Babova, O.; Occhipinti, A.; Capuzzo, A.; Maffei, M.E. Extraction of bilberry (*Vaccinium myrtillus*) antioxidants using supercritical/subcritical CO_2 and ethanol as co-solvent. *J. Supercrit. Fluids* **2016**, *107*, 358–363. [CrossRef]
74. Idham, Z.; Nasir, H.M.; Yunus, M.A.C.; Yian, L.N.; Peng, W.L.; Hassan, H.; Setapar, S.H.M. Optimisation of supercritical CO_2 extraction of red colour from roselle (*Hibiscus Sabdariffa* Linn.) calyces. *Chem. Eng. Trans.* **2017**, *56*, 871–876.
75. Pazir, F.; Kocak, E.; Turan, F.; Ova, G. Extraction of anthocyanins from grape pomace by using supercritical carbon dioxide. *J. Food Process. Preserv.* **2021**, *45*, e14350. [CrossRef]
76. Mushtaq, M.; Sultana, B.; Anwar, F.; Adnan, A.; Rizvi, S.S.H. Enzyme-assisted supercritical fluid extraction of phenolic antioxidants from pomegranate peel. *J. Supercrit. Fluids* **2015**, *104*, 122–131. [CrossRef]
77. Reategui, J.L.P.; da Fonseca Machado, A.P.; Barbero, G.F.; Rezende, C.A.; Martinez, J. Extraction of antioxidant compounds from blackberry (*Rubus* sp.) bagasse using supercritical CO_2 assisted by ultrasound. *J. Supercrit. Fluids* **2014**, *94*, 223–233. [CrossRef]
78. Solana, M.; Mirofci, S.; Bertucco, A. Production of phenolic and glucosinolate extracts from rocket salad by supercritical fluid extraction: Process design and cost benefits analysis. *J. Food Eng.* **2016**, *168*, 35–41. [CrossRef]
79. Wang, Y.; Ye, Y.; Wang, L.; Yin, W.; Liang, J. Antioxidant activity and subcritical water extraction of anthocyanin from raspberry process optimization by response surface methodology. *Food Biosci.* **2021**, *44*, 101394. [CrossRef]
80. Kang, H.J.; Ko, M.J.; Chung, M.S. Anthocyanin Structure and pH Dependent Extraction Characteristics from Blueberries (*Vaccinium corymbosum*) and Chokeberries (*Aronia melanocarpa*) in SubcriticalWater State. *Foods* **2021**, *10*, 527. [CrossRef]
81. Sharifi, A.; Mortazav, S.A.; Maskooki, A.; Niakousari, M.; Elhamirad, A.H. Optimization of Subcritical Water Extraction of Bioactive Compounds from Barberry Fruit (*Berberis vulgaris*) by Using Response Surface Methodology. *Int. J. Agri Crop. Sci.* **2013**, *6*, 89–96.

82. Teixeira, R.F.; Benvenutti, L.; Burin, V.M.; Gomes, T.M.; Ferreira, S.R.S.; Ferreira-Zielinski, A.A. An eco-friendly pressure liquid extraction method to recover anthocyanins from broken black bean hulls. *Innov. Food Sci. Emerg. Technol.* **2021**, *67*, 102587. [CrossRef]
83. Andrade, T.A.; Hamerski, F.; Fetzer, D.E.L.; Roda-Serrat, M.C.; Corazza, M.L.; Norddahl, B.; Errico, M. Ultrasound-assisted pressurized liquid extraction of anthocyanins from *Aronia melanocarpa* pomace. *Sep. Purif. Technol.* **2021**, *276*, 119290. [CrossRef]
84. Altuner, E.M.; Tokusoglu, O. The effect of high hydrostatic pressure processing on the extraction, retention and stability of anthocyanins and flavonols contents of berry fruits and berry juices. *Int. J. Food Sci. Technol.* **2013**, *48*, 1991–1997. [CrossRef]
85. Barba, F.J.; Zhu, Z.; Kouba, M.; Sant'Ana, A.S.; Orlien, V. Green alternative methods for the extraction of antioxidant bioactive compounds from winery wastes and by-products: A review. *Trends Food Sci. Technol.* **2016**, *49*, 96–109. [CrossRef]
86. Fernandes, L.; Casal, S.I.P.; Pereira, J.A.; Ramalhosa, E.; Saraiva, J.A. Optimization of high pressure bioactive compounds extraction from pansies (*Viola* × *wittrockiana*) by response surface methodology. *High Press. Res.* **2017**, *37*, 415–429. [CrossRef]
87. Briones-Labarca, V.; Giovagnoli-Vicuña, C.; Chacana-Ojeda, M. High pressure extraction increases the antioxidant potential and in vitro bio-accessibility of bioactive compounds from discarded blueberries. *CyTA J. Food* **2019**, *17*, 622–631. [CrossRef]
88. Liu, S.; Xu, Q.; Li, X.; Wang, Y.; Zhu, J.; Ning, C.; Chang, X.; Meng, X. Effects of high hydrostatic pressure on physicochemical properties, enzymes activity, and antioxidant capacities of anthocyanins extracts of wild *Lonicera caerulea* berry. *Innov. Food Sci. Emerg. Technol.* **2016**, *36*, 48–58. [CrossRef]
89. Zhang, W.; Shen, Y.; Li, Z.; Xie, X.; Gong, E.S.; Tian, J.; Si, X.; Wang, Y.; Gao, N.; Shu, C.; et al. Effects of high hydrostatic pressure and thermal processing on anthocyanin content, polyphenol oxidase and β-glucosidase activities, color, and antioxidant activities of blueberry (*Vaccinium* Spp.) puree. *Food Chem.* **2021**, *342*, 128564. [CrossRef]
90. Morata, A.; Guamis, B. Use of UHPH to obtain juices with better nutritional quality and healthier wines with low levels of SO_2. *Front. Nutr.* **2020**, 241. [CrossRef]
91. Loira, I.; Morata, A.; Bañuelos, M.A.; Puig-Pujol, A.; Guamis, B.; González, C.; Suárez-Lepe, J.A. Use of Ultra-High Pressure Homogenization processing in winemaking: Control of microbial populations in grape musts and effects in sensory quality. *Innov. Food Sci. Emerg. Technol.* **2018**, *50*, 50–56. [CrossRef]
92. Chen, T.; Li, B.; Shu, C.; Tian, J.; Zhang, Y.; Gao, N.; Cheng, Z.; Xie, X.; Wang, J. Combined effect of thermosonication and high hydrostatic pressure on bioactive compounds, microbial load, and enzyme activities of blueberry juice. *Food Sci. Technol. Int.* **2021**, *25*, 33765872. [CrossRef]
93. Zderic, A.; Zondervan, E. Polyphenol extraction from fresh tea leaves by pulsed electric field: A study of mechanisms. *Chem. Eng. Res. Des.* **2016**, *109*, 586–592. [CrossRef]
94. Buchmann, L.; Mathys, A. Perspective on Pulsed Electric Field Treatment in the Bio-based Industry. *Front. Bioeng. Biotechnol.* **2019**, *7*, 265. [CrossRef] [PubMed]
95. Lončarić, A.; Celeiro, M.; Jozinović, A.; Jelinić, J.; Kovač, T.; Jokić, S.; Babić, J.; Moslavac, T.; Zavadlav, S.; Lores, M. Green Extraction Methods for Extraction of Polyphenolic Compounds from Blueberry Pomace. *Foods* **2020**, *9*, 1521. [CrossRef]
96. Medina-Meza, I.G.; Barbosa-Cánovas, G.V. Assisted extraction of bioactive compounds from plum and grape peels by ultrasonics and pulsed electric fields. *J. Food Eng.* **2015**, *166*, 268–275. [CrossRef]
97. Gagneten, M.; Leiva, G.; Salvatori, D.; Schebor, C.; Olaiz, N. Optimization of Pulsed Electric Field Treatment for the Extraction of Bioactive Compounds from Blackcurrant. *Food Bioprocess Technol.* **2019**, *12*, 1102–1109. [CrossRef]
98. Lamanauskas, N.; Bobinaitė, R.; Šatkauskas, S.; Viškelis, P.; Pataro, G.; Ferrari, G. Pulsed electric field-assisted juice extraction of frozen/thawed blueberries. *Zemdirb. Agric.* **2015**, *102*, 59–66. [CrossRef]
99. Taiebirad, F.; Bakhshabadi, H.; Rashidzadeh, S.H. Optimization of extraction of bioactive compounds from seedless barberry fruit using pulsed electric field pretreatment. *Iran. J. Food Sci. Technol.* **2021**, *18*, 305–317.
100. Aadil, R.M.; Zeng, X.-A.; Sun, D.-W.; Wang, M.-S.; Liu, Z.-W.; Zhang, Z.-H. Combined effects of sonication and pulsed electric field on selected quality parameters of grapefruit juice. *LWT Food Sci. Technol.* **2015**, *62*, 890–893. [CrossRef]
101. Teusdea, A.C.; Bandici, L.; Kordiaka, R.; Bandici, G.E.; Vicas, S.I. The Effect of Different Pulsed Electric Field Treatments on Producing High Quality Red Wines. *Not. Bot. Horti Agrobot. Cluj-Napoca* **2017**, *45*, 540–547. [CrossRef]
102. Aadil, R.M.; Zeng, X.-A.; Han, Z.; Sahar, A.; Khalil, A.A.; Rahman, U.U.; Khan, M.; Mehmood, T. Combined effects of pulsed electric field and ultrasound on bioactive compounds and microbial quality of grapefruit juice. *J. Food Process. Preserv.* **2018**, *42*, e13507. [CrossRef]
103. Yildiz, S.; Pokhrel, P.R.; Unluturk, S.; Barbosa-Cánovas, G.V. Changes in Quality Characteristics of Strawberry Juice After Equivalent High Pressure, Ultrasound, and Pulsed Electric Fields Processes. *Food Eng. Rev.* **2021**, *13*, 601–612. [CrossRef]
104. Faisal Manzoor, M.; Ahmed, Z.; Ahmad, N.; Karrar, E.; Rehman, A.; Aadil, R.M.; Al-Farga, A.; Waheed Iqbal, M.; Rahaman, A.; Zeng, X.A. Probing the combined impact of pulsed electric field and ultra-sonication on the quality of spinach juice. *J. Food Process. Preserv.* **2021**, *45*, e15475. [CrossRef]
105. Barba, F.J.; Galanakis, C.M.; Esteve, M.J.; Frigola, A.; Vorobiev, E. Potential use of pulsed electric technologies and ultrasounds to improve the recovery of high-added value compounds from blackberries. *J. Food Eng.* **2015**, *167*, 38–44. [CrossRef]
106. Siddeeg, A.; Manzoor, M.F.; Ahmad, M.H.; Ahmad, N.; Ahmed, Z.; Khan, M.K.I.; Maan, A.A.; Mahr-Un-Nisa; Zeng, X.-A.; Ammar, A.-F. Pulsed Electric Field-Assisted Ethanolic Extraction of Date Palm Fruits: Bioactive Compounds, Antioxidant Activity and Physicochemical Properties. *Processes* **2019**, *7*, 585. [CrossRef]

107. Sotelo, K.A.G.; Hamid, N.; Oey, I.; Pook, C.; Gutierrez-Maddox, N.; Ma, Q.; Leong, S.Y.; Lu, J. Red cherries (*Prunus avium* var. Stella) processed by pulsed electric field–Physical, chemical and microbiological analyses. *Food Chem.* **2018**, *240*, 926–934. [CrossRef]
108. Bandici, L.; Vicas, S.; Bandici, G.E.; Teusdean, A.C.; Popa, D. The Effect of Pulsed Electric Field (PEF) Treatment on the Quality of Wine. In Proceedings of the 14th International Conference on Engineering of Modern Electric Systems (EMES), Oradea, Romania, 1–2 June 2017; IEEE Institute of Electrical and Electronics Engineers: Manhattan, NY, USA, 2017.
109. Jin, T.Z.; Yub, Y.; Gurtlera, J.B. Processing on microbial survival, quality change and nutritional characteristics of blueberries. *LWT Food Sci. Technol.* **2017**, *77*, 517–524. [CrossRef]
110. Leong, S.Y.; Oey, I.; Burritt, D.J. Pulsed Electric Field Technology Enhances Release of Anthocyanins from Grapes and Bioprotective Potential against Oxidative Stress. In Proceedings of the 1st World Congress on Electroporation and Pulsed Electric Fields in Biology, Medicine and Food & Environmental Technologies, Portorož, Slovenia, 6–10 September 2015; Springer: Singapore, 2015; pp. 47–50.
111. Aganovic, K.; Smetana, S.; Grauwet, T.; Toepfl, S.; Mathys, A.; Van Loey, A.; Heinz, V. Pilot scale thermal and alternative pasteurization of tomato and watermelon juice: An energy comparison and life cycle assessment. *J. Clean. Prod.* **2017**, *141*, 514–525. [CrossRef]
112. Reineke, K.; Schottroff, F.; Meneses, N.; Knorr, D. Sterilization of liquid foods by pulsed electric fields—An innovative ultra-high temperature process. *Front. Microbiol.* **2015**, *6*, 400. [CrossRef]
113. Jaeger, H.; Knorr, D. Pulsed Electric Fields Treatment in Food Technology: Challenges and Opportunities. In *Handbook of Electroporation*; Miklavčič, D., Ed.; Springer: Cham, Germany, 2017; pp. 2657–2680.
114. Lotfi, M.; Hamdami, N.; Dalvi-Isfahan, M.; Fallah-Joshaqani, S. Effects of high voltage electric field on storage life and antioxidant capacity of whole pomegranate fruit. *Innov. Food Sci. Emerg. Technol.* **2022**, *75*, 102888. [CrossRef]
115. Wang, Q.; Li, Y.; Sun, D.-W.; Zhu, Z. Enhancing Food Processing by Pulsed and High Voltage Electric Fields: Principles and Applications. *Crit. Rev. Food Sci. Nutr.* **2018**, *58*, 2285–2298. [CrossRef]
116. Boussetta, N.; Vorobiev, E. Extraction of valuable biocompounds assisted by high voltage electrical discharges: A review. *Comptes Rendus. Chim.* **2014**, *17*, 197–203. [CrossRef]
117. Rajha, H.N.; Boussetta, N.; Louka, N.; Maroun, R.G.; Vorobiev, E. Electrical, mechanical, and chemical effects of high-voltage electrical discharges on the polyphenol extraction from vine shoots. *Innov. Food Sci. Emerg. Technol.* **2015**, *31*, 60–66. [CrossRef]
118. Meini, M.-R.; Cabezudo, I.; Boschetti, C.E.; Romanini, D. Recovery of phenolic antioxidants from Syrah grape pomace through the optimization of an enzymatic extraction process. *Food Chem.* **2019**, *283*, 257–264. [CrossRef] [PubMed]
119. Lotfi, L.; Kalbasi-Ashtari, A.; Hamedi, M.; Ghorbani, F. Effects of enzymatic extraction on anthocyanins yield of saffron tepals (Crocos sativus) along with its color properties and structural stability. *J. Food Drug Anal.* **2015**, *23*, 210–218. [CrossRef] [PubMed]
120. Binaschi, M.; Duserm Garrido, G.; Cirelli, C.; Spigno, G. Biotechnological strategies to valorise grape pomace for food applications. *Chem. Eng. Trans.* **2018**, *64*, 367–372.
121. Vardakas, A.T.; Shikov, V.T.; Dinkova, R.H.; Mihalev, K.M. Optimisation of the enzyme-assisted extraction of polyphenols from saffron (*Crocus sativus* L.) tepals. *Acta Sci. Pol. Technol. Aliment.* **2021**, *20*, 359–367.
122. Serea, D.; Râpeanu, G.; Constantin, O.E.; Bahrim, G.E.; Stănciuc, N.; Croitoru, C. Ultrasound and enzymatic assisted extractions of bioactive compounds found in red grape skins băbească neagră (*vitis vinifera*) variety. *Ann. Univ. Dunarea Jos Galati Fascicle VI Food Technol.* **2021**, *45*, 9–25. [CrossRef]
123. Li, Y.; Tao, F.; Wang, Y.; Cui, K.; Cao, J.; Cui, C.; Nan, L.; Li, Y.; Yang, J.; Wang, Z. Process optimization for enzymatic assisted extraction of anthocyanins from the mulberry wine residue. *IOP Conf. Series Earth Environ. Sci.* **2020**, *559*, 012011. [CrossRef]
124. Rezende, Y.R.R.S.; Nogueira, J.P.; Silva, T.O.M.; Barros, R.G.C.; de Oliveira, C.S.; Cunha, G.C.; Gualberto, N.C.; Rajan, M.; Narain, N. Enzymatic and ultrasonic-assisted pretreatment in the extraction of bioactive compounds from Monguba (*Pachira aquatic* Aubl) leaf, bark and seed. *Food Res. Int.* **2021**, *140*, 109869. [CrossRef]
125. Araujo, N.M.P.; Pereira, G.A.; Arruda, H.S.; Prado, L.G.; Ruiz, A.L.T.G.; Eberlin, M.N.; de Castro, R.; Pastore, G.M. Enzymatic treatment improves the antioxidant and antiproliferative activities of *Adenanthera pavonina* L. seeds. *Biocatal. Agric. Biotechnol.* **2019**, *18*, 101002. [CrossRef]
126. Li, X.; Zhu, F.; Zeng, Z. Effects of different extraction methods on antioxidant properties of blueberry anthocyanins. *Open Chem.* **2021**, *19*, 138–148. [CrossRef]
127. Szymanowska, U.; Baraniak, B. Antioxidant and Potentially Anti-Inflammatory Activity of Anthocyanin Fractions from Pomace Obtained from Enzymatically Treated Raspberries. *Antioxidants* **2019**, *8*, 299. [CrossRef] [PubMed]
128. Oancea, S.; Perju, M. Influence of enzymatic and ultrasonic extraction on phenolics content and antioxidant activity of Hibiscus Sabdariffa, L. flowers. *Bulg. Chem. Commun.* **2020**, *52*, 25–29.
129. Xue, H.; Tan, J.; Li, Q.; Tang, J.; Cai, X. Ultrasound-Assisted Enzymatic Extraction of Anthocyanins from Raspberry Wine Residues: Process Optimization, Isolation, Purification, and Bioactivity Determination. *Food Anal. Methods* **2021**, *14*, 1369–1386. [CrossRef]

Review

Anthocyanins: Factors Affecting Their Stability and Degradation

Bianca Enaru, Georgiana Drețcanu, Teodora Daria Pop, Andreea Stănilă and Zorița Diaconeasa *

Faculty of Food Science and Technology, University of Agricultural Science and Veterinary Medicine, 400372 Cluj-Napoca, Romania; bianca.enaru@stud.ubbcluj.ro (B.E.); georgiana.dretcanu@stud.ubbcluj.ro (G.D.); teodora-daria.pop@student.usamvcluj.ro (T.D.P.); andreea.stanila@usamvcluj.ro (A.S.)
* Correspondence: zorita.sconta@usamvcluj.ro; Tel.: +40-751-033-871

Abstract: Anthocyanins are secondary metabolites and water-soluble pigments belonging to the phenolic group, with important functions in nature such as seed dispersal, pollination and development of plant organs. In addition to these important roles in plant life, anthocyanins are also used as natural pigments in various industries, due to the color palette they can produce from red to blue and purple. In addition, recent research has reported that anthocyanins have important antioxidant, anti-cancer, anti-inflammatory and antimicrobial properties, which can be used in the chemoprevention of various diseases such as diabetes, obesity and even cancer. However, anthocyanins have a major disadvantage, namely their low stability. Thus, their stability is influenced by a number of factors such as pH, light, temperature, co-pigmentation, sulfites, ascorbic acid, oxygen and enzymes. As such, this review aims at summarizing the effects of these factors on the stability of anthocyanins and their degradation. From this point of view, it is very important to be precisely aware of the impact that each parameter has on the stability of anthocyanins, in order to minimize their negative action and subsequently potentiate their beneficial health effects.

Keywords: anthocyanins; stability; degradation; pigments; health benefit

Citation: Enaru, B.; Drețcanu, G.; Pop, T.D.; Stănilă, A.; Diaconeasa, Z. Anthocyanins: Factors Affecting Their Stability and Degradation. *Antioxidants* **2021**, *10*, 1967. https://doi.org/10.3390/antiox10121967

Academic Editors: Agustín G. Asuero and Noelia Tena

Received: 14 October 2021
Accepted: 8 December 2021
Published: 9 December 2021

Publisher's Note: MDPI stays neutral with regard to jurisdictional claims in published maps and institutional affiliations.

Copyright: © 2021 by the authors. Licensee MDPI, Basel, Switzerland. This article is an open access article distributed under the terms and conditions of the Creative Commons Attribution (CC BY) license (https://creativecommons.org/licenses/by/4.0/).

1. Introduction

Anthocyanins are a class of natural water-soluble pigments that are part of the flavonoid family. They are very widespread in nature, found not only in the colored petals of flowers but also in the roots, stems, tubers, leaves, fruits and seeds [1,2]. This type of pigment has a strong absorption in the UV-visible region of the electromagnetic spectrum and is the main determinant of red-blue colors and their derivatives in the plant kingdom. These characteristics place them second in importance, immediately following chlorophyll pigments [3]. Anthocyanins play an important role in seed dispersal, pollination, development of plant organs, but also in their adaptation to various changes in biotic (pathogenic attacks) and abiotic (drought, lack of nutrients, high intensity light) [4] factors. Due to their chemical structure, with a central core in the form of 2-phenylbenzopyrylium or flavylium cation, anthocyanins can be classified as polyphenols and secondary metabolites [5].

Structurally, anthocyanins are found in the form of glycosides of polyhydroxy and polymethoxy derivatives of 2-phenylbenzopyrrile salts and are composed of an aglycone also called anthocyanidin and a carbohydrate residue that may be glucose, xylose, galactose, arabinose, rhamnose or rutinose. These carbohydrate residues are generally attached to the anthocyanidin skeleton through the C3 hydroxyl group in ring C [6,7]. Organic acids can be added to the sugar moieties and thus the acylation of the initial structure takes place. The most common acylation compounds include aromatic acids such as p-coumaric, sinapic, gallic, ferulic and caffeic acid, but also a number of aliphatic acids such as the malic, succinic, oxalic, tartaric and acetic acid [8,9]. It should be noted that this structure makes them dependent on the composition and conditions of the solution in which they are dissolved. Additionally, anthocyanins can perform interactions with other compounds but also with each other, where both their color and their structural balance are influenced [10]. In this way, a class of compounds comprising over 700 distinct species of anthocyanins is obtained, but only 6 of them (cyanidin, pelargonidin, delphinidin, peonidin, petunidin and malvidin) are abundantly found in nature and represent about 90% of all anthocyanins identified so far [11]. These 6 anthocyanins are found in fruits and vegetables in various percentages, for example: 50%—cyanidin, 12%—in the case of pelargonidin, delphinine, petunidin and 7%—for peonidin and malvidin. In nature, cyanidin is found predominantly in berries and other red vegetables because it appears in the form of a red-purple pigment, similar to the color magenta. Pelargonidin appears freely in the form of a red pigment but gives flowers an orange hue and fruits a red color. Delphinidin appears as a blue-reddish or purple pigment in the plant, leading to the blue color of flowers. Petunidin is a water-soluble methylated anthocyanin, dark red or purple, which is often found in blackcurrants and purple flowers. Peonidin is another methylated anthocyanin, which appears in the form of a magenta pigment, also found abundantly in berries, grapes and in red wines. Malvidin is an O-methylated anthocyanin, which appears as a purple pigment, and determines the blue color of certain flowers, but it is the major pigment in red wines [12]. Thus, the anthocyanins cyanidin, pelargonidin and delphinidin are often found in fruits, while in flowers the predominant anthocyanins are peonidin, petunidin and malvidin [13]. All these color variations of anthocyanins are represented in Figure 1.

Figure 1. The structure and color of the most common anthocyanins present in nature.

Therefore, Figure 2 shows the structural variations in anthocyanins, which occur due to differences in the type of sugar, position and number of hydroxyl or methoxyl groups in the B ring but also due to the presence or absence of acylation with both aromatic and aliphatic compounds [8]. The chemical structure of these polyphenols appears in the form of a C6 skeleton, and is pH dependent [14–16].

Anthocyanin synthesis takes place in a special segment of the flavonoid synthesis pathway, which is regulated on several levels. The first step is the conversion of phenylalanine to cinnamic acid, and subsequently, it is transformed into the main precursor of anthocyanins, 4-coumaroyl CoA, through a series of reactions catalyzed by cinnamate 4-hydroxylase (C4H) and 4-coumaroyl CoA ligase (4CL). In the next step one molecule of 4-coumaroyl CoA and three molecules of malonyl CoA will be condensed using the chalcone synthase, resulting in chalcones. Finally, after a series of enzymatic reactions, the main anthocyanins will be produced [17]. This synthesis takes place in the cytosol, and the anthocyanins obtained will be transported to vacuoles, where they will be stored as colored coalescences called anthocyanic vacuolar inclusions [18].

In addition, recent research has concluded that the synthesis and accumulation of anthocyanins can be controlled by regulating the expression of certain genes, epigenetic changes in plants, and post-translational changes in proteins that coordinate the activity of the transcription factor [17]. Furthermore, the regulatory molecules (precursors and enzymes) of biosynthetic and degradation pathways influence the amount of anthocyanin in different plant regions and in different plants [19].

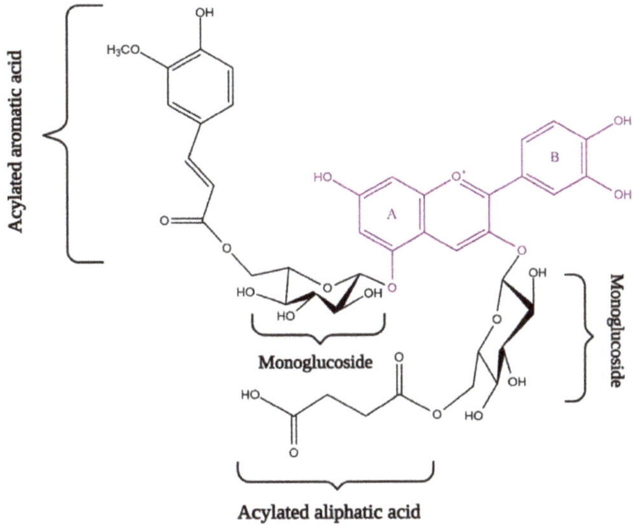

Figure 2. General chemical structure of anthocyanidin (flavylium cation) with sugar moieties and aromatic or aliphatic acid.

Both red and blue fruits and vegetables are the main sources of anthocyanins, but their anthocyanin content suffers from important variations between different species, and within the same species. The factors influencing anthocyanin content include plant variety, climate, growing area, cultivation processes, harvesting period, ripening, seasonal variability, processing and storage, light and temperature [20].

Initially these compounds were mainly used as natural dyes in the food industry, due to their wide range of colors, ranging from salmon pink to red and from purple to almost black [21]. Over the past few years, numerous studies have reported that anthocyanins exhibit anti-cancer, anti-inflammatory, antimicrobial and antioxidant properties, with a role in reducing the incidence of cardiovascular disease, diabetes and obesity [22]. Therefore, it can be stated that anthocyanins have enormous potential; they can be used both in food science and in the medical field [23].

The main disadvantage of anthocyanins is their extremely low stability, which is easily influenced by a wide range of parameters such as relative humidity, light, pH, temperature, sugars (acylated and unacylated), vitamin C, oxygen levels, sulfur dioxide or sulfites, enzymes, co-pigments and metal ions [24]. Thus, these factors and processes will be able to determine changes in the concentration and bioactivity of anthocyanins, which will affect the degree of substance acceptance by the consumer [25]. In recent years, several studies have focused on the bioavailability of anthocyanins, proving that it would be less than 1%. At the same time, it was revealed that the bioavailability of anthocyanins depends on the glycosidic radical and, therefore, non-acylated anthocyanins are absorbed more efficiently than acylated ones [16]. Thereby, the stability of anthocyanins can be modified in three ways: by polymerization, cleavage and derivatization. The cleavage of anthocyanins results in colorless compounds, polymerization causes browning, and derivatization reactions cause the production of several molecules of different colors [26].

As such, in order to benefit from all the properties of the anthocyanins mentioned above, scientists have conducted numerous studies to understand how these parameters influence the stability and more precisely how to tackle these shortcomings to increase the stability and henceforth the effectiveness of anthocyanins.

The purpose of this research is to provide an overview of anthocyanin stability. In this respect, the most important factors that influence their stability were reviewed, and an analysis was performed on how anthocyanin degradation takes place under the action of these factors. Thus, each factor will undergo discussions throughout this paper, determining the positive or negative effects it produces on anthocyanin stability. For a better understanding of their activity, a series of figures and tables are also included. It is in this way that the opportunity for future studies is presented, towards improving their stability and their intensive use both in the food industry and medical field.

2. Antioxidant Potential of Anthocyanins

Reactive oxygen species (ROS) and reactive nitrogen species (RNS) are compounds of great importance for the immune system with a role in cell signaling and other body functions. However, the production of large quantities of these compounds, which the body can no longer eliminate with the help of endogenous and exogenous antioxidants, will cause alteration of the oxidative balance of the biological system, and the occurrence of oxidative stress [27–29]. Its appearance produces oxidative changes in molecules, tissue damage and causes accelerated cell death, all of which disrupt cell function, which is the first step in the emergence of many diseases [30].

A compound can be called an antioxidant if it delays or prevents the oxidation of a substrate at low concentrations. The basic property of an antioxidant is that it helps to limit oxidative damage in the human body by preventing or detecting a chain of oxidative propagation by stabilizing the produced radical. Substances with antioxidant capacity can act through multiple mechanisms such as: hydrogen atom transfer (HAT), single electron transfer (SET) or the ability to chelate transition metals [29].

The main endogenous antioxidants are enzymes such as superoxide dismutase, glutathione peroxidase and catalase, but certain non-enzymatic compounds may also have antioxidant functions, such as bilirubin, albumin, metallothionein and uric acid. In some cases, endogenous agents fail to provide effective protection and strict control against reactive oxygen species, so there is a need to administer exogenous antioxidants in the form of nutritional supplements or pharmaceutical compounds that contain an antioxidant. The most important exogenous antioxidants are vitamins C and E, certain minerals, β-carotene and flavonoids [31].

The most relevant compounds in the flavonoid class are anthocyanins, which have a flavylium cation structure, as is shown in the next chapter. Thus, most of the functional properties of anthocyanins, such as their sensory quality and antioxidant capacity are determined by their chemical structure, which acts as an acid. It is worth mentioning that the structure and properties of these polyphenols, including their antioxidant capacity, are influenced by certain factors such as temperature, pH and solvents, factors that should be controlled to obtain relevant results in terms of their antioxidant activity [32]. In addition, even the glycosylation of anthocyanins has been shown to decrease antioxidant activity and the ability to capture free radicals, compared to the aglycone form, thereby decreasing the power of anthocyanin radicals to delocalize electrons [27]. On the other hand, anthocyanin bioavailability influences the efficacy of antioxidant activity of this compound in oxidative stress [32]. In general, anthocyanins neutralize reactive radical species by transferring a single electron or by removing the hydrogen atom from phenolic groups. The central component of the antioxidant activity of anthocyanins is represented by the oxidation of phenolic hydroxyl groups, more precisely the para- and orthophenolic groups, which have a crucial role in the formation of semiquinones and in the stabilization of one-electron oxidation products [20].

Therefore, a large amount of anthocyanins are found in fruits and vegetables, so their consumption also involves the intake of a certain amount of antioxidants that will contribute to the protection of various types of diseases caused by oxidative stress [30]. However, we must remember that these foods include a variety of different phytochemicals and vitamins that may interact with anthocyanins in a synergistic or antagonistic manner, improving or lowering their antioxidant activity [27].

Undoubtedly, measuring the antioxidant activity of biological samples and food is of crucial importance, not only for ensuring food quality but also for determining the effectiveness of food antioxidants in preventing and treating diseases based on oxidative stress [33]. To this end, several antioxidant assays have been developed to evaluate anthocyanin capacity to inhibit the oxidation process that happens naturally [32].

Measurement methods for the antioxidant capacity of foods and various products have improved considerably in recent years, so it is necessary to establish and standardize measurement tests for the antioxidant capacity of various compounds [30]. In addition, evaluation methods for the antioxidant capacity must have certain characteristics such as speed and reproducibility and can be performed with small amounts of the analyzed compounds [29].

With these in mind, the most common methods for determining antioxidant capacity, which are able to assess this anthocyanin property, will be mentioned. There are two categories of such methods, one based on hydrogen atom transfer (HAT) and the second based on single electron transfer (SET). The methods based on the HAT mechanism include oxygen radical absorbance capacity (ORAC), while the types of assays that are SET based include: ferric ion reducing antioxidant power assay (FRAP), 2,2-Diphenyl-1-picrylhydrazyl radical scavenging assay (DPPH$^\bullet$) and cupric ions (Cu^{2+}) reducing antioxidant power assay (CUPRAC). Likewise, specialist studies have reported that the method 2,2-Azinobis 3-ethylbenzthiazoline-6-sulfonic acid radical scavenging assay (ABTS$^{\bullet+}$) uses both mechanisms. At the same time, there are other types of tests that are not based on the aforementioned mechanisms that measure the sample scavenging ability for oxidants, which interact and produce a negative effect on major macromolecules in biological systems. One such test is hydrogen peroxide (H_2O_2) scavenging assay [30].

2.1. DPPH$^\bullet$ (Diphenyl-1-Picrylhydrazyl) Assay

DPPH$^\bullet$ is a common and frequently used spectrophotometric procedure for determining antioxidant capacities of components. This technique can be used on both solid and liquid samples, and is based on the ability of the free radical (DPPH$^\bullet$) to react with a hydrogen donor (AH^+) [32]. Thus, electron donation by the antioxidant takes place to neutralize the DPPH$^\bullet$ radical [33].

This free radical is a stable one, but when the delocalization of the electron takes place, a purple color is obtained, which presents an intense abortion in the UV–vis spectral region at 517 nm [29,32]. When DPPH$^\bullet$ interacts with a hydrogen donor, the reduced form, DPPH, is formed; this results in the violet color vanishing. As a result, the decrease of DPPH$^\bullet$ gives an index for estimating the test compound capacity to capture radicals [29,31].

In conclusion, this technique is a simple, fast and economical one, used successfully to determine the antioxidant activity of some substances. In addition, the DPPH$^\bullet$ assay is the oldest indirect method of determining antioxidant activity and has been used for the first time in determining the antioxidant capacity of phenolic compounds [30].

2.2. Method of Inhibition of the (2,2'-Azinobis-(3-Ethylbenozothiazoline-6-Sulfonate)) Radical Cation (ABTS$^{\bullet+}$)

ABTS is also a simple spectrophotometric method, which measures the total antioxidant capacity [33]. This technique is initiated by the reaction of ABTS with persulfate potassium, which produces a stable cation radical (ABTS$^{\bullet+}$) of blue/green color that is absorbed at a maximum wavelength of 734 nm [29]; thus, absorbance decreases for this radical when it reacts with an antioxidant [32].

As part of this test, when ABTS interacts with the antioxidant, it forms ABTS$^{\bullet+}$, which has an intense color, and the antioxidant capacity of the substance is expressed as the ability of the test compounds to decrease the color that reacts directly with the ABTS radical [30]. As a result, the level of discoloration can be reported as a percentage of ABTS$^{\bullet+}$ inhibition, which is computed as a function of antioxidant concentration and time. ABTS assay is suitable for both lipophilic and hydrophilic compounds, can be used at various pH levels and is excellent for investigating the impact of pH on antioxidant activity [29,30].

At the same time, this method is important for determining the antioxidant activity of mixtures of substances, and has a role in distinguishing between additive and synergistic effects [32]. It should also be noted that the ABTS assay is frequently correlated with DPPH assay to determine antioxidant activity because both mechanisms of these methods involve the acceptance of electrons and H• from antioxidant agents. This combination of tests is mainly used to determine anthocyanin antioxidant activity, and numerous studies have been conducted on various food matrices such as wine, pomegranate juice, blueberries and corn [32].

2.3. The FRAP (Ferric Reducing Antioxidant Power) Method

FRAP assay is a colorimetric method, which measures the ability of antioxidants to reduce the ferric ion complex (Fe^{3+}) to the ferrous complex (Fe^{2+}), which has an intense blue color [33]. Therefore, the increase of the absorption is measured at 593 nm and is related to a standard solution of ferrous ions or to a standard antioxidant solution to obtain the FRAP values [30].

Certain conditions are required in order to perform this test to determine the antioxidant capacity, namely, the FRAP test must take place in acidic conditions (pH = 3.6) to maintain the solubility of iron, the tested samples must be aqueous [32,33], and there must be an incubation time of 4 min at 37 °C to activate the antioxidant capacity of most samples [29].

In the past, the FRAP test used tripyridyltriazine (TPTZ) as an iron ion binding agent, but potassium ferricyanide is currently used. Thus, in the case of the use of potassium ferriyanide, Prussian blue is obtained as the final reaction compound that can be quantified spectrophotometrically and which shows the reducing power of the antioxidants that have been tested. Antioxidants can lead to the formation of Prussian blue by two different mechanisms. The first mechanism involves the reduction of Fe^{3+} from the solution to Fe^{2+} by binding ferricyanide, which will give the Prussian blue color, and the second mechanism is achieved by reducing ferricyanide to ferrocyanide that will then bind free Fe^{3+} [33].

The FRAP method is very easy, quick, reproducible and inexpensive, and it does not require any additional equipment. It has been used successfully on a large scale to determine the antioxidant activity of anthocyanins in various matrices, for example: in elderberry, raspberry, blackberry, red currant, carrot, cabbage, potato, onion and eggplant and even red wine [29,32].

2.4. Oxygen Radical Absorbance Capacity (ORAC) Method

ORAC assay is a fluorescence technique that involves the test sample (AH^+), a fluorescent compound such as fluorescein or protein phycoerthrin (β-PE) and a free radical generator [32]. The most common free radical generators used for this test are azo compounds such as 2,2-azobis(2-amidinopropane) chlorhydrate (ABAP), 2,2'-azobis(2,4-dimethylvaleronytril) (AMVN), α,α,-azobisizobutyronytril (AIBN) and 2,2'-azobis(2-amidinopropane) dihydrochloride (AAPH) [33].

The mechanism of this method is based on the degradation of the fluorescent compound, as the interaction with free radicals determines its oxidation. However, when antioxidant agents (AH^+) are present in the reaction, the fluorescent compound is protected from oxidative degradation, and thus, the fluorescence signal remains unaltered [32]. The ORAC test measures the capacity of antioxidants to produce hydrogen atoms, it assesses both the time effect and the degree of inhibition simultaneously [29], and hence, it is a HAT-based assay [33].

In the case of the ORAC test, certain reaction conditions are required for relevant results, the pH must be 7.4 and both fluorescein and the source of free radicals need to be kept at a constant temperature of 37 °C. The first stage of this procedure involves the thermal decomposition of AAPH—the free radical generator—which stimulates the degradation of the fluorescent substance. The introduction of the antioxidant agent, whose capacity is being tested, favors the elimination of peroxyl radicals, thus protecting the fluorescein from damage. The decrease of fluorescence is monitored at an interval of 1 min for 35 min at a wavelength of 485 nm for excitation and at 535 nm for emission. Finally, the change in fluorescence, due to the attack of free radicals but also the protection offered by antioxidants, generates a curve on the graph. Thus, the antioxidant capacity is determined from the area under the fluorescence decrease curve [29].

The use of the ORAC method is very frequent because it can be easily adapted to the type of sample, hydrophobic or hydrophilic, while solely the free radical generator needs to be changed [32]. Therefore, this assay has been widely used to assess the antioxidant content of drinks, supplements and fruits and vegetables [29]. Moreover, many authors have assessed the antioxidant activity of anthocyanins using this method [32].

2.5. CUPRAC (Cupric Ion Reducing Antioxidant Capacity) Assay

CUPRAC assay is similar to the FRAP method, a spectrophotometric technique, which measures the antioxidant capacity of a compound based on the reduction of the cupric ion (Cu^{2+}) to cuprous (Cu^+) [33]. The optimal pH for the CUPRAC test is 7.0, which is close to physiological pH (7.4) and mimics the antioxidant effect under real conditions [30].

Similar to the previously mentioned tests, this method also has a specific ligand, Neocuproine (2,9-dimethyl-1,10-phenanthroline), used to form a copper–ligand complex whose absorbance will be spectrophotometrically measured. In short, the reduction of Cu^{2+} to the Cu^+ complex occurs in the presence of an antioxidant and neocuproine complex, which has a maximum absorption peak at 450 nm. Additionally, the reaction time required to complete the reaction depends on the speed of the antioxidant, thus varying between 30 and 60 min. It is to be mentioned that flavonoid glycosides may require preliminary hydrolysis to completely reveal their antioxidant potential [33].

This method is cost-effective, fast, stable and can be used on a wide range of antioxidant agents, regardless of the chemical or hydrophobic type. However, citric acid and simple sugars, which are not true antioxidants, cannot be oxidized by the CUPRAC reagent [30].

2.6. Hydrogen Peroxide (H_2O_2) Scavenging Assay

Hydrogen peroxide (H_2O_2) is a biological compound found in air, water, plants, microorganism and food at low concentration. H_2O_2 appears in the body as a by-product of normal aerobic metabolism. Thus, under stress, H_2O_2 is produced in excess and is harmful to cells because it could be transformed into ROS, such as hydroxyl radicals, which initiate lipid peroxidation and cause damage to DNA [34,35].

Furthermore, the H_2O_2 test detects reactive oxygen uptake and is an important aspect of total antioxidant activity. This assay evaluates the uptake capacity against H_2O_2 and relies on the principle of the intrinsic absorption of this molecule at a wavelength of 230 nm. For this method, a solution of H_2O_2 (40 mM) is prepared with phosphate buffer at a pH of 7.4. Therefore, after an incubation period of 10 min, the remaining concentration of H_2O_2 is measured spectrophotometrically and compared to a control solution that includes phosphate buffer without H_2O_2. The absorbance value at 230 nm falls when the H_2O_2 concentration decreases in the presence of scavenger agents [30].

Thus, a colorimetric test was created in order to evaluate the capacity of capturing the H_2O_2 by antioxidants, especially by plant extracts. This assay is based on the reaction between H_2O_2, phenol and 4-aminoantipyrine in the presence of horseradish peroxidase (HRP) and produces a pink compound, namely quinonimine. As such, antioxidant agents will decrease chromatophore production, changing the color of the solution [36].

In conclusion, this approach is used to test the hydrogen peroxide scavenging capacities of common antioxidants such as ascorbic acid, gallic acid and tannic acid, as well as selected plant extracts [36].

Considering all these different evaluation methods for antioxidant activity, Table 1 includes the main advantages and disadvantages of each method.

Table 1. Common assay for antioxidant activity, their reaction, advantages and disadvantages.

Method	Reaction	Advantages	Disadvantages	References
DPPH$^\bullet$ assay	ArOH + DPPH$^\bullet$ → ArOH$^\bullet$ + DPPH$_2$	- can be used to test hydrophilic and lipophilic antioxidants with polar and nonpolar organic solvents	- can interact with other radicals, its stability is sensitive to certain solvents	[29,37]
ABTS$^{\bullet+}$ assay	(NH$_4$)$_2$S$_2$O$_3$ + ABTS → ABTS$^{\bullet+}$ + ArOH → ABTS + ArO$^\bullet$	- permit detection of a wide range of antioxidant compounds, reacts quickly with synthetic and natural antioxidant agents	- lack of biological relevance, because ABTS radical cation is not found naturally	[32,33,37],
FRAP assay	[Fe^{3+}-(TPTZ)$_2$]$^{3+}$ + ArOH → [Fe^{3+}-(TPTZ)$_2$]$^{2+}$ + ArO$^\bullet$ + H$^+$	- great sensitivity and precision allow it to distinguish between samples, can test a variety of biological samples such as plasma, blood, serum, saliva, tears and urine	- the tendency of the Prussian blue complex to precipitate and to stain the measuring vessel	[29,33,37]
ORAC assay	R-N=N-R + O$_2$ → N$_2$ + 2ROO$^\bullet$ + Fluorescein → Non-fluorescent products	- can be modified for the detection of lipophilic antioxidants	- non-specificity of the fluorescence compounds that can react with other samples, thus losing fluorescence	[29,32]
CUPRAC assay	Cu(Nc$_2$)$^{2+}$ + ArOH → Cu(Nc$_2$)$^+$ + ArO$^\bullet$ + H$^+$	- cheap reagents, more stable and accessible than DPPH or ABTS reagents	- antioxidant enzymes cannot be measurable	[33,37]
HPS assay	C$_{11}$H$_{13}$N$_3$O + 2H$_2$O$_2$ + ArOH → C$_{11}$H$_{11}$N$_3$OArO$^\bullet$ + 4H$_2$O$_2$	- is fast and simple	- secondary metabolites found in plants that absorb UV light may cause interference	[36]

DPPH: Diphenyl-1-Picrylhydrazyl; ABTS: 2,2′-Azinobis-(3-Ethylbenozothiazoline-6-Sulfonate; FRAP: Ferric Reducing Antioxidant Power; ORAC: Oxygen Radical Absorbance Capacity; CUPRAC: Cupric Ion Reducing Antioxidant Capacity; HPS: Hydrogen Peroxide.

In addition, for a better understanding of these methods to determine the antioxidant activity of phenolic compounds, specifically anthocyanins, Figure 3 includes all the presented tests, compounds used and their mechanism of action.

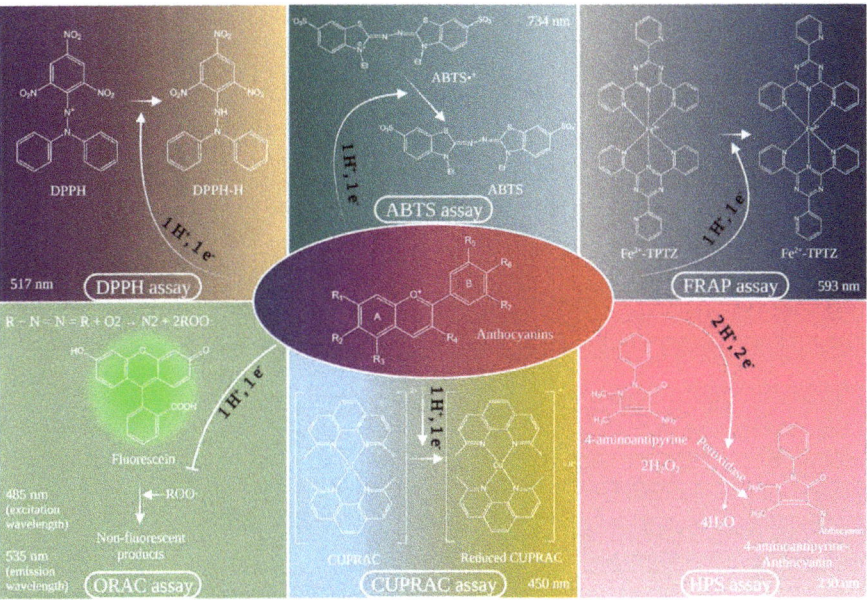

Figure 3. The action mechanism of methods for evaluating antioxidant activity.

Due to the very high antioxidant potential of anthocyanins, but also the other health benefits, it is necessary to study the factors that influence their bioavailability and implicitly the antioxidant capacity. To this end, the most important parameters that affect the bioavailability of these polyphenols will be highlighted.

3. The Influence of pH

Due to the ionic nature of the molecular structure, the color of anthocyanins is influenced by pH, as this is the first parameter that is approached. As it is known, anthocyanins are found in four different chemical forms, which depend on the pH of the solution. Thus, in an acidic environment, at pH = 1, anthocyanins are found in the form of the flavylium cation (red color) which makes them very soluble in water [12], and this form is also responsible for the production of red and purple colors. When the pH increases between 2–4, the quinoidal blue species is found abundantly, while at a pH between 5–6, carbinol pseudobase and a chalcone appear, compounds that are colorless [38]. Finally, at a pH higher than 7, the anthocyanins will be degraded according to the substituent groups. However, the four forms of anthocyanins can co-exist at a pH range of 4–6, the balance of these forms being maintained by the flavylium cation (Figure 4) [39,40]. Generally speaking, ring B substituents and the presence of additional hydroxyl or methoxyl groups are responsible for the stability of anthocyanins, which in neutral environments decrease aglycone stability. On the other hand, although in a neutral environment, aglycones are not stable, monoglycosides and diglycoside derivatives are more stable under these conditions, because sugar molecules will avoid the degradation of unstable intermediates into phenol and aldehyde acid molecules [15,26]. In short, the stability of anthocyanins increases with increasing methylation and decreases with an increasing number of hydroxyl groups in the B ring of anthocyanidins. At the same time acylation plays an important role in improving the stability of anthocyanins [41].

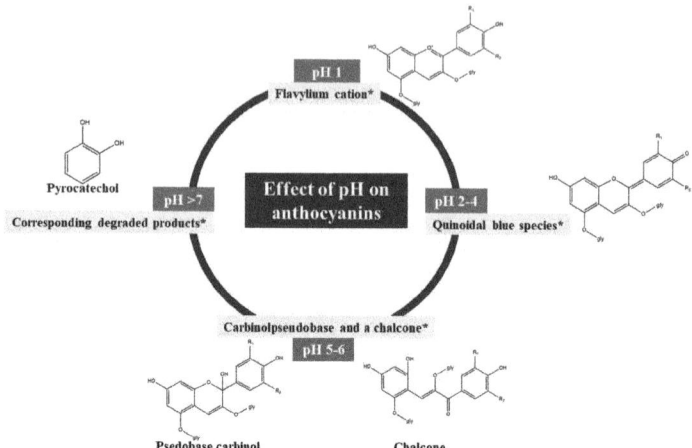

Figure 4. Anthocyanins at different pH values, * predominant chemical form [42].

4. Co-Pigmentation Effect

The following paragraph will focus on another factor that influences the stability of anthocyanins, namely co-pigmentation, this phenomenon is specific to anthocyanins, and cannot be observed in other classes of polyphenols or other non-phenolic compounds [43].

Co-pigmentation is the process by which pigments form molecular or complex associations with other colorless organic compounds or metal ions, thus producing a change or increase in color intensity [44]. For a molecule to act as a co-pigment, it must meet two requirements: to possess sufficiently extended π-conjugated systems, which are supposed to favor π–π stacking interaction, and to have hydrogen bond donor/acceptor groups, such as OH and C=O groups [45]. Thus, the solution of anthocyanins with the co-pigment will have a much more intense color than would theoretically be expected at the pH value of the medium [46]. It should be mentioned that this co-pigmentation phenomenon is pH dependent, because an increase in the pH value will cause the destruction of anthocyanins and, therefore, a decrease in the color intensity of the solution [47]. At the same time, the stability of co-pigmented complexes is influenced by thermal degradation. Therefore, at high temperatures the co-pigmentation mechanism has no visible effects or can even lead to opposite results. For example, co-pigmentation with ferulic acid, sinapic acid or heat treatment at 90 °C for 12 h or 88 °C for 2 min resulted in accelerated degradation of anthocyanins [48]. In addition, the interaction between anthocyanins and co-pigments is different depending on their nature and concentration, thus obtaining different shades and intensities of color [46]. The co-pigments are colorless in a free form, but when mixed with anthocyanins there is an interaction, which results in a hyperchromic effect and a bathochromic shift change in the absorption spectra, as pigment-free anthocyanins absorb light at a certain wavelength, and after co-pigmentation this value increases significantly. As such, the co-pigment anthocyanin complex will have a longer absorption wavelength [49,50]. The nature of these compounds can be extremely varied, therefore, pigments can be alkaloids, amino acids, nucleotides, organic acids, flavonoids, metals or even other anthocyanins [39,51]. The efficiency of the co-pigmentation effect depends on the steric arrangement of the substance (the best co-pigmentation effect occurs in the case of a planar structure) and its size, in order to achieve the stabilization of the flavylium ion [46].

Studies reveal that the main mechanism used to stabilize the color of anthocyanins is co-pigmentation with other substances. This phenomenon is possible due to the electron-rich structure of the co-pigments, so it is associated with flavylium cations that are poor in electrons, thus producing stabilization. Furthermore, this association protects the flavylium ion from the nucleophilic attack of water by generating steric hindrance [45]. The interaction between anthocyanins and co-pigments can be achieved in several ways, depending on the nature of the co-pigment. As such, an association between anthocyanins and colorless co-pigments is called inter-molecular co-pigmentation. An intra-molecular association appears when an interaction exists between the chromophore and nonchromophoric parts of the same molecule. There is also an inter-molecular interaction between anthocyanidin nuclei or between anthocyanidin nuclei and aromatic acyl groups called self-association [10]. If the co-pigment is metallic, a complexation occurs, and there are also more complex cases of co-pigmentation when anthocyanins form compounds at the same time with several molecules such as aglycones, sugars, co-pigments and protons. A special case of co-pigmentation occurs when anthocyanins form an interaction with another phenolic compound, this interaction is transitory due to the lack of chemical bonds. This type of bond is the result of chemical phenomena known as charge–transfer complex formation or π-π interactions and occurs when the interacting compounds are electrically charged differently. Therefore, in our case, because the flavylium ion has a positive charge, it is a good candidate for the formation of complexes by electronic charge transfer with rich electronic substrates [39,45].

4.1. Metallic Interaction

A type of metal co-pigmentation is used in the food industry, but there is an increased risk of contaminating products with metals, leading to their becoming toxic to consumers. However, the use of metals to stabilize the color of anthocyanins could be a successful method as long as the metals used are not a risk to the health of the individual and they are essential minerals in a balanced diet [39]. Thus, in this type of co-pigmentation, the metal cation has the ability to modify the absorption spectrum of the pyrylium ring by affecting the distribution of unlocated electrons. The strongest color effects can be observed when co-pigmentation is conducted with positively charged alkaline earth metals or with poor metals (+2, +3) [13]. Among anthocyanins, the only compounds derived from cyanidin, delphinidin and petunidin are capable of metal chelation, due to the free hydroxyl groups in the B ring. The most common metals that can form complexes with anthocyanins are copper (Cu), iron (Fe), magnesium (Mg), tin (Sn) and potassium (K) [45,52]. In addition, some authors believe that the blue color of plants is the effect of the interaction of anthocyanins and metals. For example, the interaction between anthocyanins and molybdenum is thought to be responsible for the stabilization of the blue color in Hindu cabbage tissue [42]. Current research has found that the interaction between o-di-hydroxyl anthocyanins and Fe (III) or Mg (II) ions placed in a solution with pH = 5 is responsible for the formation of blue color in plants, which is possible if the stoichiometric ratio between anthocyanin and Fe (III) is 1:6 [39,53].

Finally, this process can reduce the concentration of free ions in the cell, thus minimizing the possible harmful effects of metal toxicity on plants. According to research, it has been shown that toxic metals that form complexes with anthocyanins will be sequestered in vacuoles, thus the ions of these metals are accumulated in non-photosynthetic tissues, which are less sensitive and, therefore, important processes in plant growth and development will not be affected. At the same time, they play a role in the correct assimilation of carbon. These properties of anthocyanins have important biotechnological implications as plants that have a large amount of anthocyanins can be used to phyto-remedy soils contaminated with toxic metals [54].

4.2. Self-Association

Anthocyanins can form bonds with other anthocyanins, this type of interaction being called self-association. The obtained complex does not present a degree of association as strong as co-pigmentation, as a higher concentration of anthocyanins is necessary in order to produce and observe this phenomenon. However, there are some stronger self-association interactions than others in the case of neutral species, but they can be destabilized due to the rejection between positively charged flavylium cations and negatively charged anionic bases [55]. In addition, it has been observed that anthocyanin molecules are arranged vertically in chiral (helical) aggregates, with left geometry. Furthermore, an important role in strengthening the structure is played by the hydrogen bonds between the sugars attached to the anthocyanins; however, the position and size of the sugar are responsible for the overall alignment of the complex [46,55]. This process has been observed during the aging of wine and is supposed to play a role in determining the color of aged red wines. However, future research is necessary to discover the mechanism responsible for this process [52].

4.3. Intramolecular and Intermolecular Co-Pigmentation

Intramolecular interactions are performed by stacking the hydrophobic acyl moiety covalently bound to sugar and the flavylium nucleus, thus reducing the hydrolysis of anthocyanins [56], and intermolecular interactions occur due to van der Waals forces between the planar polarizable nucleus of the anthocyanin and colorless co-pigment (Figure 5) [57]. Intermolecular co-pigmentation frequently occurs in red wines, due to the already mentioned van der Waals interactions, but also due to hydrophobic effects and hydrogen bonds, which achieve a non-covalent interaction between anthocyanin molecules and other molecules (co-pigment). Thus, these interactions impart the wines with more shades of purple and a darker color, through the two occurring actions, namely, the hyperchromic effect and the bathochromic one. At the same time, it is assumed that this color change is the result of changing the conformation of the anthocyanin chromatophore, but also avoiding the hydration reaction, thus determining an increase in the proportion between the flavylium cation (red color) and the quinonoidal base (purple color) [58].

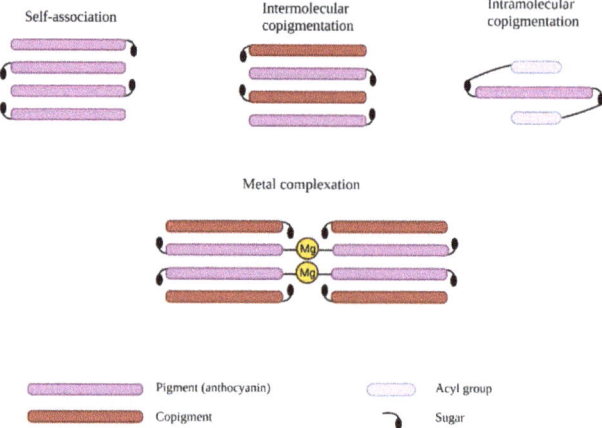

Figure 5. Types of anthocyanin co-pigmentation.

This interaction plays an important role in preventing whitening during light and heat treatments, which is necessary to obtain certain products [59]. Additionally, inter-molecular co-pigmentation plays an important role in the case of anthocyanin–metal complexes that we have already discussed and which are a special case [49], but unlike metal complexation, intermolecular interaction occurs in all major anthocyanidins [52].

Finally, intermolecular co-pigmentation took place in vacuolar fluids but also in extra vacuolar ones, and this is the main mechanism responsible for the multitude of colors that appear in flowers, considering the limited number of existing anthocyanins [46].

On the other hand, intra-molecular co-pigmentation requires an interaction between the anthocyanidin backbone and a co-pigment forming part of the anthocyanin itself. This phenomenon occurs due to the alignment of the aromatic acyl fragments of the acylated anthocyanins and the anthocyanidin core structure, the structural characteristics of anthocyanins are the main determinants of the effectiveness of this interaction. In addition, the sugar from the anthocyanin structure acts as a spacer between the anthocyanin skeleton and the acyl moiety, thus allowing the molecule to fold in a way that the aromatic acyl group will interact with the π-system of the planar pyrylium ring protecting it against the nucleophilic water attack [43,46]. Due to the formation of sandwich structures (the anthocyanin core will be incorporated between two acyl portions), an efficient stabilization of anthocyanins pigments will be achieved for those having two or more aromatic acyl portions necessary to achieve this complex [43].

As anthocyanins from flowers are most often found in acylate form, intramolecular interactions occur that stabilize the color of the flowers even at neutral pH values. This phenomenon takes place not only in flowers but also in the case of other edible plants that have a high content of acylated anthocyanins, for example, black carrot [52]. Thus, it is assumed that due to the bonds formed in intramolecular co-pigmentation, this interaction is more efficient and strong than intermolecular co-pigmentation in stabilizing the color of anthocyanins [52]. Finally, due to the molecular interactions between anthocyanins and other uncolored phenolic components, it has been shown that all types of co-pigmentation reactions have a beneficial effect on improving the stability of anthocyanins [50].

5. Temperature

As is well known, heat processing is one of the most commonly used methods for preserving and extending the shelf life of food and ensuring food safety. Thus, depending on the functional parameters of the food and the shelf life desired by the manufacturer, heat treatment can take place at temperatures between 50–180 °C. Due to this processing under high temperature for certain periods of time, foods can undergo changes in color, the amount of anthocyanins but also their antioxidant capacity [25]. Due to heat processing, anthocyanins can undergo a multitude of mechanisms such as glycosylation, nucleophilic attack of water, cleavage and polymerization that will cause the loss of this pigment and their degradation [57]. Therefore, temperature is another factor that affects the stability of the molecular structure of anthocyanins, so with increasing temperature the degradation of these compounds occurs [60] and determines the browning of products in the presence of oxygen [42]. As such, both the color intensity determined by monomeric anthocyanins and their amount decreases depending on time/temperature, while the amount of brown pigments/polymer fraction increased. It has been proven that in all food pigments, including anthocyanins, stability decreases with increasing temperature. At the same time, current research has reported that the stability of these compounds is closely related to their chemical structure, where the sugar fraction is an important factor [61].

According to a study by Turker and colleagues in 2004, anthocyanins stored at different temperatures, in an acylated form, have greater stability than nonacylated anthocyanins [62]. Another study that analyses the influence of storage temperature on anthocyanin content and their half-life was performed by Hellström and colleagues in 2013. The current study shows how storage temperature affects the stability of anthocyanins, one of the most important quality criteria in berry juices. Therefore, the half-life ($t_{\frac{1}{2}}$) of anthocyanins was considerably shorter at room temperature than in cold storage in all juices examined. In this respect, in the long-term efficient conservation of anthocyanins, it is recommended to avoid storing them at room temperature [63].

On the other hand, research on grape extracts reported that heat treatment at 35 °C reduced the total anthocyanin content to less than half, compared to the same grape extract that was subjected to a heat treatment of only 25 °C. In addition, the color of anthocyanins changed from red to orange up to temperatures of 40 °C, a phenomenon observed in an environment with acidic pH, which did not influence this change [64].

Thermal processes that require high temperatures such as bleaching and pasteurization at 95 °C degrees Celsius for 3 min to obtain blueberry puree, caused a loss of 43% of total monomeric anthocyanins, compared to the amount observed in fresh fruit before heat treatment. It should be noted that while the amount of anthocyanins decreased, the values of polymer colors increased from 1% to 12%. This finding implies that heat labile parameters can promote anthocyanin pigment degradation, and it backs up the theory that pigment destruction in juice processing is caused by endogenous enzymes in fruits [65]. Although, so far, it has been shown that heat treatment has a negative effect on the amount of anthocyanins in fruits and vegetables, it can still have a beneficial effect. Polyphenols, so implicitly anthocyanins, are enzymatically degraded by polyphenol oxidase, but this enzyme can be thermally inactivated [65,66]. As a consequence, in the food processing sector, mild heat treatment of raw materials, such as blanching, can reduce anthocyanin oxidation by polyphenol oxidase [12]. Therefore, a short-term heat treatment has been shown to improve the stability of anthocyanins by inhibiting native enzymes that are harmful to anthocyanins [66,67]. At the same time, thermal degradation can be prevented by decreasing the pH value of the anthocyanin solution. Furthermore, the decrease in oxygen concentration, a parameter that will be discussed further, has been shown to protect anthocyanins from thermal degradation [52,68].

6. Oxygen

Due to the unsaturated chemical structure of anthocyanins, these compounds are susceptible to reaction with molecular oxygen [69]. Therefore, oxygen is another important factor that influences the stability of anthocyanins, having a role in their degradation, as the presence of oxygen can accelerate the degradation process of anthocyanins in two ways: by a direct oxidative mechanism or by the action of oxidizing enzymes [65]. This factor causes a detrimental effect on anthocyanins and numerous studies have reported that the stability of these compounds increases when stored under vacuum, nitrogen or argon, compared to their storage in an atmosphere containing predominantly oxygen. Therefore, the concentration of anthocyanin dropped in all of the atmospheres studied; where high oxygen produced a greater decline [61]. As mentioned in the previous chapter, removing oxygen from the anthocyanin solution prevents thermal degradation. In fact, high temperature and the presence of oxygen have proven to be the most harmful combination of all the factors that influence the stability of these compounds [70]. For many years, it has been known that filling bottles completely with hot grape juice may postpone the color degeneration from purple to dull brown. Other juices containing anthocyanins have yielded similar results [69]. In addition, because anthocyanins react with oxygen radicals, such as peroxyradicals, they have an antioxidant character, rendering these compounds beneficial against cardiovascular diseases [52,71].

On the other hand, some investigators concluded that if food is stored in an environment enriched with 60–100% oxygen and at a low temperature, during the onset (0–7 days) of cold storage, the content of phenols and anthocyanins will increase. This effect, however, reduces with time spent in storage [70].

7. Ascorbic Acid

Ascorbic acid is recognized as performing a significant role as an antioxidant in the human body [72]. This compound is crucial in food processing because any change in vitamin C level shows that the quality of the food has deteriorated after processing and during storage [73]. Ascorbic acid is found in various fruits and vegetables, and, in addition, it is added as an antioxidant in many types of foods to increase their nutritional value. It is also known that the presence of ascorbic acid in an environment containing anthocyanins causes their faster degradation and loss of color, thus suggesting a direct interaction between the two molecules [69,74]. At the same time, the presence of oxygen in the environment will favor faster degradation of anthocyanins by ascorbic acid, thus determining the formation of the polymeric pigment and the whitening of the anthocyanin pigment. The specific process of degradation is yet unknown; however, adding ascorbic acid to anthocyanins increases the rate of decomposition of both molecules. The presumed mechanisms by which this phenomenon would occur are the direct condensation of ascorbic acid with anthocyanins, or the formation of hydrogen peroxide and the oxidative cleavage of the pyrylium ring by peroxides [75]. The loss of anthocyanin pigments by the second mechanism, namely the oxidative cleavage of the pyrylium ring, is due to the ability of anthocyanins to act as a molecular oxygen activator, the result of this reaction being free radicals [76].

On the other hand, electrophilic molecules, such as bisulphites, hydrogen peroxide, and ascorbic acid are believed to target anthocyanin nucleophilic sites. It has been proposed that ascorbic acid causes reciprocal and irreversible degradation of both pigment and micronutrients. This process is distinct from bisulfited bleaching, which is reversible and pH sensitive. Therefore, this phenomenon is a significant barrier to the use of anthocyanins-based colorants in the food sector, particularly in juices and drinks that are frequently fortified with vitamin C [74]. However, the effect of ascorbic acid on anthocyanins is a complex one, depending on several factors but also on the matrix [67]. For example, in an environment with H_2O_2, concentrations between 60–80 mg/L of ascorbic acid favored the stabilization of anthocyanins in pomegranate juice, while in cherry juice a concentration of 80 mg/L produced rapid degradation of anthocyanins [77]. Therefore, when both oxygen and ascorbic acid levels are high, the degrading impact is most evident. Copper ions are known to speed up the processes [61].

8. Light

Plants are exposed to light, which stimulates anthocyanin synthesis and accumulation, hence, light is another important parameter in the stability of these compounds [69]. However, anthocyanins are affected by light in two ways. Light is required for anthocyanins production, but it also speeds up their breakdown [70]. In addition, the amount of molecular oxygen present affects the rate of light-induced breakdown. It should be mentioned that when pigments are exposed to fluorescent light, the most intense anthocyanin loss occurs [61]. These compounds are responsible for determining the colors orange, red and blue in many plants, such as grapes and berries, as anthocyanins are good absorbers of visible light. Therefore, the color is mainly influenced by the aglycon's B-ring substitution pattern, as opposed to the flavan structure glycosylation pattern, which impacts color production to a smaller extent [61]. Over the years, studies have been conducted on the impact of light on the stability of anthocyanin extracts from various biological sources. Thus, in the case of grape juice that was stored at a 20 °C temperature in the dark, it was observed that about 30% of the anthocyanin amount was destroyed. Nevertheless, placing the identical samples in the presence of light at the same temperature and duration removed nearly half of the total pigments [78].

Light, in addition to the above-mentioned information, also affects the antioxidant activity of anthocyanins. Thus, a significant decrease in anthocyanin content and antioxidant activity was observed in mulberry fruit extracts that were stored at room temperature and exposed to fluorescent light for 10 h. Thus, the longer the exposure of the extract to light, the more the anthocyanin content and their antioxidant activity will decrease [79].

In order to minimize the negative impact of light on anthocyanins, packaging can be made of materials that can block light from the visible spectrum and especially from the ultraviolet field of the spectrum, thus creating a protective barrier. In addition, glycosylation, acylation and co-pigmentation play an important role in increasing the stability of anthocyanins to light [61].

9. Sulfites

Although widely used in the storage of fruit derived foods, sulfites and sulfates can cause the loss of anthocyanin pigment. This discoloration is caused by forming colorless sulfur derivative structures when these sulfur compounds are added in positions 2 or 4 [70]. A quantity of sulfur dioxide will be released from the anthocyanins when heated, and it is in this way that the specific color of these compounds can be partially recovered. Acidification to a low pH also regenerates anthocyanins by releasing SO_2. For instance, sulfur dioxide is widely used in the fruit and vegetable preservation industry, as an inhibitor of microbial growth and enzymatic and non-enzymatic browning [61]. On the other hand, high sulfite concentrations (more than 10 g/kg) trigger permanent anthocyanin degradation [70].

Sulfur dioxide has been used for food preservation since the 18th century, and has been shown to be an excellent choice for low-pH foods such as wine [80]. Therefore, SO_2 is intensely used in the wine industry, due to its antiseptic properties against yeasts and bacteria and its antioxidant properties, making it one of the most versatile and efficient additives [81,82]. Its molecular form may cross the cell membrane of bacteria, disrupting the action of cell enzymes and proteins, and thus regulating microbial development [80].

These properties of sulfur dioxide play a role in stopping unwanted fermentations, such as malolactic and acetic fermentation, and limit the activity of oxidase that is endogenous in grapes or may come from fungal infection. In addition, SO_2 causes pigment bleaching and eliminates unpleasant odors that are obtained from oxidation [81].

Furthermore, it acts as a solvent during winemaking, allowing grape solid components such as stems, seeds and skins, to be extracted. Additionally, SO_2 is utilized in another three stages of wine making. First, it is used during the prefermentation process, in the grapes or must, with the primary goal of avoiding oxidation. However, this phenomenon does not occur by direct elimination of oxygen from edibles, but by binding to the precursors involved in oxidative reactions and compounds resulting from oxidation [80,82]. Second, after the fermentation procedures are ended and before the aging or storage phases, it is used to limit microbial development that might affect the wines, and last, shortly before bottling, the wines are stabilized with SO_2 to avoid any changes or accidents in the bottles [82].

Therefore, it can be said that this treatment with sulphur dioxide is one of great importance in wine technology. However, the use of this agent is strictly controlled, because high doses of SO_2 can cause organoleptic changes of the final product (such as unwanted flavors in sulphur gas or reducing products, hydrosulphate and mercaptans). At the same time, the ingestion of this compound produces negative effects on human health [82]. In a number of individuals, it causes a variety of symptoms associated with allergic responses. As a result, the maximum SO_2 content in wines authorized by law has gradually been decreased to 150 and 200 mg/L for red and white wines, respectively. Thus, at present minimizing or eliminating the use of SO_2 in vinification is being sought, through research on innovative technologies (Figure 6) [80].

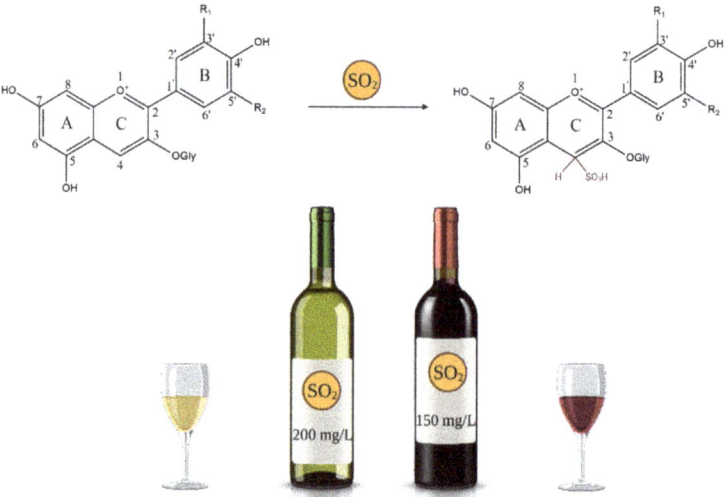

Figure 6. Content of SO_2 in red and white wines, and anthocyanin reaction with SO_2.

10. Enzymes

The most common enzymes that degrade anthocyanins are glycosidases, peroxidases (phenol oxidases) and phenolases (polyphenol oxidases) [42]. The collective name for these enzymes is anthocyanases [69]. These enzymes may be produced by the plant and are present in its tissues, or may occur as an effect of microbial contamination. It is worth noting that glycosidases directly affect anthocyanins, while peroxidases and phenolases have indirect effects on their stability [83].

Therefore, glycosidases break the covalent bond between the glycosyl residue and the aglycone of an anthocyanin, and thus unstable anthocyanidin are formed [42]. This process further affects the color of the compound [83]. On the one hand, polyphenol oxidase and peroxidase, which are plant-specific enzymes, once removed from cellular compartments during anthocyanin extraction, will accelerate the degradation of these compounds [67]. On the other hand, polyphenol oxidase is the enzyme that will catalyze two types of reactions: o-hydroxylation of monophenols in o-diphenols and oxidation of o-diphenols in o-quinones in the presence of molecular oxygen. Quinones are highly reactive electrophilic compounds that can covalently alter nucleophiles like anthocyanins to produce brown pigments known as melanin. Although anthocyanins are weak substrates for this enzyme, polyphenol oxidase reacts directly with the compounds, causing their degradation by a cooxidation reagent of enzymatically generated oquinones and/or secondary oxidation products of quinone [66]. It is assumed that the first enzyme that affects the stability of anthocyanins is β-glucosidase, which forms anthocyanidins that can be further oxidized by polyphenol oxidase and/or peroxidase [83]. Following the action of these enzymes, the solubility of anthocyanins decreases and their transformation into colorless compounds takes place, thus losing the color intensity of these pigments [69].

11. Encapsulation of Anthocyanins

The first step required before encapsulating anthocyanins in various systems is their extraction from natural sources. Thus, solvents that do not interfere with physical and antioxidant properties of anthocyanins are used. For the food applications of these compounds small organic alcohols are recommended, because they can efficiently permeate plant tissue, exert great affinity, and have a high solubility for anthocyanins. The most used and efficient solvents for the extraction of anthocyanins are acidified organic solvents, such as water, acetone, ethanol and methanol. Their acidification is achieved with weaker organic acids, such as acetic acid, but hydrochloric acid can also be used, with the role of improving the stability of anthocyanins [84].

As already mentioned, anthocyanins have multiple health benefits, but their effectiveness is limited by their low stability, which is influenced by the factors previously described such as pH, temperature, light, oxygen and others. Furthermore, due to their low absorption and rapid metabolization in the organism, these compounds have very low bioavailability and the best strategy to overcome these disadvantages is to create anthocyanin-encapsulation systems [85,86].

Encapsulation can be defined as the process by which a bioactive agent of nature: solid, liquid or gaseous is covered by a polymeric material or introduced into a matrix in order to protect it from harmful environmental factors. Through this process, certain parameters can be controlled such as place and time of delivery for maximum efficiency [85,87]. The size of vesicles produced by this method varies, ranging from 1 to 1000 μm [88].

In recent years, a great number of techniques for anthocyanin encapsulation have been developed and optimized in micro- and nanocarriers. There are several methods of obtaining compounds that contain anthocyanins, the most common are spray-drying, freeze-drying or electro-spinning/spraying. At the same time, lipid-based carriers can be obtained for anthocyanins, such as liposomes and emulsions [85]. Some of the following methods of encapsulating anthocyanins and their characteristics are of paramount importance and are presented in Figure 7.

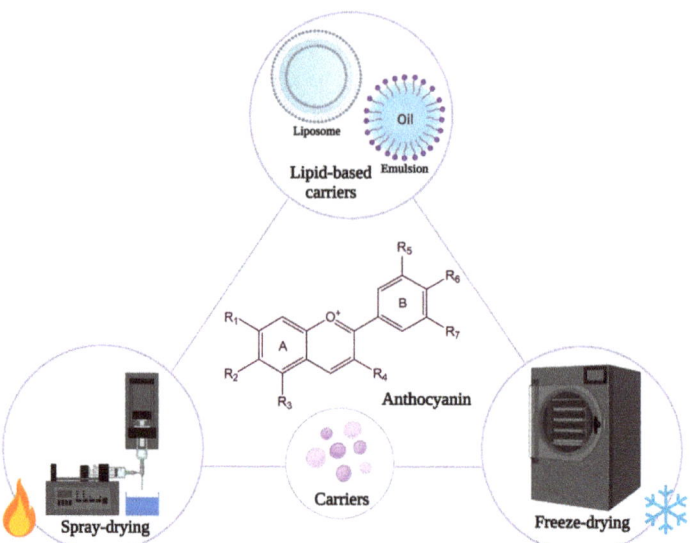

Figure 7. Methods for anthocyanins encapsulation.

11.1. Spray-Drying

According to studies, spray-drying is the oldest and the most frequent method used for microencapsulating anthocyanins. For the encapsulation of anthocyanins by this technique, certain steps are necessary: first the fluid feed is atomized through a high-pressure nozzle into a drying chamber, into a solution, emulsion or suspension. The next step involves the evaporation of the solvent used by heating the environment where the sprayed drops are found to 150–220°. As the final step, a filter or cyclone is used to separate and recover the powdered product from the air [85,88].

This method is quick, adaptable, cost-effective and simple to scale up, with high encapsulation efficiency and relatively good storage stability [88]. In addition, various types of encapsulating agents can be used such as lipids, proteins and polysaccharides depending on the type of incorporated compound [89]. A crucial role for the efficiency of encapsulating compounds by this method, is the choice of a suitable wall material for microencapsulation by spray drying. However, the range of compounds that can be chosen for this technique is limited, because the agent must meet certain criteria such as: low viscosity at high concentrations, acceptable solubility, film-forming capacity and emulsifying properties [88]. Bearing these criteria in mind, the most used compounds are polysaccharides for spray-drying encapsulation of anthocyanins, because they have low viscosity, desirable solubility and adequate emulsification characteristics, but also high capacity to retain volatile compounds. Therefore, they are widely used in microencapsulation of polyphenolic compounds [85].

Following this technique, non-uniform particles of various sizes are obtained, so in spray-drying the size and morphology of the carries cannot be controlled. To this drawback is added the fact that the high temperatures used in the process can degrade anthocyanins [88].

11.2. Freeze-Drying

This technique uses an opposite procedure to spray-drying, using low temperatures for dehydration. Therefore, freeze-drying, also known as lyophilization due to the low temperatures used, is suitable for encapsulating high temperature sensitive compounds, such as anthocyanins. [88,89]. The steps of this technique include freezing, sublimation (primary drying), desorption (secondary drying) and finally the storage of the resulting dry material [90].

The advantages of freeze-drying rely on the fact that it is a simple process, which takes place in the absence of air and at a low temperature and, as a result, the obtained compounds are resistant to oxidation or chemical modification [90]. This technique is used to improve thermal and color stability especially for anthocyanins, using different wall materials [89]. However, the freeze-drying method also has disadvantages such as high costs due to the vacuum technology needed, and a long period of time for dehydration, about 20 h [89,90].

11.3. Liposomes

Liposomes are phospholipid vesicles of a spherical shape composed of one or more concentric lipid bilayers comprising an aqueous space [86]. In recent years, methods and systems for obtaining liposomes have advanced considerably, but there are several basic steps in this process. The first step is lipid drying from organic solvents, then dispersion of lipids in an aqueous medium. Afterwards, liposomes need purification, and are followed by post-processing steps such as sonication or extrusion [91]. Thus, depending on the method of production, the compounds used and the environmental conditions, liposomes can have different sizes and structures [90].

Due to their good ability to encapsulate and protect hydrophilic substances, liposomes are used to encapsulate anthocyanins. Therefore, by incorporating these polyphenols, they are protected from degrading environmental factors. At the same time, liposomes can increase absorption and bioavailability of anthocyanins due to their biocompatibility, amphiphilicity, nontoxicity and non-immunogenicity, [88,89]. All these advantages have contributed to the growing interest in these vesicles, having versatile applications in the biomedical, food and cosmetic industries [89].

Despite the many advantages presented, they can undergo oxidation processes due to unsaturated fatty acids in the membrane component, leading to the formation of hydroperoxides. The high costs of manufacturing and the fact that there is no standardization of methods of production, require future research and add to the estimated shortcomings [88,89].

11.4. Emulsion

Emulsions consist of at least two immiscible liquids, most commonly water and oil, and surfactants, in which one of the liquids is dispersed in small spherical droplets into the other. They can be found in different forms depending on the liquid that is dispersed, so we have simple emulsions such as oil in water (O/W) or water in oil (W/O), and double emulsions, oil in water in oil (O/W/O) or water in oil in water (W/O/W) [85,89].

Likewise, a double emulsion of water in oil in water (W/O/W) is a suitable method for the encapsulation of hydrophilic substances, such as anthocyanins [88]. As such, anthocyanins can easily be incorporated into the dispersed phase of an emulsion, thus protecting them from harmful environmental factors and against degradation. At the same time, this encapsulation system controls the release of anthocyanins at a certain time or place and in addition improves their bioavailability in the gastrointestinal tract [89].

Although they manage to effectively encapsulate anthocyanins, the main disadvantage of emulsions is that they are thermodynamically unstable systems, and tend to break down with time [89].

12. Conclusions and Future Perspectives

The aim of the present review has been to gather the scientific information of the antioxidant capacity of anthocyanins and its evaluation methods, and the main parameters that influence anthocyanin stability and encapsulation methods. There is a wide variety of factors that can affect their stability and, implicitly, their bioavailability. Since anthocyanins are the most common and widely used category of water-soluble natural colors, a large number of studies have been conducted to improve their stability, in order for them to be used in different fields, from the food industry to the cosmetics industry. Their main use is as food dyes, as they impart a wide range of colors and can replace widely-used synthetic dyes nowadays. The latter can exhibit toxicity and adverse health effects. In addition to their main use, as pigments, anthocyanins also have important antioxidant, anticancer, antitumor, antimutagenic and antidiabetic characteristics. To get the most out of these anthocyanin properties, it is important to know how their stability is influenced by pH, temperature, co-pigmentation, oxygen, ascorbic acid, light, sulfites or enzymes.

In order to extend their efficiency, new directions of research involve the incorporation of anthocyanins into targeted delivery systems, and, therefore, the load of the compound is projected by the external environment that could influence their stability. Thus, various systems such as liposomes, microcapsules or emulsions have been developed to increase their bioavailability, so, implicitly the effects produced by anthocyanins would be stronger. However, future studies are needed to focus on the action of each factor that influences stability and how to reduce the negative impact they have on it.

Author Contributions: Conceptualization, Z.D.; writing—original draft preparation, B.E.; writing—review and editing, G.D., T.D.P. and A.S.; funding acquisition, Z.D. All authors have read and agreed to the published version of the manuscript.

Funding: This research was funded by the Romanian National Authority for Scientific Research (UEFISCDI), grant numbers PN-III-P1-1.1-TE-2019-0960/178TE/2020.

Institutional Review Board Statement: Not applicable.

Informed Consent Statement: Not applicable.

Acknowledgments: We would like to thank Mihaela Mihai for English language editing.

Conflicts of Interest: The authors declare no conflict of interest.

Abbreviations

4CL	4-coumaroyl CoA ligase
AAPH	2,2′-azobis(2-amidinopropane) dihydrochloride
ABAP	2,2-azobis(2-amidinopropane) chlorhydrate
ABTS$^{\bullet+}$	2,2-Azinobis 3-ethylbenzthiazoline-6-sulfonic acid radical scavenging
AH$^+$	Hydrogen donor
AIBN	$\alpha, \alpha,$-azobisizobutyronytril
AMVN	2,2′-azobis(2,4-dimethylvaleronytril)
ANCs	Anthocyanins
ArOH	Antioxidant agent
C4H	Cinnamate 4-hydroxylase
CUPRAC	Cupric ions reducing antioxidant capacity
DPPH$^{\bullet}$	2,2-Diphenyl-1-picrylhydrazyl radical scavenging
FRAC	Ferric ion reducing antioxidant power
H_2O_2	Hydrogen peroxide
HAT	Hydrogen atom transfer
HRP	Horseradish peroxidase
O/W	Oil in water emulsion
O/W/O	Oil in water in oil double emulsion
ORAC	Oxygen radical absorbance capacity
RNS	Reactive nitrogen species
ROS	Reactive oxygen species
SET	Single electron transfer
TPTZ	Tripyridyltriazine
W/O	Water in oil emulsion
W/O/W	Water in oil in water double emulsion

References

1. Liu, P.; Li, W.; Hu, Z.; Qin, X.; Liu, G. Isolation, purification, identification, and stability of anthocyanins from Lycium ruthenicum Murr. *LWT* **2020**, *126*, 109334. [CrossRef]
2. Williams, C.A.; Grayer, R.J. Anthocyanins and other flavonoids. *Nat. Prod. Rep.* **2004**, *21*, 539–573. [CrossRef] [PubMed]
3. Sinopoli, A.; Calogero, G.; Bartolotta, A. Computational aspects of anthocyanidins and anthocyanins: A review. *Food Chem.* **2019**, *297*, 124898. [CrossRef] [PubMed]
4. Šamec, D.; Karalija, E.; Šola, I.; Vujčić Bok, V.; Salopek-Sondi, B. The Role of Polyphenols in Abiotic Stress Response: The Influence of Molecular Structure. *Plants* **2021**, *10*, 118. [CrossRef] [PubMed]
5. Brooks, M.S.-L.; Celli, G.B. *Anthocyanins from Natural Sources: Exploiting Targeted Delivery for Improved Health*; Royal Society of Chemistry: Londone, UK, 2019.
6. Oancea, S.; Draghici, O. pH and thermal stability of anthocyanin-based optimised extracts of Romanian red onion cultivars. *Czech J. Food Sci.* **2013**, *31*, 283–291. [CrossRef]
7. Diaconeasa, Z.; Ayvaz, H.; Rugină, D.; Leopold, L.F.; Stănilă, A.; Socaciu, C.; Tăbăran, F.; Luput, L.; Mada, D.C.; Pintea, A.; et al. Melanoma Inhibition by Anthocyanins Is Associated with the Reduction of Oxidative Stress Biomarkers and Changes in Mitochondrial Membrane Potential. *Plant Foods Hum. Nutr.* **2017**, *72*, 404–410. [CrossRef]
8. Bakowska-Barczak, A. Acylated Anthocyanins as Stable, Natural Food Colorants: A Review. *Pol. J. Food Nutr. Sci.* **2005**, *14*, 107–116.
9. Saha, S.; Singh, J.; Paul, A.; Sarkar, R.; Khan, Z.; Banerjee, K. Anthocyanin Profiling Using UV-Vis Spectroscopy and Liquid Chromatography Mass Spectrometry. *J. AOAC Int.* **2020**, *103*, 23–39. [CrossRef]
10. Fernandes, A.; Brás, N.F.; Mateus, N.; de Freitas, V. A study of anthocyanin self-association by NMR spectroscopy. *New J. Chem.* **2015**, *39*, 2602–2611. [CrossRef]
11. Wallace, T.C.; Giusti, M.M. Anthocyanins—Nature's Bold, Beautiful, and Health-Promoting Colors. *Foods* **2019**, *8*, 550. [CrossRef]
12. Khoo, H.E.; Azlan, A.; Tang, S.T.; Lim, S.M. Anthocyanidins and anthocyanins: Colored pigments as food, pharmaceutical ingredients, and the potential health benefits. *Food Nutr. Res.* **2017**, *61*, 1361779. [CrossRef]
13. Morata, A.; López, C.; Tesfaye, W.; González, C.; Escott, C. Anthocyanins as Natural Pigments in Beverages. In *Value-Added Ingredients and Enrichments of Beverages*; Academic Press: Cambridge, MA, USA, 2019; Volume 14, pp. 383–428. [CrossRef]
14. Aura, A.-M.; Martin-Lopez, P.; O'Leary, K.A.; Williamson, G.; Oksman-Caldentey, K.-M.; Poutanen, K.; Santos-Buelga, C. In vitro metabolism of anthocyanins by human gut microflora. *Eur. J. Nutr.* **2005**, *44*, 133–142. [CrossRef]
15. Fleschhut, J.; Kratzer, F.; Rechkemmer, G.; Kulling, S.E. Stability and biotransformation of various dietary anthocyanins in vitro. *Eur. J. Nutr.* **2006**, *45*, 7–18. [CrossRef]
16. Yousuf, B.; Gul, K.; Wani, A.A.; Singh, P. Health Benefits of Anthocyanins and Their Encapsulation for Potential Use in Food Systems: A Review. *Crit. Rev. Food Sci. Nutr.* **2015**, *56*, 2223–2230. [CrossRef] [PubMed]
17. Bendokas, V.; Skemiene, K.; Trumbeckaite, S.; Stanys, V.; Passamonti, S.; Borutaite, V.; Liobikas, J. Anthocyanins: From plant pigments to health benefits at mitochondrial level. *Crit. Rev. Food Sci. Nutr.* **2019**, *60*, 3352–3365. [CrossRef] [PubMed]
18. Flamini, R.; Mattivi, F.; De Rosso, M.; Arapitsas, P.; Bavaresco, L. Advanced Knowledge of Three Important Classes of Grape Phenolics: Anthocyanins, Stilbenes and Flavonols. *Int. J. Mol. Sci.* **2013**, *14*, 19651–19669. [CrossRef]
19. Belwal, T.; Singh, G.; Jeandet, P.; Pandey, A.; Giri, L.; Ramola, S.; Bhatt, I.D.; Venskutonis, P.R.; Georgiev, M.; Clément, C.; et al. Anthocyanins, multi-functional natural products of industrial relevance: Recent biotechnological advances. *Biotechnol. Adv.* **2020**, *43*, 107600. [CrossRef]
20. Mattioli, R.; Francioso, A.; Mosca, L.; Silva, P. Anthocyanins: A Comprehensive Review of Their Chemical Properties and Health Effects on Cardiovascular and Neurodegenerative Diseases. *Molecules* **2020**, *25*, 3809. [CrossRef]
21. Torskangerpoll, K.; Andersen, Ø.M. Colour stability of anthocyanins in aqueous solutions at various pH values. *Food Chem.* **2005**, *89*, 427–440. [CrossRef]
22. Yang, P.; Yuan, C.; Wang, H.; Han, F.; Liu, Y.; Wang, L.; Liu, Y. Stability of Anthocyanins and Their Degradation Products from Cabernet Sauvignon Red Wine under Gastrointestinal pH and Temperature Conditions. *Molecules* **2018**, *23*, 354. [CrossRef] [PubMed]
23. Zhao, X.; Sheng, F.; Zheng, J.; Liu, R. Composition and Stability of Anthocyanins from Purple Solanum tuberosum and Their Protective Influence on Cr(VI) Targeted to Bovine Serum Albumin. *J. Agric. Food Chem.* **2011**, *59*, 7902–7909. [CrossRef] [PubMed]
24. Sharma, R.J.; Gupta, R.C.; Singh, S.; Bansal, A.K.; Singh, I.P. Stability of anthocyanins- and anthocyanidins-enriched extracts, and formulations of fruit pulp of Eugenia jambolana ('jamun'). *Food Chem.* **2016**, *190*, 808–817. [CrossRef]
25. Aprodu, I.; Milea, Ş.A.; Enachi, E.; Râpeanu, G.; Bahrim, G.; Stănciuc, N. Thermal Degradation Kinetics of Anthocyanins Extracted from Purple Maize Flour Extract and the Effect of Heating on Selected Biological Functionality. *Foods* **2020**, *9*, 1593. [CrossRef]
26. Weber, F.; Larsen, L.R. Influence of fruit juice processing on anthocyanin stability. *Food Res. Int.* **2017**, *100*, 354–365. [CrossRef]
27. Smeriglio, A.; Barreca, D.; Bellocco, E.; Trombetta, D. Chemistry, Pharmacology and Health Benefits of Anthocyanins: Anthocyanins and Human Health. *Phytother. Res.* **2016**, *30*, 1265–1286. [CrossRef]
28. Brainina, K.; Stozhko, N.; Vidrevich, M. Antioxidants: Terminology, Methods, and Future Considerations. *Antioxidants* **2019**, *8*, 297. [CrossRef]
29. Santos-Sánchez, N.F.; Salas-Coronado, R.; Villanueva-Cañongo, C.; Hernández-Carlos, B. *Antioxidant Compounds and Their Antioxidant Mechanism*; IntechOpen: London, UK, 2019.

30. Gulcin, İ. Antioxidants and antioxidant methods: An updated overview. *Arch. Toxicol.* **2020**, *94*, 651–715. [CrossRef] [PubMed]
31. Pisoschi, A.M.; Negulescu, G.P. Methods for Total Antioxidant Activity Determination: A Review. *Biochem. Anal. Biochem.* **2011**, *1*, 106. [CrossRef]
32. Tena, N.; Martín, J.; Asuero, A.G. State of the Art of Anthocyanins: Antioxidant Activity, Sources, Bioavailability, and Therapeutic Effect in Human Health. *Antioxidants* **2020**, *9*, 451. [CrossRef]
33. Munteanu, I.G.; Apetrei, C. Analytical Methods Used in Determining Antioxidant Activity: A Review. *Int. J. Mol. Sci.* **2021**, *22*, 3380. [CrossRef]
34. Ma, X.; Li, H.; Dong, J.; Qian, W. Determination of hydrogen peroxide scavenging activity of phenolic acids by employing gold nanoshells precursor composites as nanoprobes. *Food Chem.* **2011**, *126*, 698–704. [CrossRef]
35. Haida, Z.; Hakiman, M. A comprehensive review on the determination of enzymatic assay and nonenzymatic antioxidant activities. *Food Sci. Nutr.* **2019**, *7*, 1555–1563. [CrossRef]
36. Fernando, C.D.; Soysa, P. Optimized enzymatic colorimetric assay for determination of hydrogen peroxide (H_2O_2) scavenging activity of plant extracts. *MethodsX* **2015**, *2*, 283–291. [CrossRef]
37. Bibi Sadeer, N.; Montesano, D.; Albrizio, S.; Zengin, G.; Mahomoodally, M.F. The Versatility of Antioxidant Assays in Food Science and Safety—Chemistry, Applications, Strengths, and Limitations. *Antioxidants* **2020**, *9*, 709. [CrossRef] [PubMed]
38. Kang, H.-J.; Ko, M.-J.; Chung, M.-S. Anthocyanin Structure and pH Dependent Extraction Characteristics from Blueberries (*Vaccinium corymbosum*) and Chokeberries (*Aronia melanocarpa*) in Subcritical Water State. *Foods* **2021**, *10*, 527. [CrossRef]
39. Castañeda-Ovando, A.; Pacheco-Hernández, M.D.L.; Páez-Hernández, M.E.; Rodríguez, J.A.; Galán-Vidal, C.A. Chemical studies of anthocyanins: A review. *Food Chem.* **2009**, *113*, 859–871. [CrossRef]
40. Pina, F.; Melo, M.J.; Laia, C.A.T.; Parola, A.J.; Lima, J.C. Chemistry and applications of flavylium compounds: A handful of colours. *Chem. Soc. Rev.* **2011**, *41*, 869–908. [CrossRef]
41. Fei, P.; Zeng, F.; Zheng, S.; Chen, Q.; Hu, Y.; Cai, J. Acylation of blueberry anthocyanins with maleic acid: Improvement of the stability and its application potential in intelligent color indicator packing materials. *Dye. Pigment.* **2020**, *184*, 108852. [CrossRef]
42. Riaz, M.; Zia-Ul-Haq, M.; Saad, B. Biosynthesis and Stability of Anthocyanins. In *Anthocyanins and Human Health: Biomolecular and Therapeutic Aspects*; Springer: Cham, Switzerland, 2016; pp. 71–86. [CrossRef]
43. Kammerer, D.R. Anthocyanins. In *Handbook on Natural Pigments in Food and Beverages*; Elsevier: Amsterdam, The Netherlands, 2016; pp. 61–80. [CrossRef]
44. Boulton, R. The Copigmentation of Anthocyanins and Its Role in the Color of Red Wine: A Critical Review. *Am. J. Enol. Vitic.* **2001**, *52*, 67–87.
45. Trouillas, P.; Sancho-Garcia, J.-C.; Freitas, V.; Gierschner, J.; Otyepka, M.; Dangles, O. Stabilizing and Modulating Color by Copigmentation: Insights from Theory and Experiment. *Chem. Rev.* **2016**, *116*, 4937–4982. [CrossRef]
46. Escribano-Bailon, M.T.; Santos-Buelga, C. Anthocyanin Copigmentation—Evaluation, Mechanisms and Implications for the Colour of Red Wines. *Curr. Org. Chem.* **2012**, *16*, 715–723. [CrossRef]
47. Jamei, R.; Babaloo, F. Stability of blueberry (*Cornus mas*—Yulyush) anthocyanin pigment under pH and co-pigment treatments. *Int. J. Food Prop.* **2017**, *20*, 2128–2133. [CrossRef]
48. Tan, C.; Dadmohammadi, Y.; Lee, M.C.; Abbaspourrad, A. Combination of copigmentation and encapsulation strategies for the synergistic stabilization of anthocyanins. *Compr. Rev. Food Sci. Food Saf.* **2021**, *20*, 3164–3191. [CrossRef]
49. Andersen, Ø.M.; Jordheim, M. Chemistry of Flavonoid-Based Colors in Plants. In *Comprehensive Natural Products II*; Elsevier: Amsterdam, The Netherlands, 2010; pp. 547–614. [CrossRef]
50. Molaeafard, S.; Jamei, R.; Marjani, A.P. Co-pigmentation of anthocyanins extracted from sour cherry (*Prunus cerasus* L.) with some organic acids: Color intensity, thermal stability, and thermodynamic parameters. *Food Chem.* **2020**, *339*, 128070. [CrossRef]
51. Oliveira, J.; Azevedo, J.; Teixeira, N.; Araújo, P.; de Freitas, V.; Basílio, N.; Pina, F. On the Limits of Anthocyanins Co-Pigmentation Models and Respective Equations. *J. Agric. Food Chem.* **2021**, *69*, 1359–1367. [CrossRef]
52. Rein, M. Copigmentation Reactions and Color Stability of Berry Anthocyanins. Ph.D. Dissertation, University of Helsinki, Helsinki, Finland, 8 April 2005; 87p.
53. Cortez, R.; Luna-Vital, D.A.; Margulis, D.; Gonzalez De Mejia, E. Natural Pigments: Stabilization Methods of Anthocyanins for Food Applications. *Compr. Rev. Food Sci. Food Saf.* **2017**, *16*, 180–198. [CrossRef]
54. Landi, M.; Tattini, M.; Gould, K.S. Multiple functional roles of anthocyanins in plant-environment interactions. *Environ. Exp. Bot.* **2015**, *119*, 4–17. [CrossRef]
55. Houghton, A.; Appelhagen, I.; Martin, C. Natural Blues: Structure Meets Function in Anthocyanins. *Plants* **2021**, *10*, 726. [CrossRef]
56. Mateus, N.; de Freitas, V. Anthocyanins as Food Colorants. In *Anthocyanins*; Winefield, C., Davies, K., Gould, K., Eds.; Springer: New York, NY, USA, 2008; pp. 284–304. [CrossRef]
57. Rodriguez-Amaya, D.B. Update on natural food pigments—A mini-review on carotenoids, anthocyanins, and betalains. *Food Res. Int.* **2019**, *124*, 200–205. [CrossRef]
58. Zhao, X.; Ding, B.-W.; Qin, J.-W.; He, F.; Duan, C.-Q. Intermolecular copigmentation between five common 3-O-monoglucosidic anthocyanins and three phenolics in red wine model solutions: The influence of substituent pattern of anthocyanin B ring. *Food Chem.* **2020**, *326*, 126960. [CrossRef]

59. Malien-Aubert, C.; Dangles, O.; Amiot, M.J. Color Stability of Commercial Anthocyanin-Based Extracts in Relation to the Phenolic Composition. Protective Effects by Intra- and Intermolecular Copigmentation. *J. Agric. Food Chem.* **2001**, *49*, 170–176. [CrossRef]
60. Laleh, G.H.; Frydoonfar, H.; Heidary, R.; Jameei, R.; Zare, S. The Effect of Light, Temperature, pH and Species on Stability of Anthocyanin Pigments in Four Berberis Species. *Pak. J. Nutr.* **2005**, *5*, 90–92. [CrossRef]
61. Remini, H.; Dahmoune, F.; Sahraoui, Y.; Madani, K.; Kapranov, V.; Kiselev, E. Recent Advances on Stability of Anthocyanins. *RUND J. Agron. Anim. Ind.* **2018**, *13*, 257–286. [CrossRef]
62. Turker, N.; Aksay, S.; Ekiz, H.I. Effect of Storage Temperature on the Stability of Anthocyanins of a Fermented Black Carrot (*Daucus carota* var. L.) Beverage: Shalgam. *J. Agric. Food Chem.* **2004**, *52*, 3807–3813. [CrossRef]
63. Hellström, J.; Mattila, P.; Karjalainen, R. Stability of anthocyanins in berry juices stored at different temperatures. *J. Food Compos. Anal.* **2013**, *31*, 12–19. [CrossRef]
64. West, M.E.; Mauer, L.J. Color and Chemical Stability of a Variety of Anthocyanins and Ascorbic Acid in Solution and Powder Forms. *J. Agric. Food Chem.* **2013**, *61*, 4169–4179. [CrossRef] [PubMed]
65. Patras, A.; Brunton, N.; O'Donnell, C.; Tiwari, B. Effect of thermal processing on anthocyanin stability in foods; mechanisms and kinetics of degradation. *Trends Food Sci. Technol.* **2010**, *21*, 3–11. [CrossRef]
66. Deylami, M.Z.; Rahman, R.A.; Tan, C.P.; Bakar, J.; Olusegun, L. Effect of blanching on enzyme activity, color changes, anthocyanin stability and extractability of mangosteen pericarp: A kinetic study. *J. Food Eng.* **2016**, *178*, 12–19. [CrossRef]
67. He, J. Isolation of Anthocyanin Mixtures from Fruits and Vegetables and Evaluation of Their Stability, Availability and Biotransformation in the Gastrointestinal Tract. Ph.D. Thesis, The Ohio State University, Columbus, OH, USA, 2008.
68. Sipahli, S.; Mohanlall, V.; Mellem, J. Stability and degradation kinetics of crude anthocyanin extracts from *H. sabdariffa*. *Food Sci. Technol.* **2017**, *37*, 209–215. [CrossRef]
69. Sikorski, Z. Fennema's Food Chemistry (Fifth Edition)—Edited by Srinivasan Damodaran and Kirk L. Parkin. *J. Food Biochem.* **2018**, *42*, e12483. [CrossRef]
70. Cavalcanti, R.N.; Santos, D.T.; Meireles, M.A.A. Non-thermal stabilization mechanisms of anthocyanins in model and food systems—An overview. *Food Res. Int.* **2011**, *44*, 499–509. [CrossRef]
71. Van Hung, P. Phenolic Compounds of Cereals and Their Antioxidant Capacity. *Crit. Rev. Food Sci. Nutr.* **2016**, *56*, 25–35. [CrossRef]
72. Sheraz, M.; Khan, M.; Ahmed, S.; Kazi, S.; Ahmad, I. Stability and Stabilization of Ascorbic Acid. *Househ. Pers. Care Today* **2015**, *10*, 20–25.
73. Tewari, S.; Sehrawat, R.; Nema, P.K.; Kaur, B.P. Preservation effect of high pressure processing on ascorbic acid of fruits and vegetables: A review. *J. Food Biochem.* **2017**, *41*, e12319. [CrossRef]
74. Farr, J.E.; Giusti, M.M. Investigating the Interaction of Ascorbic Acid with Anthocyanins and Pyranoanthocyanins. *Molecules* **2018**, *23*, 744. [CrossRef]
75. Levy, R.; Okun, Z.; Shpigelman, A. The Influence of Chemical Structure and the Presence of Ascorbic Acid on Anthocyanins Stability and Spectral Properties in Purified Model Systems. *Foods* **2019**, *8*, 207. [CrossRef]
76. Socaciu, C. *Food Colorants: Chemical and Functional Properties*; CRC Press: Boca Raton, FL, USA, 2008.
77. Ozkan, M. Degradation of anthocyanins in sour cherry and pomegranate juices by hydrogen peroxide in the presence of added ascorbic acid. *Food Chem.* **2002**, *78*, 499–504. [CrossRef]
78. Amogne, N.Y.; Ayele, D.W.; Tsigie, Y.A. Recent advances in anthocyanin dyes extracted from plants for dye sensitized solar cell. *Mater. Renew. Sustain. Energy* **2020**, *9*, 23. [CrossRef]
79. Aramwit, P.; Bang, N.; Srichana, T. The properties and stability of anthocyanins in mulberry fruits. *Food Res. Int.* **2010**, *43*, 1093–1097. [CrossRef]
80. Christofi, S.; Malliaris, D.; Katsaros, G.; Panagou, E.; Kallithraka, S. Limit SO_2 content of wines by applying High Hydrostatic Pressure. *Innov. Food Sci. Emerg. Technol.* **2020**, *62*, 102342. [CrossRef]
81. Gabriele, M.; Gerardi, C.; Lucejko, J.J.; Longo, V.; Pucci, L.; Domenici, V. Effects of low sulfur dioxide concentrations on bioactive compounds and antioxidant properties of Aglianico red wine. *Food Chem.* **2018**, *245*, 1105–1112. [CrossRef]
82. Pozo-Bayon, M.A.; Monagas, M.; Bartolomé, B.; Moreno-Arribas, M.V. Wine Features Related to Safety and Consumer Health: An Integrated Perspective. *Crit. Rev. Food Sci. Nutr.* **2012**, *52*, 31–54. [CrossRef] [PubMed]
83. Marszałek, K.; Woźniak, Ł.; Kruszewski, B.; Skąpska, S. The Effect of High Pressure Techniques on the Stability of Anthocyanins in Fruit and Vegetables. *Int. J. Mol. Sci.* **2017**, *18*, 277. [CrossRef] [PubMed]
84. Nistor, M.; Diaconeasa, Z.; Frond, A.D.; Stirbu, I.; Socaciu, C.; Pintea, A.; Rugina, D. Comparative efficiency of different solvents for the anthocyanins extraction from chokeberries and black carrots, to preserve their antioxidant activity. *Chem. Pap.* **2021**, *75*, 813–822. [CrossRef]
85. Sharif, N.; Khoshnoudi-Nia, S.; Jafari, S.M. Nano/microencapsulation of anthocyanins; a systematic review and meta-analysis. *Food Res. Int.* **2020**, *132*, 109077. [CrossRef]
86. Enaru, B.; Socaci, S.; Farcas, A.; Socaciu, C.; Danciu, C.; Stanila, A.; Diaconeasa, Z. Novel Delivery Systems of Polyphenols and Their Potential Health Benefits. *Pharmaceuticals* **2021**, *14*, 946. [CrossRef]
87. Robert, P.; Fredes, C. The Encapsulation of Anthocyanins from Berry-Type Fruits. Trends in Foods. *Molecules* **2015**, *20*, 5875–5888. [CrossRef]

88. Mohammadalinejhad, S.; Kurek, M. Microencapsulation of Anthocyanins—Critical Review of Techniques and Wall Materials. *Appl. Sci.* **2021**, *11*, 3936. [CrossRef]
89. Tarone, A.G.; Cazarin, C.B.B.; Junior, M.R.M. Anthocyanins: New techniques and challenges in microencapsulation. *Food Res. Int.* **2020**, *133*, 109092. [CrossRef] [PubMed]
90. Ozkan, G.; Franco, P.; De Marco, I.; Xiao, J.; Capanoglu, E. A review of microencapsulation methods for food antioxidants: Principles, advantages, drawbacks and applications. *Food Chem.* **2019**, *272*, 494–506. [CrossRef] [PubMed]
91. Trucillo, P.; Campardelli, R.; Reverchon, E. Liposomes: From Bangham to Supercritical Fluids. *Processes* **2020**, *8*, 1022. [CrossRef]

Review

Emerging Non-Thermal Technologies for the Extraction of Grape Anthocyanins

Antonio Morata *, Carlos Escott, Iris Loira, Carmen López, Felipe Palomero and Carmen González

enotecUPM, Chemistry and Food Technology Department, Technical University of Madrid UPM, 28040 Madrid, Spain; carlos.escott@gmail.com (C.E.); iris.loira@upm.es (I.L.); carmen.lopez@upm.es (C.L.); felipe.palomero@upm.es (F.P.); carmen.gchamorro@upm.es (C.G.)
* Correspondence: antonio.morata@upm.es

Abstract: Anthocyanins are flavonoid pigments broadly distributed in plants with great potential to be used as food colorants due to their range of colors, innocuous nature, and positive impact on human health. However, these molecules are unstable and affected by pH changes, oxidation and high temperatures, making it very important to extract them using gentle non-thermal technologies. The use of emerging non-thermal techniques such as High Hydrostatic Pressure (HHP), Ultra High Pressure Homogenization (UHPH), Pulsed Electric Fields (PEFs), Ultrasound (US), irradiation, and Pulsed Light (PL) is currently increasing for many applications in food technology. This article reviews their application, features, advantages and drawbacks in the extraction of anthocyanins from grapes. It shows how extraction can be significantly increased with many of these techniques, while decreasing extraction times and maintaining antioxidant capacity.

Keywords: non-thermal technologies; grapes; wine; anthocyanins; HHP; UHPH; PEFs; US; irradiation

1. Introduction

Anthocyanins are flavonoid pigments responsible for the color of many fruits, flowers and vegetable tissues. Extensive details on their properties and features can be found in the literature [1–9]. They have been extensively studied for their potential applications as natural colorants [10–12] as they are innocuous and safe molecules, but also for their positive impact on health due to their antioxidant properties [13,14] and their effect on the gut microbiome [15]. Anthocyanin color depends on the substitution pattern in the B-ring and the acylation patterns, both of which affect the electron density and the observed color, ranging in grapes from red orange (brownish red) to bluish red (purple), with typical ranges from 518 nm of maximum absorption for cyanidin to 528 nm for malvidin [16]. Acylation normally increases the maximum absorption (e.g., malvidin 528 nm to coumaroyl malvidin 535 nm) (Table 1). The color of anthocyanins is also affected by low pH, which increases the color intensity by the hyperchromic effect, shifting the equilibria to increase the amount of pyrilium cation. Additionally, anthocyanins can also undergo SO_2 bleaching and co-pigmentation processes that produce bluish red pigments by bathochromic shifts in the maximum wavelength of absorbance [16].

In most grape varieties, anthocyanins are located in the exocarp (skins) (Figure 1A,B), which are the layers of cells in the outer surface of the berry; only a few varieties also have anthocyanins in the pulp [16]. The skins have a thicker cell wall than the pulp to protect the berry mechanically and against rot and pests.

Table 1. Molecular structure, substitution pattern in the B-ring (R1 and R2), acylation patterns (R3), maximum λ (nm), and color of grape's anthocyanins.

Anthocyanin	R1	R2	R3	λ max [1]	Colour	$[M]^+$/Aglycon (m/z) [2]
Delphinidin-3-O-gucoside	-OH	-OH	-H	526.9	Red	465/303
Cyanidin-3-O-glucoside	-OH	-H	-H	518.4	Orange-red	449/287
Petunidin-3-O-glucoside	-OH	-OCH$_3$	-H	528.1	Red	479/317
Peonidin-3-O-glucoside	-OCH$_3$	-H	-H	518.4	Orange-red	463/301
Malvidi-3-O-glucoside	-OCH$_3$	-OCH$_3$	-H	529.3	Red	493/331
Delphinidin-3-O-(6-O-acetyl)-glucoside	-OH	-OH	-COCH$_3$	529.3	Red	507/303
Cyanidin-3-O-(-6-O-acetyl)-glucoside	-OH	-H	-COCH$_3$	520.8	Red	491/287
Petunidin-3-O-(-6-O-acetyl)-glucoside	-OH	-OCH$_3$	-COCH$_3$	530.5	Bluish-red	521/317
Peonidin-3-O-(-6-O-acetyl)-glucoside	-OCH$_3$	-H	-COCH$_3$	520.8	Red	505/301
Malvidin-3-O-(-6-O-acetyl)-glucoside	-OCH$_3$	-OCH$_3$	-COCH$_3$	530.5	Bluish-red	535/331
Delphinidin-3-O-(6-O-p-coumaroyl)-glucoside	-OH	-OH	-COCH=CHC$_6$H$_4$-OH	534.2	Bluish-red	611/303
Cyanidin-3-O-(-6-O-p-coumaroyl)-glucoside	-OH	-H	-COCH=CHC$_6$H$_4$-OH	525.7	Red	595/287
Petunidin-3-O-(-6-O-p-coumaroyl)-glucoside	-OH	-OCH$_3$	-COCH=CHC$_6$H$_4$-OH	535.4	Bluish-red	625/317
Peonidin-3-O-(-6-O-p-coumaroyl)-glucoside	-OCH$_3$	-H	-COCH=CHC$_6$H$_4$-OH	524.5	Red	609/301
Malvidin-3-O-(-6-O-p-coumaroyl)-glucoside	-OCH$_3$	-OCH$_3$	-COCH=CHC$_6$H$_4$-OH	535.4	Bluish-red	639/331
Malvidin-3-O-(-6-O-caffeoyl)-glucoside	-OCH$_3$	-OCH$_3$	-COCH=CHC$_6$H$_3$-(OH)$_2$	536.6	Bluish-red	655/331
Molecular structure			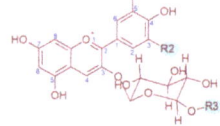			

[1] Obtained experimentally with HPLC-DAD-ESI/MS; [2] From [17].

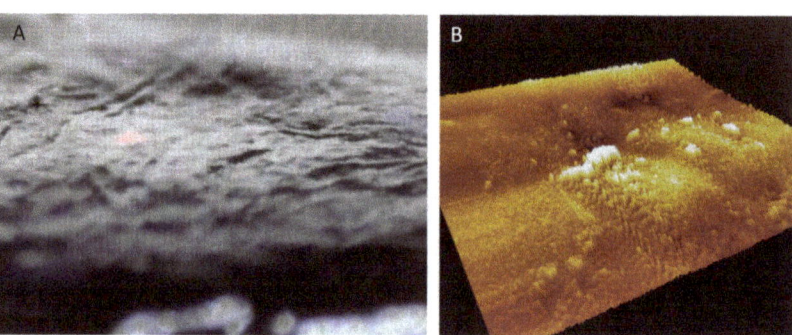

Figure 1. (**A**) Red grape skin (exocarp) *Vitis vinifera* L. Tempranillo variety by 60 μm optical camera built-in part of the AFM. (**B**) 3D Topography of the same skin by atomic force microscopy.

The structure and shape of the cells in the berries are flat cells in the skin and large polyhedral cells in the pulp (Figure 2A). Anthocyanins are located in the cells of the skin, inside the vacuole (Figure 2B). To extract the anthocyanins and to keep enough color (not only in red winemaking, but also in red juice production), it is necessary to disaggregate the cell wall polysaccharides, mainly the pectins (Figure 2C). In conventional winemaking, depolymerization of the cell wall and separation of polysaccharide fibers is achieved during maceration by means of soaking, fermentation temperature and mechanical treatments (i.e., punch downs, pump overs, délestage) [18].

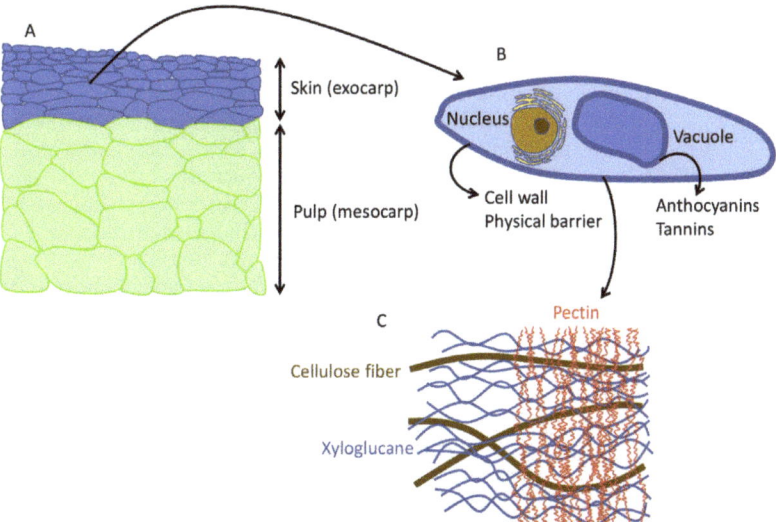

Figure 2. (**A**) Red grape section with flat colored cells in the skins and polyhedral cells in the pulp. (**B**) Skin cells shape and structure. (**C**) Cell wall fiber components.

In addition, cryomacerations (cold soak) by heat exchanger cooling or dry ice can be used to preferentially extract anthocyanins and aroma compounds in the absence of fermentation [19]. This is advantageous because it reduces the extraction of tannins whose solubility is lower in the absence of alcohol and better reduces astringency in young wines and juices. Another powerful technology to quickly degrade cell wall pectins and promote the extraction of anthocyanins, tannins and aroma compounds is the use of pectolytic enzymes [20], especially endo-polygalacturonases that break the pectin sequence depolymerizing the cell wall and releasing the pigments in the juice.

Currently emerging non-thermal technologies are increasingly being used to improve the extraction of bioactive compounds, such as anthocyanins, from food products, but also to decrease or eliminate spoilage or pathogenic microorganisms and sometimes even to inactivate oxidative enzymes [21,22]. Among them, continuous and discontinuous high-pressure technologies (i.e., High Hydrostatic Pressure (HHP) and Ultra High Pressure Homogenization (UHPH) [23–25], Pulsed Electric Fields (PEFs) [26–29], Ultrasound (US) [30–34] and β-irradiation [35–37].

This review is focused on the features of emerging non-thermal technologies that make them suitable for the extraction of anthocyanins from grape skins while protecting their natural coloring and antioxidant properties.

2. Use of High-Pressure Technologies to Extract Anthocyanins

The use of high-pressure technologies is growing exponentially in the food industry. A PubMed search using the keywords high, pressure and food yields 40,077 research articles in the period 1970–2021 with 36,920 since 2000. Although several technologies can be found, research with continuous (Ultra)-High Pressure Homogenization processes (UHPH and HPH) and discontinuous High Hydrostatic Pressure (HHP) technologies stand out. All of them share gentle food processing as they are non-thermal treatments with low impact on food quality, sensory constituents and nutraceutical components [25,38–41]. HHP and UHPH technologies are industrially implemented and several brands compete in the market. In batch technologies, the leading companies are Hiperbaric (https://www.hiperbaric.com/es/ (accessed on 2 November 2021)) and Avure (https://www.jbtc.com/es/north-america/foodtech/products-and-solutions/brands/avure-technologies (accessed on 2 November 2021)). In UHPH (continuous processing), the most effective technology is the one developed by Ypsicon (https://www.ypsicon.com/ (accessed on 2 November 2021)). These technologies have specific characteristics that will be described separately below.

2.1. High Hydrostatic Pressure (HHP)

HHP involves the application of high pressures to the food by means of a fluid (hydrostatic), which is usually water. The fluid is pumped into a high-strength steel vessel containing the food product, where pressures above 100 MPa, commonly in the range of 400–600 MPa, are reached during processing [42]. HHP treatments consist of pressurizing the food product in this pressure range for 2–10 min. The main effect of this is the destruction of the cell walls and membranes of microorganisms, but plant and animal tissue cells are also similarly affected. HHP can be considered a non-thermal technology because, even when compression produces adiabatic heating, this is quite moderate and ranges between 2 and 3 °C/100 MPa. This slight increase in temperature can be controlled by cooling the vessel or lowering the temperature of the food at the inlet. Moreover, HHP processing is not able to affect covalent bonds, so pigments, aroma and flavors are usually protected [24,38,42].

The effect of HHP on plant tissues is to damage the integrity of the cell walls, resulting in small pores or fissures that can facilitate the extraction of metabolites from the cell wall. HHP has been used to enhance the extraction of anthocyanins from grapes [23,24] and grape pomace [26]. The extraction of anthocyanins by HHP has been increased compared to controls in the range of 23–82% (Table 2). The extraction of phenolic compounds (tannins) is also increased in grapes, with a total polyphenol index of +26% [24], and the antioxidant activity of the extracts is higher than in controls [23,26]. HHP is a powerful technology to extract anthocyanins from plant tissues, and specifically from grape skin, preserving or enhancing the antioxidant capacity of the extracts, working at low temperatures (<30 °C, at 550 MPa for 10 min [24]), even under refrigeration. Furthermore, anthocyanin extraction has been reported to be selective depending on the acylation pattern and the methoxilation ratio on the B ring [23]. In grapes, anthocyanin migration from the skin to the pulp and seeds is observed after HHP treatment (Figure 3), which is evidence of cell wall poration and anthocyanin migration into the berry under the effect of pressure [24]. HHP causes the intensification of mass transfer phenomena, thus affecting cell permeability and molecule diffusion [43]. However, the external shape and structure of the berry are completely preserved (Figure 3). It is noticeable that similar extractions can be achieved in the range of 200–550 MPa, so it is possible to use milder HHP conditions, making the process cheaper while working at lower temperatures by adiabatic compression. Lower temperatures help to protect the anthocyanins during extraction and later in the ongoing process [44], probably reducing the risk of oxidation that can occur under thermal conditions.

The processing of grapes by HHP can be done with whole grapes (Figure 3), but it is also possible to process the separated skins that can be obtained from by-products such as pomace. Additionally, the stability of anthocyanins can be improved by using additives such as ethanol or other preservatives. HHP processing of berries also helps to sanitize them by easily removing yeasts and highly reducing bacterial loads [21,24]. This helps to obtain healthier anthocyanin extracts with reduced microbial loads, which facilitates the implantation of starters if these extracts are subsequently used in fermented foods [45]. Additionally, gentle extraction together with inactivation of microorganisms and higher antioxidant activity can reduce the use of antioxidants such as sulfites in the extracts, and also in the subsequent use of these extracts in food products [21,46].

Figure 3. External shape and appearance of control and pressurized grapes (200 MPa, 10 min), and details of the internal structure showing colored pulp and seeds in HHP-processed grapes.

Table 2. Emerging non-thermal technologies, processing conditions, and effects on the extraction of anthocyanin.

Emerging Non-Thermal Technology	Processing Mode	Product/Conditions	Effect	Extraction of Anthocyanins	Reference
HHP	Discontinuous	Grape skins 70 °C, 600 MPa	↑extraction ↑antioxidant activity	+23%	[23]
		Grapes <30 °C, 200–550 MPa, 10 min	↑extraction Migration of anthocyanins to pulp and seeds	+80%	[24]
		Grape by-products 70 °C, 600 MPa	↑extraction Phenols +50% ↑antioxidant activity ×3	+41%	[26]
UHPH	Continuous	Grape juice 300 MPa, 78 °C, <0.2 s	↑extraction ↑antioxidant activity	+2.6%	[47]
PEF	Continuous	Grapes 3 kV/cm, 50 pulses	↑extraction ↑juice yield +5%	×3	[48]
	Discontinuous	Grape by-products Exponential decay pulses, 70 °C, 30 kV/cm, 10 kJ/kg, 30 pulses, 2 Hz, 15 s	↑extraction Phenols +50% ↑antioxidant activity ×4	+77%	[26]
	Discontinuous	Mazuelo grapes Exponential decay pulses, <30 °C, 2, 5 and 10 kV/cm, 0.4, 1.8, and 6.7 kJ/kg, 50 pulses, 1 Hz	↑extraction Phenols +20−31%	+20.3, 28.6 and 41.8% after 120 h	[49]
	Discontinuous	Pinot noir grapes Square wave bipolar pulses, 1.5 kV/cm, 15 and 70 kJ/kg, 50 Hz, pulse width of 20 μs, pulse numbers of 243 and 1033	↑extraction ↑Phenols ↑bioprotective capacity	+43−74% after 2 days	[50]
	Continuous	Cabernet sauvignon grapes Square pulses, width of 3 μs, collinear chamber, 2, 5 and 7 kV/cm; 0.56, 3.67, 6.76 kJ/kg, <23 °C, 50 pulses, 122 Hz Flow 118 kg/h. Average residence time 0.41 s	↑extraction Phenols +14−36%	+18−45% after 24 h	[51]
	Continuous	Garnacha grapes Square pulses, width of 3 μs, collinear chamber, 4.3 kV/cm, 60 μs Flow 1900 kg/h. Average residence time 0.41 s	↑extraction Phenols +23%	+25% after 7 days	[52]
	Continuous	Merlot grapes >30 kV/cm, 4.7–49.4 kJ/L Flow 500 kg/h	↑extraction ↑Phenols +23−162% Shortening of cold macerations	+17−636% after 24 h	[53]

Table 2. Cont.

Emerging Non-Thermal Technology	Processing Mode	Product/Conditions	Effect	Extraction of Anthocyanins	Reference
	Continuous	Rondinella grapes Square-wave pulses at 1.5 kV/cm, 1, 5 and 10 µs, 2, 10 and 20 kJ/kg, 400 Hz Flow 250 L/h	↑extraction ↑Phenols +37%	+30% color intensity after fermentation	[54]
	Continuous	Grenache grapes Pulses of 5 kV/cm, 63.4 kJ/kg. 1800 µS (45 pulses of 40 µS) Average residence time 0.38 s. Temperature < 32 ± 2 Flow 120 kg/h	↑extraction ↑Phenols ×1.6	×2.2 after 24 h	[55]
Ultrasounds	Discontinuous	Tannat grape pomace. US bath: 15–60 °C, 0–100 W, 5–50 min	↑extraction ↑Phenols +50%	+50%	[56]
	Discontinuous	Red grape pomace US bath: 25 °C, 160 W, 40 kHz, 30 min, 0.4 W/mL, 720 J/mL	↑extraction	+59% after 5 min	[33]
	Discontinuous	Monastrell grapes US bath: 18 °C, 40 kHz, 280 W, 90 min	↑extraction ↑Phenols +9%	+8% first day of maceration	[57]
	Discontinuous	Wine lees Sonifier Cell Disruptor Model 450, high gain horn of $\frac{3}{4}''$ of diameter. Time of sonication 30–90 s	≈extraction Lower time	33% of the control time	[58]
	Discontinuous	Grape pomace Moldova variety ultrasonic transducer coupled with a function generator	↑extraction	+18% from 12.5 to 25 kHz	[59]
β-irradiation	Continuous	Tempranillo grapes 10-MeV, 50-kW Rhodotron accelerator, scan frequency of 100 Hz oses of 0 (control), 0.5, 1 and 10 kGy	↑extraction	+71% at 10 kGy	[37]

2.2. Ultra High Pressure Homogenization (UHPH)

UHPH consists of the continuous pressurization of a fluid to 200 MPa or more, through a special valve, and its subsequent release at low pressure (usually atmospheric pressure) [25,41,60,61]. Typical processing conditions are the use of 300 MPa with valve residence times less than 0.2 s. The process can be heat-assisted by using upstream heat exchangers, which greatly increases efficiency. The short processing time, even when high temperatures are used in the valve, produces a very gentle treatment with high nutritional and sensory quality [25]. The preservation of antioxidant activity [62], the control of oxidative enzymes such as polyphenol oxidases (PPOs) [62–64], the preservation of delicate aromatic molecules such as terpenes [64], and the absence of thermal markers have been observed in the processing of grape juices by UHPH [64].

The UHPH process and the passing through the valve produce high impact forces and intense shear stresses, and the result is a significant nanofragmentation of plant tissues with removal of microorganisms, including spores depending on the temperature in the valve, inactivation of enzymes and nanofragmentation of colloidal particles. The consequence is increased extraction by cell disruption and improved bioaccessibility [65]. The mechanical effect is highly dependent on the valve design, and the antimicrobial effect with mild impact on the residence time and the design of the heat exchangers upstream and downstream of the valve [25] (Figure 4). One of the most effective designs is the one developed by Ypsicon [66].

There is a size requirement concerning the maximum size of colloidal particles in the grape juice before pressurization due to the cross-sections in the fluidic components of the pump and valve. Particles in the fluid should be less than 500 µm to avoid clogging (Figure 4) [25]. After the treatment, the particles are fragmented in the range of 100–500 nm [25,64]. When grape juice, which has many colloidal constituents with a polyhedral appearance (Figure 5A), is processed by UHPH, a finer structure can be observed (Figure 5B) without large fragments [47].

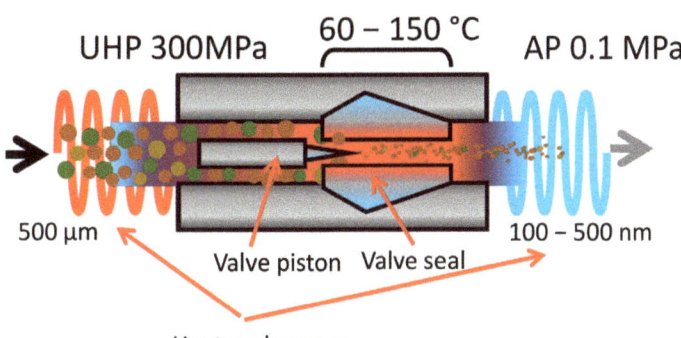

Figure 4. Scheme of the structure and components of a UHPH-Ypsicon valve. Intense impact and shear stresses together with the help of heating produce: pasteurization/sterilization, nano-fragmentation, enzyme inactivation, nano-coating and nano-encapsulation.

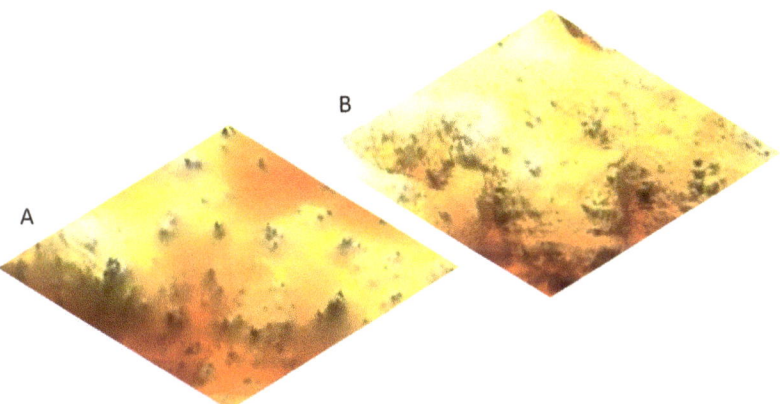

Figure 5. (**A**) Atomic Force Microscopy (AFM) topography of the surface of a dried red grape juice showing polyhedral granules, which are the colloidal particles of the juice (i.e., plant cell fragments and fibers). (**B**) The same dried red grape juice by AFM after UHPH treatment, with smaller granules and a flatter surface (no large polyhedral granules).

3. Pulsed Electric Fields (PEFs) in the Extraction of Anthocyanins

Like the previous ones (HHP and UHPH), PEFs have become a global technology with numerous applications in food processing, preservation and stabilization [67–71]. PEF is based on the use of high intensity electric fields (3–40 kV/cm) for a very short time (milli-micro seconds). Food is processed by PEF when it passes through two electrodes. The Electric Field Strength (E) is the voltage (kV) divided by the distance between the electrodes (cm), i.e., $E = V/d$. PEF systems are currently available on an industrial scale for food processing in the range of 50–10,000 L/h for fluids and 1–70 tonnes/h for solids such as French fries. The effect of PEFs is the poration of cells at the nanoscale, which affects the selective permeability [72]. These pores are difficult to observe by electronic microscopy. However, the pores produce various effects depending on size and number, tending to increase cell permeability, thus facilitating the extraction of cell compounds (e.g., anthocyanins and many others), the entry of compounds and the temporal or definitive inactivation of cells depending on the intensity [73]. Pulses can be applied in several modalities. The main parameters are the pulse shape (i.e., squared, exponential, sinusoidal), the polarity (i.e., monopolar or bipolar), the number of pulses and the pulse duration (Figure 6). The intensity and effectiveness of the treatments depend on the above parameters with squared bipolar pulses being more effective and the number of pulses making the process more powerful. Even when the pulse duration also improves the efficacy, it should be kept at a low value because it affects the temperature of the food by ohmic heating.

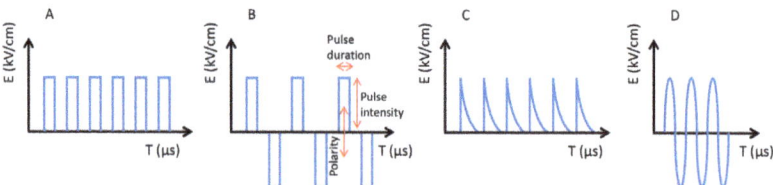

Figure 6. Types of pulses. (**A**) Squared monopolar. (**B**) Squared bipolar. (**C**) Exponentially decaying. (**D**) Sinusoidal.

Plant cells need lower intensities than microorganisms, especially bacteria, depending on size and shape. To induce permeabilization in plant cells (size 40–200 µm), E must be 1–2 kV/cm, while in microorganisms (size 1–10 µm), 12–20 kV/cm are required [74]. Therefore, to extract bioactive molecules from vegetal tissues, less than 5 kV/cm is necessary, however, for microbial inactivation, E should normally be higher than 10 kV/cm. When plant cells are pored (i.e., grape skins), the consequence is an increased extraction of biomolecules such as anthocyanins, tannins and aroma compounds (Figure 7).

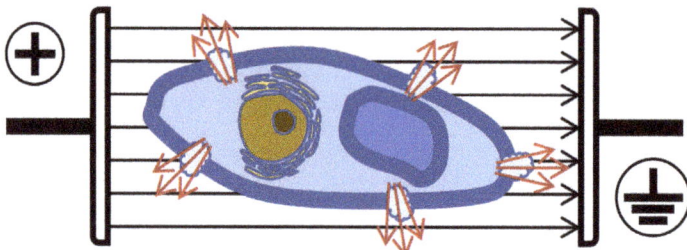

Figure 7. Electroporation and cell permeabilization.

At pilot and industrial scale, several works have demonstrated the efficiency of PEFs to increase the extraction of anthocyanins, and other phenols at low temperature while preserving their antioxidant capacity (Table 1). Currently grapes or by-products (grape pomace) can be processed continuously at a flow rate of several hundreds to a few tonnes of kg per hour (118 kg/h, [51]; 500 kg/h [53], 1900 kg/h [52]). Usually, the crushed gape is pumped by a progressive cavity pump [51] or a peristaltic pump [52] and later processed in a collinear chamber by applying exponentially decay pulses or, more frequently, squared pulses of an electric field strength ranging from 2 to 10 kV/cm [26,28,48,49,52,53]. Anthocyanin extraction increases in the range of 17–100% (Table 1) depending on processing conditions and post-maceration time. The temperature is increased by only 2–15 °C [55], therefore it is easy to work at room temperature or under refrigerated conditions. In addition to improved anthocyanin and phenol extraction, PEFs can be used for gentle non-thermal pasteurization of the must, thus improving the implantation of non-*Saccharomyces* starters [55] and potentially reducing the use of SO_2. The effect of PEFs on the extraction of phenolic compounds from seeds has also been reported and should be considered in winemaking processes [75,76].

4. Ultrasounds (USs) in the Extraction of Anthocyanins

Ultrasounds (USs) are mechanic waves with a frequency above 20 kHz, which is not perceptible to the human ear (typically in the range 20 Hz–20 kHz) [77]. It is a key technology for obtaining bioactive compounds (e.g., anthocyanins), like the others described above, as it can be considered a sustainable 'green' extraction method [78] as it does not use organic solvents and is gentle to heat-sensitive molecules [79]. The compression and rarefaction of the products produced by the US waves produce the successive reduction in size and expansion of the bubbles formed by cavitation (Figure 8). When these bubbles collapse, large amounts of energy are released, reaching localized temperatures of 5000 °K and pressures of 200 MPa [80]. These phenomena are responsible for the depolymerization of biostructures [80] and facilitate the extraction of molecules from plant tissues. Depolymerization can occur by bubble collapse, cavitation or degradation of the polymer by impact with radicals formed during sonication [81]. Depolymerization of cell wall polysaccharides accelerates the release of anthocyanins from the skin cells in grapes (Figure 9). The extraction of anthocyanins in water within a few minutes and the increase in temperature due to the cavitation effect can be observed. High power ultrasounds with the best extraction potential are considered to be in the range of 20–25 kHz [78].

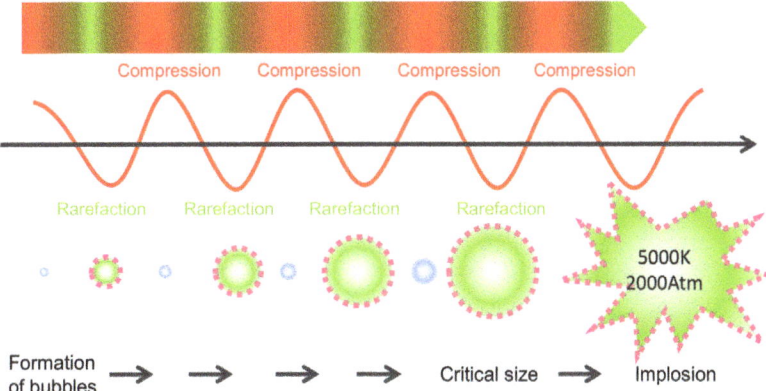

Figure 8. Implosion of bubbles and cavitation produced by alternative compression-rarefaction effects generated by US waves [22].

Figure 9 shows the application of USs on grape berries by means of a sonotrode and reveals, after a few minutes, how the anthocyanins are extracted to the surrounding media (water) due to the depolymerization of the cell walls of the grape skins. Additionally, the heating effect produced by cavitation can be observed, which in this case is about 5 °C in the center of the flask according to infrared thermography.

There are several systems for applying USs to plant tissues with the aim of favoring the extraction of compounds: Ultrasound baths, sonotrodes, sonoplates. However, on an industrial scale, the most effective system is the use of continuous tubular exchangers on the external surface of which sonoplates are distributed to apply US waves during the flow of the mash or liquid through the exchanger. For a better distribution of the sonoplates on the exchange surface, the section is usually hexagonal instead of circular (Figure 10).

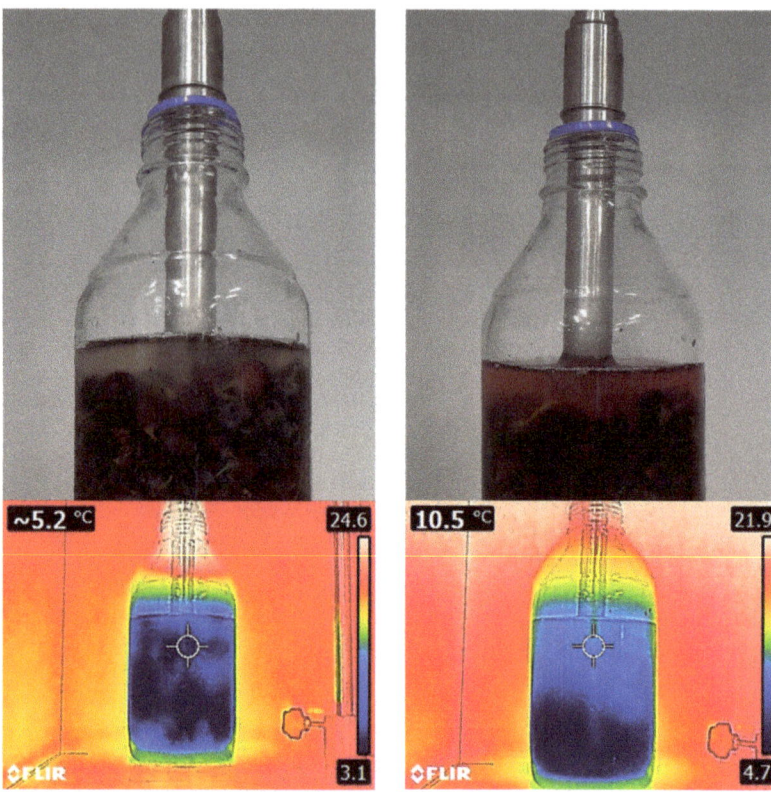

Figure 9. Use of USs in the extraction of grape anthocyanins and effect on temperature measured with an infrared camera. Left: before ultrasonication, right: after US treatment.

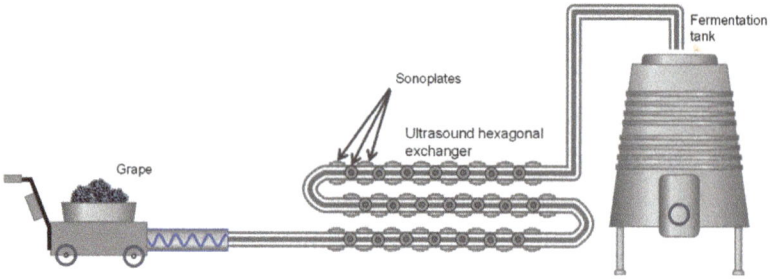

Figure 10. Cavitation cells arranged in a hexagonal tubular exchanger with the sonoplates for applying US waves [22].

This technology has been used to process Tempranillo grapes, achieving the same anthocyanin content in a final wine after only continuous US treatment and 72 h of skin maceration as in the control wine [82]. In discontinuous treatment at the laboratory scale, the USs have been shown to increase the extraction of anthocyanins and phenols by more than 50% compared to controls [56]. US can also be applied continuously after the application of pectolytic enzymes at industrial level, increasing color intensity by 18% and total polyphenols by 21% in wines [57]. The use of US-assisted extraction can be improved by optimizing other physicochemical parameters (temperature, ethanol and time), thus reaching a maximum of 6.26 mg/mL under the best conditions of 45.14 °C, 52.3% ethanol and 24.5 min [83]. USs can be used to improve extraction and/or reduce extraction time in grapes [56,57,82], and by-products such as pomace [33] and lees [58]. USs have been applied to *Vitis vinifera* L. varieties Cabernet Franc [84], Tempranillo [82], Tannat [56], and Monastrell [57]. The influence of the US frequency has also been analyzed, considering the values of 12.5, 25, and 37.5 kHz, as well as from 12.5 to 25 kHz, the extraction of anthocyanins increased by 18% in grape pomace, however, the higher the frequency, the lower the extraction [59].

5. Effect of E-Beam Irradiation in the Extraction of Anthocyanins

Electron beam (e-beam) irradiation or β-irradiation involves the use of accelerated electrons at high energy, typically 10 MeV [21,37,85], to process foods and eliminate microorganisms, allowing pasteurization (1–5 kGy) or sterilization (>10 kGy) depending on the dose [21,37,85]. The irradiation dose is measured in Grays (Gy) or kGrays (kGy). One Gy is defined as the absorption of 1 Joule of energy per kg of irradiated mass. Irradiation is widely used to preserve food in more than 55 countries and is considered a safe technology approved by WHO, FAO and IAEA [86]. This technique is cheap on a large scale, environmentally friendly and time efficient [87]. e-Beam irradiation is a complex technology that requires expensive irradiation accelerators and large-scale facilities. e-Beam irradiation can be applied on an industrial scale in a continuous process [21]. The radiation dose can be monitored and verified by placing radiochromic dosimeters on the treated food (Figure 11). This can be used to verify the real dose received by the food at every width. e-Beam irradiation can be considered a gentle non-thermal technology with temperature increments of less than 5 °C at doses up to 10 kGy [21]. In addition, e-beam irradiation has been proposed as an alternative to sulphites in wine preservation [88] and has demonstrated its ability to delay browning in plant foods [89]. However, some negative effects have been observed such as loss of aroma [37] and reduction in vitamin C content [85], due to free radical-mediated oxidation [90].

Figure 11. Red grapes in plastic bags after e-beam irradiation. The white arrows indicate the location of radiochromic dosimeters.

Even when the external appearance of the grapes after irradiation remains unchanged (Figure 12A), the release of some juice in the bags can be observed, especially at high doses (10 kGy). This leakage of juice from the grapes shows the weakening of the plant tissues due to irradiation. The main effect of e-beam irradiation on plant tissues is the fragmentation of fibrillar polymers such as pectins and other polysaccharides, promoting the release and extraction of cell components, including anthocyanins. It has been reported that the molecular weight of pectins can be reduced by 90% using doses of 3–10 kGy [91]. The effect on grapes is the increased extraction of phenols and anthocyanins [37], which can be observed in the more intense color of the running juice, especially when grapes are processed at 10 kGy (Figure 12B). Up to 1 kGy, the anthocyanin extraction is low and not too high compared to the controls, however at 10 kGy, the anthoyanins extracted in the running juice were 125 mg/L compared with 72 mg/L on average in the controls (+71%) [37]. Better antioxidant and sensory properties and higher phenol content have also been observed in grapes processed up to 2 kGy [92]. With blueberries, the use of e-beam irradiation at doses below 3 kGy has demonstrated to be a gentle processing that does not affect monomeric anthocyanin and antioxidant activity [93]. Protection of anthocyanins, color, phenols and antioxidant activity has also been observed in strawberries processed at 1 kGy [94].

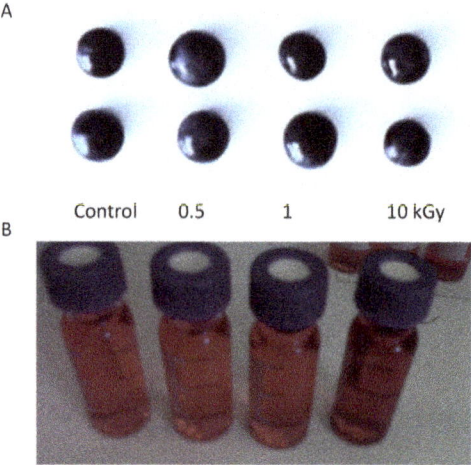

Figure 12. Effect of e-beam irradiation on the external appearance of grapes (**A**). Running juice from grapes processed by e-beam irradiation at various doses (**B**).

6. Pulsed Light

Pulsed light (PL) treatment involves the application of high-intensity, low-duration pulses of radiation in the 200 nm (UV) to 2500 nm (IR) range [95,96]. The intensity can be higher than 10^6 fold that of sunlight at sea level in the summer midday and the duration ranges from micro to milliseconds. Detailed parameters for pulsed light processing conditions have recently been revised [97]. The temperature rise after standard PL treatments is usually less than 5 °C, so it can be considered a mild non-thermal technology that can be used in delicate foods [96].

The ability to extract anthocyanins and phenols from plant tissues is lower than some of the previous techniques and the literature reports inconclusive results. Non-significant differences have been found in the anthocyanin content of wines made from PL-processed grapes compared to controls [98]. Temperature increases in grape skins of 2–3 °C after pulsed light treatments have also been reported without severe damage to the skin surface observed by AFM scanning [99]. Furthermore, PL processing of strawberries at 4–8 J/cm^2 does not affect the quality and antioxidant capacity [100].

7. Conclusions

The use of emerging non-thermal technologies is a successful tool for the extraction of anthocyanins from grapes, increasing the yield, accelerating the process and preserving the antioxidant capacity. Many of these techniques can be applied in continuous flow (UHPH, PEF, US, Irradiation and PL), which is suitable for industrial processing. Most of these techniques can work at room temperature or even using refrigerated crushed grapes, although temperature is always a synergistic parameter. These technologies can be used for the extraction of anthocyanins from grapes, and also from by-products such as pomace, generating high-value pigments from them. Additionally, emerging technologies can be used to improve the winemaking process by increasing the extraction of anthocyanins and phenolic compounds in maceration and controlling oxidations.

Author Contributions: Conceptualization, A.M.; methodology, A.M. and C.G.; writing—original draft preparation, A.M., C.E., I.L., C.L., F.P.; writing—review and editing, A.M., C.E., I.L., C.L., F.P. and C.G.; funding acquisition, A.M. All authors have read and agreed to the published version of the manuscript.

Funding: This study was funded by Ministerio de Ciencia, Innovación y Universidades project: RTI2018-096626-B-I00.

Conflicts of Interest: The authors declare no conflict of interest.

References

1. Mazza, G.; Francis, D.F.J. Anthocyanins in grapes and grape products. *Crit. Rev. Food Sci. Nutr.* **2009**, *35*, 341–371. [CrossRef] [PubMed]
2. Bridle, P.; Timberlake, C.F. Anthocyanins as natural food colours—Selected aspects. *Food Chem.* **1997**, *58*, 103–109. [CrossRef]
3. Harborne, J.B.; Williams, C.A. Anthocyanins and other flavonoids. *Nat. Prod. Rep.* **2001**, *18*, 310–333. [CrossRef] [PubMed]
4. He, F.; Liang, N.-N.; Mu, L.; Pan, Q.-H.; Wang, J.; Reeves, M.J.; Duan, C.-Q. Anthocyanins and Their Variation in Red Wines I. Monomeric Anthocyanins and Their Color Expression. *Molecules* **2012**, *17*, 1571–1601. [CrossRef] [PubMed]
5. Khoo, H.; Azlan, A.; Tang, S.; Lim, S. Anthocyanidins and anthocyanins: Colored pigments as food, pharmaceutical ingredients, and the potential health benefits. *Food Nutr. Res.* **2017**, *61*, 1361779. [CrossRef]
6. Escribano-Bailón, M.T.; Rivas-Gonzalo, J.C.; García-Estévez, I. Wine Color Evolution and Stability. In *Red Wine Technology*; Academic Press: New York, NY, USA, 2019; pp. 195–205.
7. Morata, A.; López, C.; Tesfaye, W.; González, C.; Escott, C. Anthocyanins as natural pigments in beverages. In *Value-Added Ingredients and Enrichments of Beverages*; Academic Press: New York, NY, USA, 2019; pp. 383–428.
8. Santos-Buelga, C.; González-Paramás, A.M. Anthocyanins. In *Encyclopedia of Food Chemistry*; Academic Press: Cambridge, MA, USA, 2019; pp. 10–21.
9. Mazza, G.; Miniati, E. *Anthocyanins in Fruits, Vegetables, and Grains*, 1st ed.; CRC Press: Boca Raton, FL, USA, 2018; ISBN 9781351069700.
10. Jackman, R.L.; Yada, R.Y.; Tung, M.A.; Speers, R.A. Anthocyanins as Food Colorants—A Review. *J. Food Biochem.* **1987**, *11*, 201–247. [CrossRef]
11. Mateus, N.; de Freitas, V. Anthocyanins as Food Colorants. In *Anthocyanins*; Winefield, C., Davies, K., Gould, K., Eds.; Springer: New York, NY, USA, 2008; pp. 284–304.
12. Roy, S.; Rhim, J.-W. Anthocyanin food colorant and its application in pH-responsive color change indicator films. *Crit. Rev. Food Sci. Nutr.* **2020**, *61*, 2297–2325. [CrossRef]
13. Tena, N.; Martín, J.; Asuero, A.G. State of the Art of Anthocyanins: Antioxidant Activity, Sources, Bioavailability, and Therapeutic Effect in Human Health. *Antioxidants* **2020**, *9*, 451. [CrossRef]
14. Speer, H.; D'Cunha, N.M.; Alexopoulos, N.I.; McKune, A.J.; Naumovski, N. Anthocyanins and Human Health—A Focus on Oxidative Stress, Inflammation and Disease. *Antioxidants* **2020**, *9*, 366. [CrossRef] [PubMed]
15. Hair, R.; Sakaki, J.R.; Chun, O.K. Anthocyanins, Microbiome and Health Benefits in Aging. *Molecules* **2021**, *26*, 537. [CrossRef]
16. Morata, A.; Escott, C.; Loira, I.; Manuel Del Fresno, J.; González, C.; Suárez-Lepe, J.A. Influence of *Saccharomyces* and non-*Saccharomyces* yeasts in the formation of pyranoanthocyanins and polymeric pigments during red wine making. *Molecules* **2019**, *24*, 4490. [CrossRef]
17. Monagas, M.; Núñez, V.; Bartolomé, B.; Gómez-Cordovés, C. Anthocyanin-derived pigments in Graciano, Tempranillo, and Cabernet Sauvignon wines produced in Spain. *Am. J. Enol. Vitic.* **2003**, *54*, 163–169.
18. Morata, A.; González, C.; Tesfaye, W.; Loira, I.; Suárez-Lepe, J.A. Maceration and fermentation: New technologies to increase extraction. In *Red Wine Technology*; Elsevier: Amsterdam, The Netherlands, 2019; pp. 35–49.
19. Busse-Valverde, N.; Gómez-Plaza, E.; López-Roca, J.M.; Gil-Muñoz, R.; Bautista-Ortín, A.B. The Extraction of Anthocyanins and Proanthocyanidins from Grapes to Wine during Fermentative Maceration Is Affected by the Enological Technique. *J. Agric. Food Chem.* **2011**, *59*, 5450–5455. [CrossRef] [PubMed]
20. Río Segade, S.; Pace, C.; Torchio, F.; Giacosa, S.; Gerbi, V.; Rolle, L. Impact of maceration enzymes on skin softening and relationship with anthocyanin extraction in wine grapes with different anthocyanin profiles. *Food Res. Int.* **2015**, *71*, 50–57. [CrossRef]

21. Morata, A.; Loira, I.; Vejarano, R.; González, C.; Callejo, M.J.; Suárez-Lepe, J.A. Emerging preservation technologies in grapes for winemaking. *Trends Food Sci. Technol.* **2017**, *67*, 36–43. [CrossRef]
22. Morata, A.; Loira, I.; Guamis, B.; Raso, J.; del Fresno, J.M.; Escott, C.; Bañuelos, M.A.; Álvarez, I.; Tesfaye, W.; González, C.; et al. Emerging technologies to increase extraction, control microorganisms, and reduce SO_2. In *Chemistry and Biochemistry of Winemaking, Wine Stabilization and Aging*; IntechOpen: Rijeka, Croatia, 2020; pp. 1–20. ISBN 978-1-83962-576-3.
23. Corrales, M.; García, A.F.; Butz, P.; Tauscher, B. Extraction of anthocyanins from grape skins assisted by high hydrostatic pressure. *J. Food Eng.* **2009**, *90*, 415–421. [CrossRef]
24. Morata, A.; Loira, I.; Vejarano, R.; Bañuelos, M.A.; Sanz, P.D.; Otero, L.; Suárez-Lepe, J.A. Grape processing by High Hydrostatic Pressure: Effect on microbial populations, phenol extraction and wine quality. *Food Bioprocess Technol.* **2014**, *8*, 277–286. [CrossRef]
25. Morata, A.; Guamis, B. Use of UHPH to obtain juices with better nutritional quality and healthier wines with low levels of SO_2. *Front. Nutr.* **2020**, *7*, 598286. [CrossRef] [PubMed]
26. Corrales, M.; Toepfl, S.; Butz, P.; Knorr, D.; Tauscher, B. Extraction of anthocyanins from grape by-products assisted by ultrasonics, high hydrostatic pressure or pulsed electric fields: A comparison. *Innov. Food Sci. Emerg. Technol.* **2008**, *9*, 85–91. [CrossRef]
27. López, N.; Puértolas, E.; Condón, S.; Álvarez, I.; Raso, J. Effects of pulsed electric fields on the extraction of phenolic compounds during the fermentation of must of Tempranillo grapes. *Innov. Food Sci. Emerg. Technol.* **2008**, *9*, 477–482. [CrossRef]
28. Puértolas, E.; López, N.; Condón, S.; Álvarez, I.; Raso, J. Potential applications of PEF to improve red wine quality. *Trends Food Sci. Technol.* **2010**, *21*, 247–255. [CrossRef]
29. Puértolas, E.; Saldaña, G.; Álvarez, I.; Raso, J. Experimental design approach for the evaluation of anthocyanin content of rosé wines obtained by pulsed electric fields. Influence of temperature and time of maceration. *Food Chem.* **2011**, *126*, 1482–1487. [CrossRef]
30. Ghafoor, K.; Choi, Y.H.; Jeon, J.Y.; Jo, I.H. Optimization of Ultrasound-Assisted Extraction of Phenolic Compounds, Antioxidants, and Anthocyanins from Grape (*Vitis vinifera*) Seeds. *J. Agric. Food Chem.* **2009**, *57*, 4988–4994. [CrossRef]
31. Tiwari, B.K.; Patras, A.; Brunton, N.; Cullen, P.J.; O'Donnell, C.P. Effect of ultrasound processing on anthocyanins and color of red grape juice. *Ultrason. Sonochem.* **2010**, *17*, 598–604. [CrossRef]
32. Liazid, A.; Barbero, G.F.; Azaroual, L.; Palma, M.; Barroso, C.G. Stability of Anthocyanins from Red Grape Skins under Pressurized Liquid Extraction and Ultrasound-Assisted Extraction Conditions. *Molecules* **2014**, *19*, 21034–21043. [CrossRef] [PubMed]
33. Bonfigli, M.; Godoy, E.; Reinheimer, M.A.; Scenna, N.J. Comparison between conventional and ultrasound-assisted techniques for extraction of anthocyanins from grape pomace. Experimental results and mathematical modeling. *J. Food Eng.* **2017**, *207*, 56–72. [CrossRef]
34. Tan, J.; Li, Q.; Xue, H.; Tang, J. Ultrasound-assisted enzymatic extraction of anthocyanins from grape skins: Optimization, identification, and antitumor activity. *J. Food Sci.* **2020**, *85*, 3731–3744. [CrossRef] [PubMed]
35. Ayed, N.; Yu, H.L.; Lacroix, M. Improvement of anthocyanin yield and shelf-life extension of grape pomace by gamma irradiation. *Food Res. Int.* **1999**, *32*, 539–543. [CrossRef]
36. Ayed, N.; Yu, H.L.; Lacroix, M. Using gamma irradiation for the recovery of anthocyanins from grape pomace. *Radiat. Phys. Chem.* **2000**, *57*, 277–279. [CrossRef]
37. Morata, A.; Bañuelos, M.A.; Tesfaye, W.; Loira, I.; Palomero, F.; Benito, S.; Callejo, M.J.; Villa, A.; González, M.C.; Suárez-Lepe, J.A. Electron Beam Irradiation of wine grapes: Effect on microbial populations, phenol extraction and wine quality. *Food Bioprocess Technol.* **2015**, *8*, 1845–1853. [CrossRef]
38. Bermúdez-Aguirre, D.; Barbosa-Cánovas, G.V. An Update on High Hydrostatic Pressure, from the Laboratory to Industrial Applications. *Food Eng. Rev.* **2010**, *3*, 44–61. [CrossRef]
39. Yamamoto, K. Food processing by high hydrostatic pressure. *Biosci. Biotechnol. Biochem.* **2017**, *81*, 672–679. [CrossRef] [PubMed]
40. Huang, H.W.; Hsu, C.P.; Wang, C.Y. Healthy expectations of high hydrostatic pressure treatment in food processing industry. *J. Food Drug Anal.* **2020**, *28*, 1–13. [CrossRef] [PubMed]
41. Zamora, A.; Guamis, B. Opportunities for Ultra-High-Pressure Homogenisation (UHPH) for the Food Industry. *Food Eng. Rev.* **2014**, *7*, 130–142. [CrossRef]
42. Buzrul, S. High hydrostatic pressure treatment of beer and wine: A review. *Innov. Food Sci. Emerg. Technol.* **2012**, *13*, 1–12. [CrossRef]
43. Martín, J.; Asuero, A.G. High hydrostatic pressure for recovery of anthocyanins: Effects, performance, and applications. *Sep. Purif. Rev.* **2021**, *50*, 159–176. [CrossRef]
44. Tiwari, B.K.; O'Donnell, C.P.; Cullen, P.J. Effect of non thermal processing technologies on the anthocyanin content of fruit juices. *Trends Food Sci. Technol.* **2009**, *20*, 137–145. [CrossRef]
45. Bañuelos, M.A.; Loira, I.; Escott, C.; Del Fresno, J.M.; Morata, A.; Sanz, P.D.; Otero, L.; Suárez-Lepe, J.A. Grape processing by High Hydrostatic Pressure: Effect on use of non-*Saccharomyces* in must fermentation. *Food Bioprocess Technol.* **2016**, *9*, 1769–1778. [CrossRef]
46. Christofi, S.; Malliaris, D.; Katsaros, G.; Panagou, E.; Kallithraka, S. Limit SO_2 content of wines by applying High Hydrostatic Pressure. *Innov. Food Sci. Emerg. Technol.* **2020**, *62*, 102342. [CrossRef]
47. Vaquero, C.; Escott, C.; Loira, I.; Guamis, B.; Del Fresno, J.M.; Quevedo, J.M.; Gervilla, R.; De Lamo, S.; Ferrer-Gallego, R.; González, C.; et al. Cabernet sauvignon red wine processing by UHPH without SO_2. Colloidal structure, microbial and oxidation control, colour protection and sensory quality of wine. 2021; submitted.

48. Tedjo, W.; Eshtiaghi, M.N.; Knorr, D. Einsatz nicht-thermischer Verfahren zur Zellpermeabilisierung von Weintrauben und Gewinnung von Inhaltsstoffen. *Fluss. Obs.* **2002**, *69*, 578–585.
49. López, N.; Puértolas, E.; Condón, S.; Álvarez, I.; Raso, J. Application of pulsed electric fields for improving the maceration process during vinification of red wine: Influence of grape variety. *Eur. Food Res. Technol.* **2008**, *227*, 1099–1107. [CrossRef]
50. Leong, S.Y.; Burritt, D.J.; Oey, I. Evaluation of the anthocyanin release and health-promoting properties of Pinot Noir grape juices after pulsed electric fields. *Food Chem.* **2016**, *196*, 833–841. [CrossRef]
51. Puértolas, E.; López, N.; Saldaña, G.; Álvarez, I.; Raso, J. Evaluation of phenolic extraction during fermentation of red grapes treated by a continuous pulsed electric fields process at pilot-plant scale. *J. Food Eng.* **2010**, *98*, 120–125. [CrossRef]
52. Luengo, E.; Franco, E.; Ballesteros, F.; Álvarez, I.; Raso, J. Winery Trial on Application of Pulsed Electric Fields for Improving Vinification of Garnacha Grapes. *Food Bioprocess Technol.* **2013**, *7*, 1457–1464. [CrossRef]
53. Leong, S.Y.; Treadwell, M.; Liu, T.; Hochberg, M.; Sack, M.; Mueller, G.; Sigler, J.; Silcock, P.; Oey, I. Influence of Pulsed Electric Fields processing at high-intensity electric field strength on the relationship between anthocyanins composition and colour intensity of Merlot (*Vitis vinifera* L.) musts during cold maceration. *Innov. Food Sci. Emerg. Technol.* **2020**, *59*, 102243. [CrossRef]
54. Comuzzo, P.; Voce, S.; Grazioli, C.; Tubaro, F.; Marconi, M.; Zanella, G.; Querzè, M. Pulsed Electric Field processing of red grapes (cv. *Rondinella*): Modifications of phenolic fraction and effects on wine evolution. *Foods* **2020**, *9*, 414. [CrossRef] [PubMed]
55. Vaquero, C.; Loira, I.; Raso, J.; Álvarez, I.; Delso, C.; Morata, A. Pulsed Electric Fields to Improve the Use of Non-*Saccharomyces* Starters in Red Wines. *Foods* **2021**, *10*, 1472. [CrossRef]
56. González, M.; Barrios, S.; Budelli, E.; Pérez, N.; Lema, P.; Heinzen, H. Ultrasound assisted extraction of bioactive compounds in fresh and freeze-dried *Vitis vinifera* cv Tannat grape pomace. *Food Bioprod. Process.* **2020**, *124*, 378–386. [CrossRef]
57. Osete-Alcaraz, A.; Bautista-Ortín, A.B.; Ortega-Regules, A.E.; Gómez-Plaza, E. Combined Use of Pectolytic Enzymes and Ultrasounds for Improving the Extraction of Phenolic Compounds During Vinification. *Food Bioprocess Technol.* **2019**, *12*, 1330–1339. [CrossRef]
58. Romero-Díez, R.; Matos, M.; Rodrigues, L.; Bronze, M.R.; Rodríguez-Rojo, S.; Cocero, M.J.; Matias, A.A. Microwave and ultrasound pre-treatments to enhance anthocyanins extraction from different wine lees. *Food Chem.* **2019**, *272*, 258–266. [CrossRef] [PubMed]
59. Dranca, F.; Oroian, M. Kinetic Improvement of Bioactive Compounds Extraction from Red Grape (*Vitis vinifera* Moldova) Pomace by Ultrasonic Treatment. *Foods* **2019**, *8*, 353. [CrossRef] [PubMed]
60. Patrignani, F.; Lanciotti, R. Applications of High and Ultra High Pressure Homogenization for Food Safety. *Front. Microbiol.* **2016**, *7*, 1132. [CrossRef] [PubMed]
61. Comuzzo, P.; Calligaris, S. Potential Applications of High Pressure Homogenization in Winemaking: A Review. *Beverages* **2019**, *5*, 56. [CrossRef]
62. Loira, I.; Morata, A.; Bañuelos, M.A.; Puig-Pujol, A.; Guamis, B.; González, C.; Suárez-Lepe, J.A. Use of Ultra-High Pressure Homogenization processing in winemaking: Control of microbial populations in grape musts and effects in sensory quality. *Innov. Food Sci. Emerg. Technol.* **2018**, *50*, 50–56. [CrossRef]
63. Marszałek, K.; Woźniak, Ł.; Kruszewski, B.; Skąpska, S. The Effect of High Pressure Techniques on the Stability of Anthocyanins in Fruit and Vegetables. *Int. J. Mol. Sci.* **2017**, *18*, 277. [CrossRef]
64. Bañuelos, M.A.; Loira, I.; Buenaventura, G.; Escott, C.; del Fresno, J.M.; Codina-Torrella, I.; Quevedo, J.M.; Gervilla, R.; Rodríguez Chavarría, J.M.; de Lamo, S.; et al. White wine processing by UHPH without SO_2. Elimination of microbial populations and effect in oxidative enzymes, colloidal stability and sensory quality. *Food Chem.* **2020**, *332*, 127417. [CrossRef]
65. Bevilacqua, A.; Petruzzi, L.; Perricone, M.; Speranza, B.; Campaniello, D.; Sinigaglia, M.; Corbo, M.R. Nonthermal Technologies for Fruit and Vegetable Juices and Beverages: Overview and Advances. *Compr. Rev. Food Sci. Food Saf.* **2018**, *17*, 2–62. [CrossRef]
66. Guamis López, B.; Trujillo Mesa, A.J.; Ferragut Péréz, V.; Quevedo Terré, J.M.; Lopez Pedemonte, T.; Buffa Dunat, M.N. Continuous System and Procedure of Sterilization and Physical Stabilization of Pumpable Fluids by Means of Ultra-High Pressure Homogenization. Patent Number ES2543365T3, 18 August 2015. Available online: https://patents.google.com/patent/ES25433 65T3/en (accessed on 2 November 2021).
67. Barbosa-Canovas, G.V.; Pothakamury, U.R.; Gongora-Nieto, M.M.; Swanson, B.G. *Preservation of Foods with Pulsed Electric Fields*; Taylor, S.L., Ed.; Elsevier: San Diego, CA, USA, 1999.
68. Soliva-Fortuny, R.; Balasa, A.; Knorr, D.; Martín-Belloso, O. Effects of pulsed electric fields on bioactive compounds in foods: A review. *Trends Food Sci. Technol.* **2009**, *20*, 544–556. [CrossRef]
69. Toepfl, S.; Siemer, C.; Saldaña-Navarro, G.; Heinz, V. Overview of Pulsed Electric Fields Processing for Food. In *Emerging Technologies for Food Processing*; Sun, D.-W., Ed.; Academic Press: London, UK, 2014; pp. 93–114.
70. Barba, F.J.; Parniakov, O.; Pereira, S.A.; Wiktor, A.; Grimi, N.; Boussetta, N.; Saraiva, J.A.; Raso, J.; Martin-Belloso, O.; Witrowa-Rajchert, D.; et al. Current applications and new opportunities for the use of pulsed electric fields in food science and industry. *Food Res. Int.* **2015**, *77*, 773–798. [CrossRef]
71. Gabrić, D.; Barba, F.; Roohinejad, S.; Gharibzahedi, S.M.T.; Radojčin, M.; Putnik, P.; Kovačević, D.B. Pulsed electric fields as an alternative to thermal processing for preservation of nutritive and physicochemical properties of beverages: A review. *J. Food Process Eng.* **2018**, *41*, e12638. [CrossRef]
72. Barba, F.J.; Grimi, N.; Vorobiev, E. New Approaches for the Use of Non-conventional Cell Disruption Technologies to Extract Potential Food Additives and Nutraceuticals from Microalgae. *Food Eng. Rev.* **2014**, *7*, 45–62. [CrossRef]

73. Angersbach, A.; Heinz, V.; Knorr, D. Effects of pulsed electric fields on cell membranes in real food systems. *Innov. Food Sci. Emerg. Technol.* **2000**, *1*, 135–149. [CrossRef]
74. Heinz, V.; Alvarez, I.; Angersbach, A.; Knorr, D. Preservation of liquid foods by high intensity pulsed electric fields—basic concepts for process design. *Trends Food Sci. Technol.* **2001**, *12*, 103–111. [CrossRef]
75. Arcena, M.R.; Leong, S.Y.; Then, S.; Hochberg, M.; Sack, M.; Mueller, G.; Sigler, J.; Kebede, B.; Silcock, P.; Oey, I. The effect of pulsed electric fields pre-treatment on the volatile and phenolic profiles of Merlot grape musts at different winemaking stages and the sensory characteristics of the finished wines. *Innov. Food Sci. Emerg. Technol.* **2021**, *70*, 102698. [CrossRef]
76. Martin, M.E.; Grao-Cruces, E.; Millan-Linares, M.C.; Montserrat-de la Paz, S. Grape (*Vitis vinifera* L.) Seed Oil: A Functional Food from the Winemaking Industry. *Foods* **2020**, *9*, 1360. [CrossRef] [PubMed]
77. Kumar, K.; Srivastav, S.; Sharanagat, V.S. Ultrasound assisted extraction (UAE) of bioactive compounds from fruit and vegetable processing by-products: A review. *Ultrason. Sonochem.* **2021**, *70*, 105325. [CrossRef]
78. Chemat, F.; Rombaut, N.; Sicaire, A.G.; Meullemiestre, A.; Fabiano-Tixier, A.S.; Abert-Vian, M. Ultrasound assisted extraction of food and natural products. Mechanisms, techniques, combinations, protocols and applications. A review. *Ultrason. Sonochem.* **2017**, *34*, 540–560. [CrossRef]
79. Tiwari, B.K. Ultrasound: A clean, green extraction technology. *TrAC Trends Anal. Chem.* **2015**, *71*, 100–109. [CrossRef]
80. Chemat, F.; Zill-E-Huma; Khan, M.K. Applications of ultrasound in food technology: Processing, preservation and extraction. *Ultrason. Sonochem.* **2011**, *18*, 813–835. [CrossRef] [PubMed]
81. Grönroos, A.; Pirkonen, P.; Ruppert, O. Ultrasonic depolymerization of aqueous carboxymethylcellulose. *Ultrason. Sonochem.* **2004**, *11*, 9–12. [CrossRef]
82. Gómez-Plaza, E.; Jurado, R.; Iniesta, J.A.; Bautista-Ortín, A.B. High power ultrasounds: A powerful, non-thermal and green technique for improving the phenolic extraction from grapes to must during red wine vinification. *BIO Web Conf.* **2019**, *12*, 1–5.
83. Ghafoor, K.; Hui, T.; Choi, Y.H. Optimization of ultrasonic-assisted extraction of total anthocyanins from grape peel using response surface methodology. *J. Food Biochem.* **2011**, *35*, 735–746. [CrossRef]
84. El Darra, N.; Grimi, N.; Maroun, R.G.; Louka, N.; Vorobiev, E. Pulsed electric field, ultrasound, and thermal pretreatments for better phenolic extraction during red fermentation. *Eur. Food Res. Technol.* **2012**, *236*, 47–56. [CrossRef]
85. Elias, M.I.; Madureira, J.; Santos, P.M.P.; Carolino, M.M.; Margaça, F.M.A.; Cabo Verde, S. Preservation treatment of fresh raspberries by e-beam irradiation. *Innov. Food Sci. Emerg. Technol.* **2020**, *66*, 102487. [CrossRef]
86. Farkas, J.; Mohácsi-Farkas, C. History and future of food irradiation. *Trends Food Sci. Technol.* **2011**, *22*, 121–126. [CrossRef]
87. Lung, H.M.; Cheng, Y.C.; Chang, Y.H.; Huang, H.W.; Yang, B.B.; Wang, C.Y. Microbial decontamination of food by electron beam irradiation. *Trends Food Sci. Technol.* **2015**, *44*, 66–78. [CrossRef]
88. Błaszak, M.; Nowak, L.; Lachowicz, S.; Migdał, W.; Ochmian, I. E-Beam Irradiation and Ozonation as an Alternative to the Sulphuric Method of Wine Preservation. *Molecules* **2019**, *24*, 3406. [CrossRef]
89. Fernandes, Â.; Antonio, A.L.; Oliveira, M.B.P.P.; Martins, A.; Ferreira, I.C.F.R. Effect of gamma and electron beam irradiation on the physico-chemical and nutritional properties of mushrooms: A review. *Food Chem.* **2012**, *135*, 641–650. [CrossRef]
90. Wong, P.Y.Y.; Kitts, D.D. Factors influencing ultraviolet and electron beam irradiation-induced free radical damage of ascorbic acid. *Food Chem.* **2001**, *74*, 75–84. [CrossRef]
91. Gamonpilas, C.; Buathongjan, C.; Sangwan, W.; Rattanaprasert, M.; Weizman, K.C.; Klomtun, M.; Phonsatta, N.; Methacanon, P. Production of low molecular weight pectins via electron beam irradiation and their potential prebiotic functionality. *Food Hydrocoll.* **2021**, *113*, 106551. [CrossRef]
92. Li, F.; Chen, G.; Zhang, B.; Fu, X. Current applications and new opportunities for the thermal and non-thermal processing technologies to generate berry product or extracts with high nutraceutical contents. *Food Res. Int.* **2017**, *100*, 19–30. [CrossRef]
93. Kong, Q.; Wu, A.; Qi, W.; Qi, R.; Carter, J.M.; Rasooly, R.; He, X. Effects of electron-beam irradiation on blueberries inoculated with *Escherichia coli* and their nutritional quality and shelf life. *Postharvest Biol. Technol.* **2014**, *95*, 28–35. [CrossRef]
94. Yoon, Y.-S.; Kim, J.-K.; Lee, K.-C.; Eun, J.-B.; Park, J.-H. Effects of electron-beam irradiation on postharvest strawberry quality. *J. Food Process. Preserv.* **2020**, *44*, e14665. [CrossRef]
95. Gómez-López, V.M.; Ragaert, P.; Debevere, J.; Devlieghere, F. Pulsed light for food decontamination: A review. *Trends Food Sci. Technol.* **2007**, *18*, 464–473. [CrossRef]
96. Santamera, A.; Escott, C.; Loira, I.; Del Fresno, J.M.; González, C.; Morata, A. Pulsed light: Challenges of a non-thermal sanitation technology in the winemaking industry. *Beverages* **2020**, *6*, 1–16. [CrossRef]
97. Gómez-López, V.M.; Pataro, G.; Tiwari, B.K.; Gozzi, M.; Meireles, M.A.A.; Shaojin, W.; Buenaventura, G.; Pan, Z.; Ramaswamy, H.S.; Sastry, S.; et al. Guidelines on reporting treatment conditions for emerging technologies in food processing. *Crit. Rev. Food Sci. Nutr.* **2021**, 1–25. [CrossRef]
98. Escott, C.; Vaquero, C.; del Fresno, J.M.; Bañuelos, M.A.; Loira, I.; Han, S.-y.; Bi, Y.; Morata, A.; Suárez-Lepe, J.A. Pulsed Light Effect in Red Grape Quality and Fermentation. *Food Bioprocess Technol.* **2017**, *10*, 1540–1547. [CrossRef]
99. Escott, C.; López, C.; Loira, I.; González, C.; Bañuelos, M.A.; Tesfaye, W.; Suárez-Lepe, J.A.; Morata, A. Improvement of must fermentation from late harvest cv. Tempranillo grapes treated with pulsed light. *Foods* **2021**, *10*, 1416. [CrossRef]
100. Avalos-Llano, K.R.; Martín-Belloso, O.; Soliva-Fortuny, R. Effect of pulsed light treatments on quality and antioxidant properties of fresh-cut strawberries. *Food Chem.* **2018**, *264*, 393–400. [CrossRef]

Review

Metabolomics of Chlorophylls and Carotenoids: Analytical Methods and Metabolome-Based Studies

María Roca and Antonio Pérez-Gálvez *

Food Phytochemistry Department, Instituto de la Grasa (CSIC), Building 46, 41013 Sevilla, Spain; mroca@ig.csic.es
* Correspondence: aperez@ig.csic.es; Tel.: +34-954611550

Abstract: Chlorophylls and carotenoids are two families of antioxidants present in daily ingested foods, whose recognition as added-value ingredients runs in parallel with the increasing number of demonstrated functional properties. Both groups include a complex and vast number of compounds, and extraction and analysis methods evolved recently to a modern protocol. New methodologies are more potent, precise, and accurate, but their application requires a better understanding of the technical and biological context. Therefore, the present review compiles the basic knowledge and recent advances of the metabolomics of chlorophylls and carotenoids, including the interrelation with the primary metabolism. The study includes material preparation and extraction protocols, the instrumental techniques for the acquisition of spectroscopic and spectrometric properties, the workflows and software tools for data pre-processing and analysis, and the application of mass spectrometry to pigment metabolomics. In addition, the review encompasses a critical description of studies where metabolomics analyses of chlorophylls and carotenoids were developed as an approach to analyzing the effects of biotic and abiotic stressors on living organisms.

Keywords: antioxidants; carotenoids; chlorophylls; extraction methods; novel analytical technologies; metabolomics; mass spectrometry; metabolism; pathways; pigments

1. Introduction

Metabolomics is an essential approach that allows for the acquisition of knowledge regarding the actual composition of complex mixtures of extracts from tissues of plant or animal origin. The development of metabolomics is only feasible with a holistic methodology, applying a multifaceted and interdependent sequence of experiments, techniques, and computational tools [1]. Accordingly, the successful application of metabolomics depends on the successful selection or development of extraction protocols; the arrangement of the suitable analytical platform for analyses; the implementation of software for data gathering, handling, and analysis of results, where an expert-curated learning attitude is fundamental; and, finally, the application of statistics to extract the information within a biological context [2]. Nevertheless, the importance of metabolomics lies in the information regarding the physiology of an organism, tissue, cell, etc. Indeed, metabolomics is a source that reflects a biochemical state or activity.

This review is focused on the metabolomics of chlorophylls and carotenoids, which was named "pigmentomic", as a tool for exploring their antioxidant features within the secondary plant metabolism. The antioxidant properties of both families of pigments have been deeply investigated and recently reviewed [3]. To gain an idea of the present impact of this topic, we performed a reference search in the Web of Science (ISI Web of Knowledge) databases, introducing "metabolomic*" and "chlorophyll*" as topics, and a total of 380 results were obtained (Figure 1). Moreover, when the topics "metabolomic*" and "carotenoid*" are selected, 499 results arise. However, the interesting point of both surveys is their time evolution, as half of the manuscripts were published in 2019 or thereafter, a clear signal of the exponential growth rate of metabolomics studies focusing on chlorophylls and carotenoids. This review includes metabolomics and metabonomics studies, as the difference between both terms is author dependent, and each term was defined as a subset of the other. It can be assumed that in metabolomics (stated by Fiehn and collaborators in 2001) [4], studies are necessary to identify and quantify all endogenous metabolites, while the metabonomics assessment (created by Nicholson et al. in 1999) [5] aims to identify a metabolite fingerprint. In a broad sense, a metabolomics strategy utilizes a mixture of separation techniques, such as HPLC or GC-MS, while in metabonomics studies, the use of NMR spectroscopy is more frequent.

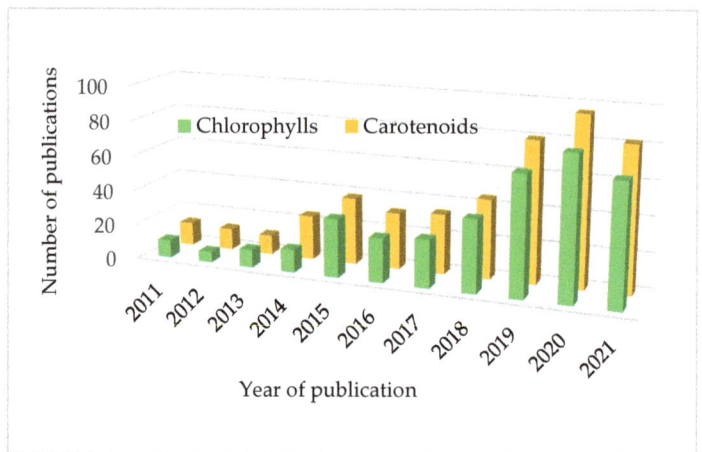

Figure 1. Number of publications since 2011 in the Web of Science (ISI Web of Knowledge) databases, introducing "metabolomic*" and "chlorophyll*" (green series) and "metabolomic*" and "carotenoid*" (orange series) as topics. The year 2021 does not cover the whole year and takes into account publications from January to September only.

We summarize the current understanding of how metabolomics describes fluctuations in chlorophylls and carotenoids, which perform essential functions and actions in photosynthetic organisms and animals that incorporate them through diet. Their involvement in plant biochemistry as key network components suggests that chlorophylls and carotenoids are key compounds involved in significant metabolic pathways. The review starts with a general description of the application of techniques for sample preparation and the acquisition of extracts suitable for analysis; a picture of the analytical platform and technologies applied for the identification and quantification of the pigment profile; and the workflow for data analysis, including software tools and the application of metabolite databases and statistics. Next, we compile the works where the metabolomics of chlorophylls and/or carotenoids has proved extremely valuable in the recognition or tuning of metabolic pathways correlated with responses to different abiotic and biotic factors; physiologic and biologic studies; and even applications to animal health. In this sense, the aim of the present review is to strengthen the potentiality of the metabolomics studies of chlorophylls and carotenoids. Moving on from an analytical determination, metabolomics is a powerful tool for comprehensive research, with multiple and diverse applications, as will be shown in this review.

2. Biochemistry of Chlorophylls and Carotenoids
2.1. Chlorophylls

Chlorophylls comprise a homogeneous group of more than 100 different structures with a unique configuration in nature. Their primary function is associated with photosynthesis, being functionals during the charge separation in the reaction centers or transferring energy in the harvesting complex. Unfortunately, this essential role has masked, other actions of chlorophylls in nature, including the interrelation with the general metabolism, and has led to the underestimation of their physiologic functions.

Chlorophylls are tetrapyrroles with an additional fifth isocyclic ring (Figure 2). They are coordinated generally with a central atom of magnesium, although this can be substituted by hydrogen or other divalent cations. In parallel, the propionic acid at $C17^3$ is esterified with a phytyl chain ($C_{20}H_{39}$), but different chlorophyll structures arise from esterification with multiple alcohols, and they can even occur in a non-esterified form (as pheophorbides). Chemically, depending on the degree of unsaturation of the macrocycle, chlorophylls could be classified as chlorin type (chlorophyll *a* and *b* among others), porphyrin type (chlorophyll *c*), or bacteriochlorin type (as certain bacteriochlorophylls), which are responsible for a complex array of different chlorophyll metabolites. Moreover, during natural (senescence or ripening) metabolism or food processing or storage, chlorophylls can be oxidized to form new chlorophylls. Among the most common are 13^2-hydroxy-chlorophylls, which are formed if the hydroxyl group is introduced at $C13^2$. In addition, $C15^1$-hydroxylactone-chlorophylls are formed if a lactone group is formed at $C15^1$, and pyroderivatives are formed if the carboxymethoxy group at $C13^2$ is lost.

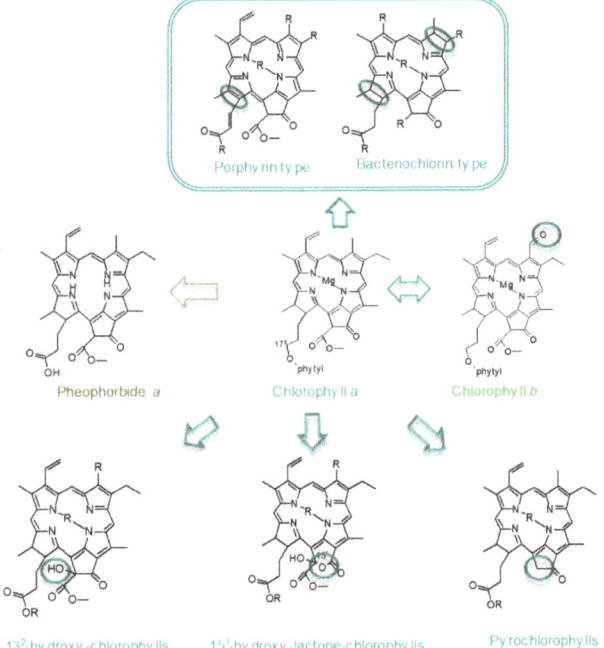

Figure 2. Main chlorophyll structures present in organisms or food due to natural metabolism or during processing or storage.

Biochemically, the chlorophyll metabolism is divided among synthesis, the chlorophyll cycle, and degradation, and it is independently regulated. Few interesting reviews have detailed the complete set of biochemical reactions, enzymes, and genes implicated in their metabolism [6–9], and, consequently, we only delineate the main reactions for a general outlook (Figure 3) in this review. Chlorophyll synthesis is initiated from the amino acid metabolism, specifically from aminolevulinic or glutamic acids, depending on the researcher. Different condensation, reductions, and decarboxylations generate protoporphyrin IX, which is the first colored chlorophyll metabolite. This point of the route is a hotspot, as it is where the branch toward heme metabolism occurs if Fe-chelatase inserts Fe in the tetrapyrrole, or, similarly, where the branch toward the chlorophyll metabolism occurs if Mg-chelatase catalyzes the reaction. Following Mg-protoporphyrin IX and after several reactions, protochlorophyllide a is formed. This compound is an interesting metabolite because the subsequent reaction is light dependent in angiosperms and thus responsible for the etiolated plants in dark conditions. After several reactions, chlorophylls a and b are synthesized, with the functional capacity of interconversion through a plastic chlorophyll cycle [9]. Such flexibility in the chlorophyll metabolism is based on the capacity to modify the relative amounts of chlorophyll a and chlorophyll b depending on the light intensity, modifying the proportion of antenna complexes and, consequently, the photosynthetic apparatus. While chlorophyll synthesis is completely developed in the chloroplast, the catabolic reactions start in the green organelle but run through the cytosol, finishing in the vacuole (Figure 3). Chlorophyll a is degraded to pheophorbide a in two reactions, liberating the magnesium atom and de-esterifying the phytol chain. Recently, it was demonstrated that phytol yielded from chlorophyll catabolism is essential for tocopherol synthesis [10]. Next, the macrocycle is oxygenolytically opened to form the first linear chlorophyll catabolite, the so-called phyllobilins due to their resemblance to the heme-derived bilins. At present, more than 40 different phyllobilins have been described [11] with unknown functions, although an antioxidant role has been assigned to them. After reduction, a fluorescent chlorophyll catabolite (FCC) is produced and exported from the chloroplast to the cytosol. FCCs could be modified in the cytosol and imported into the vacuole, where the acidic pH promotes isomerization to non-fluorescent chlorophyll catabolites (NCCs). Although a phyllobilin database for *Arabidopsis thaliana* [12] is already available, a complete database containing all phyllobilins identified at present in multiple species is necessary.

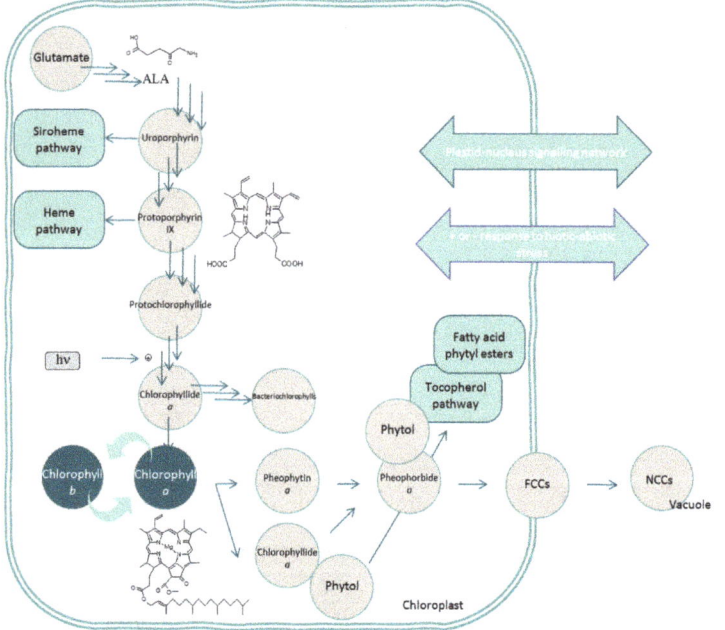

Figure 3. Brief description of the biosynthesis and catabolism of chlorophylls and routes related to other phytochemicals.

As previously stated, in addition to their key role in photosynthesis, chlorophyll compounds are implicated in different physiological actions and biochemical reactions. The photodynamic properties of several chlorophyll metabolites allow them to be implicated in the ROS response [13] and, consequently, as shown below, on different mechanisms, such as defense, stress, and cell death. However, multiple pieces of evidence demonstrate the antioxidant properties of chlorophylls [3]. Another example of the superficial valuation of chlorophylls is the simple determination of chlorophylls as a simple symptom of senescence. If we bear in mind the fact that the physical presence of chlorophylls *a* and *b* is necessary for the assembly of the photosynthetic apparatus, it can be understood that organisms named stay-greens (with a deficiency in senescence) have, in many cases, been identified as mutants in chlorophyll degradation genes.

2.2. Carotenoids

Carotenoids are a family of naturally occurring yellow, red, and orange pigments chemically derived from isoprenoids that group together ca. 1200 compounds [14]. Carotenoids are lipophilic compounds synthesized in plastids. In chloroplasts, carotenoids have an essential role in photosynthesis, assisting in harvesting light energy by transferring it to the chlorophylls and protecting the photosynthetic apparatus by quenching triplet excited states of chlorophyll molecules, singlet oxygen, and carboxy radicals [15]. Additionally, they are precursors to phytohormones and other signaling compounds [16,17]. These functions in photosynthesis, photoprotection, and key metabolic pathways make carotenoids essential metabolites. However, the biosynthesis of secondary taxon-specific carotenoids also occurs in chromoplasts, and it is linked with other roles and actions, such as antioxidant activity not being related to photosynthesis and carotenoids serving as intermediates in plant-animal interactions by furnishing flowers and fruits with fragrances and colors [18,19]. Carotenoids with specific structural arrangements are precursors for vitamin A, which has a direct impact on the function of these pigments in human nutrition [20]. Their action as antioxidants and other not yet fully understood activities in mammals have prompted evidence for their role in human health [13]. Furthermore, there is a commercial demand for carotenoids for the food, pharmacy, and cosmetics industries [21]. Altogether, this explains the enormous interest in carotenoid biosynthesis and the possibility of manipulating and engineering the carotenoid biosynthetic pathway to answer fundamental research questions and identify practical applications [22].

Carotenoid biosynthesis (Figure 4) starts with a series of isoprene condensations to yield phytoene, a substrate that undergoes desaturation and isomerization steps (yielding a group of intermediates) to form lycopene. These initial steps configure the basic structure that characterizes plant carotenoids: the typical C40 skeleton with a central polyene system that condenses the physicochemical properties of these pigments and conditions and the subsequent enzymatic processes that continue the route [23,24]. From this point, cyclization and subsequent oxygenation of the cyclic intermediates emerge as the breakthrough to the origin of a considerable diversity of carotenoid structures [25]. Different combinations of cyclic arrangements (type β and type ε) at one or both ends of the polyene system generate the first branch in the route, while the introduction of hydroxyl, keto, epoxide, etc. functions produces the classification of carotenoids in carotenes (pure hydrocarbons) and xanthophylls (oxygenated products of carotenes).

At this point, the main issue to consider in metabolome-based studies of carotenoids is the site of carotenoid biosynthesis and accumulation and the structural features that characterize the precursors, products, and catabolites of this family of natural pigments. First, carotenoids are synthesized in plastids and chloro- and chromo-plasts, meaning that compartmentation approaches can be successfully used to focus metabolomics studies and specifically analyze how the pathway operates in this separate location a priori without unexpected competition. Second, the structure of the carotenoids is the premise of solving the analytical challenge of their identification and quantification, while it includes the possibility of expanding the analysis to both parent compounds and metabolic products.

Carotenoids present a common basic skeleton (Figure 4), the polyene chain, and a combination of cyclic/linear arrangements at the ends of the skeleton with the introduction of oxygen functions at specific carbon atoms expands the number of structural blends [26]. Moreover, these structural features seem to correspond exclusively to this family of natural pigments. However, the correct identification is only feasible through the acquisition of several layers of information from different technologies (UV-visible spectrophotometry, mass spectrometry, nuclear magnetic resonance, and circular dichroism) combined with a variety of derivatization processes and a comparison with reference standards. The presence of geometric isomers, a frequent feature of carotenoids, complicates the identification task and requires the introduction of secure workflow models and a combination of analytical techniques for successful classification [27–30].

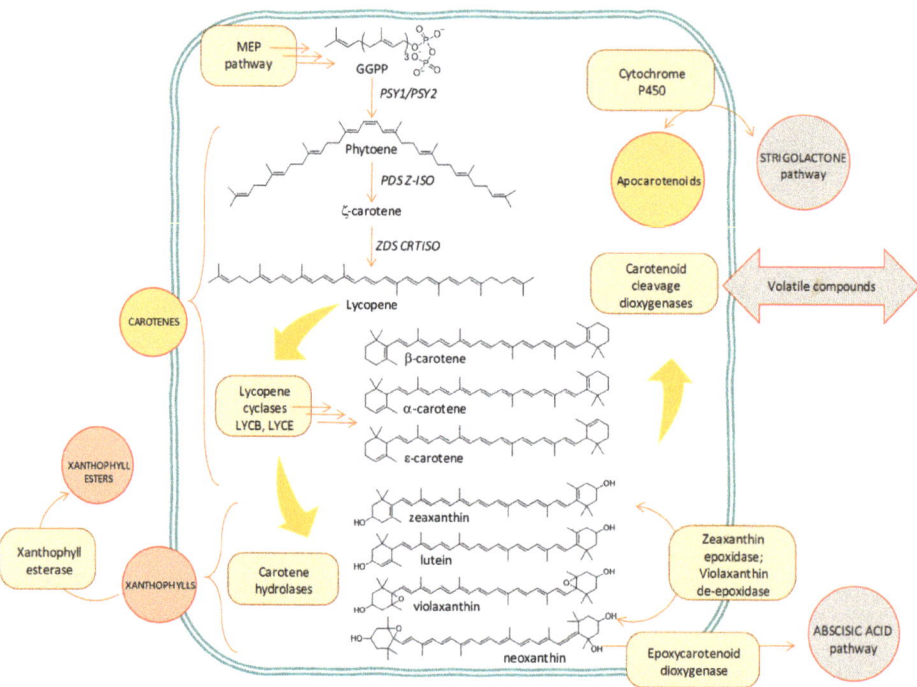

Figure 4. Scheme of the carotenoid biosynthesis route including catabolism to abscisic acid, volatile compounds, and strigolactones. MEP: methylerythritol phosphate; GGPP: geranylgeranyl pyrophosphate; PSY: phytoene synthase; PDS: phytoene desaturase; Z-ISO: ζ-carotene isomerase; ZDS: ζ-carotene desaturase; CRTISO: carotene isomerase.

3. The Praxis of Metabolomics: Essential Steps and Challenges for the Experimental Design

When working with chlorophylls and carotenoids, as with other phytochemicals, different metabolomic approaches can be developed (Figure 5). If the goal of a study can be solved with observations and the quantification of a rather limited number of metabolites, which are chosen based on previous literature reports or self-experience, targeted metabolomics is performed. If we encounter a study without a previous hypothesis, then we aim to obtain a global picture of the metabolome, measuring as many metabolites as possible, which means that untargeted metabolomics are suitable. Using this strategy, when samples are classified based on their metabolite profile, without identification of the individual peaks, fingerprinting is carried out. On the contrary, when as many compounds as possible are identified and subsequently quantified, metabolite profiling is carried out.

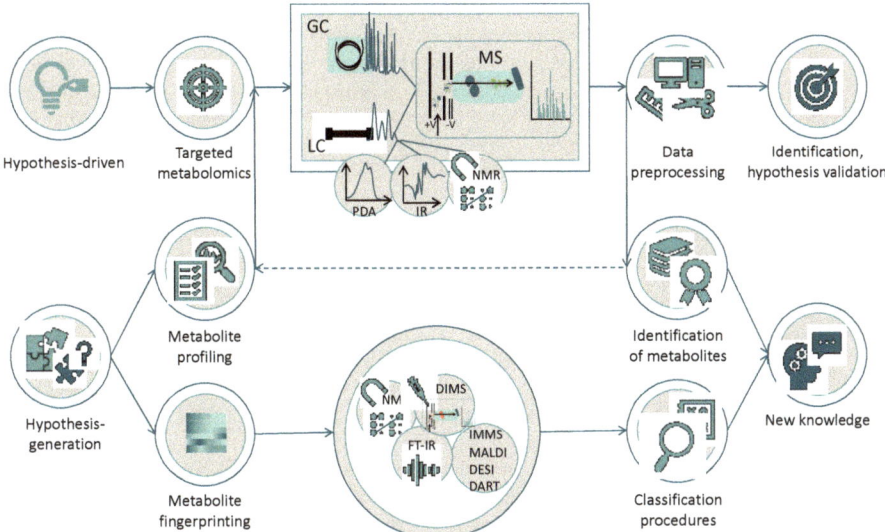

Figure 5. An outlook of the planning of a metabolomics study including the selection of the workflow (hypothesis-driven or hypothesis generation), examples of instrumental techniques, and data preprocessing and interpretation.

3.1. Material Preparation and Extraction Protocols

In metabolomics studies, frequent potential sources of bias are as follows: an unclear selection of the development stage at harvesting time, a lack of references to provide guidance on the light period and harvest duration, and a lack of a record of environmental variables and growth conditions [31]. This is crucial when working with chlorophylls and carotenoids, because the type of sample handling and applied treatments are critical to avoiding alterations to the metabolites. Additionally, it should be kept in mind that fluxes and accumulation rates are different depending on the class of metabolites (chlorophylls or carotenoids) or the metabolic process (for example, metabolites involved in the photosynthesis, antioxidant activity, catabolism of degradation products during ripening, and tissue senescence) that is focused on [32].

If the analysis of metabolites does not need the pre-processing of the tissue, then the direct flash freezing of the sample in liquid nitrogen stops metabolic conversions, and the frozen sample is homogenized into a fine powder to enhance and standardize metabolite extraction. A significant research effort has been made to refine the protocols for specific chlorophyll and carotenoid extraction, minimizing the sources of errors and increasing the reliability of the data [33]. The experimental design of most protocols aims to reduce the processing time while increasing the efficiency of the extraction. In addition, factors such as economic viability and sustainability have been introduced in the experimental design of those protocols. Therefore, different "green extraction techniques" can be applied, such as supercritical fluid extraction, microwave-assisted extraction, ultrasound-assisted extraction, pulsed electric field extraction, and extraction assisted by enzymes. These techniques have been applied mainly for carotenoid extractions, although several assays were developed for chlorophyll extractions [34]. Supercritical fluid extraction presents several advantages, such as its high purity of the extraction, simplicity, safety, and moderate temperatures [35,36]. On the contrary, it is essential to optimize the temperature and pressure conditions for a specific sample. Better results seem to be obtained when taking advantage of microwave irradiation and when applying microwave-assisted extraction [37,38]. The direct generation of heat within the matrix increases the recovery of the pigments. The studies using ultrasound-assisted extraction showed a significant reduction in the extraction time and an increase in the pigment extraction yields [39]. Pulsed electric field extraction was also used to improve pigments extractions [40,41], but its effectivity depends greatly on the intensity, amplitude, duration, number, and frequency of repetitions. However, besides the excellent results obtained with these protocols, the application of ionic solvents could be considered the most modern extraction methodology at present. An ionic solvent can be defined as compounds completely composed of ions with a melting point below 100 °C. However, additional steps of purification are required when similar structures and/or polarities are present. Therefore, an additional improvement is the set-up of the liquid-liquid extraction process using aqueous solutions of tensioactive ionic liquids and vegetable oil as an alternative to the conventional extraction processes [42], with excellent results for chlorophylls and carotenoids. However, these modern extraction protocols require a considerable amount of time to be generalized, while solvent extraction techniques are the universal protocol applied to obtain chlorophylls and/or carotenoid extracts. The high recovery and stability of the extracted compounds should be poised, and several different solvents are suitable to achieve this aim (methanol, ethanol, acetone, and mixtures at different ratios of water vs. organic solvent(s) at an acidic pH), with the help of sonication and vortex mixing. The selection of the solvent should be made considering the wide range of the polarity of compounds if untargeted metabolomics are pursued, while some solvent mixtures could be tested to extract those compounds of interest selectively for targeted metabolomics. In this case, the appearance of sample matrix effects in the subsequent instrumental analysis diminishes, while interference due to the matrix during analysis and quantification could be a serious issue in the case of metabolite profiling, which requires an almost complete extraction of metabolites.

3.2. Technologies: Instrumental Techniques for the Acquisition of Spectroscopic and Spectrometric Data

Once the extract is ready for analysis, the instrumental technique performs the acquisition of a set of data, whose complexity is related to the selected strategy for the metabolomic study (Figure 5). Fingerprinting is typically performed with ^{1}H-NMR, ignoring the problem of making individual assignments of peaks [43]. Here, the main issue is to work with signals that are typically evident as multiple peaks, hindering the analysis of data. To overcome this problem, the acquisition of ^{13}C-NMR spectra with modern probes and systems purposely created to increase the sensitivity has been noted [44]. With these approaches, the aim is to find a group of marker compounds, which are inferred from shifts of different nuclei that characterize skeletons, aromatic rings, heteroatoms, and typical structural arrangements but are not fully identified. Subsequently, statistical analysis is conducted to classify the samples and draw conclusions based on discrimination, aggrupation, or differentiation of selected variables [45,46]. Technical improvements were made in the last two decades to make definitively the combination of NMR spectroscopy with LC a successful arrangement [47,48] if metabolite profiling or targeted metabolomics is the strategy of the metabolomic study. Metabolite profiling and targeted metabolomics make use of GC, while LC can be applied to targeted and untargeted metabolomics. These techniques of analysis are coupled with one or several detection systems to achieve both the compound separation and detection, collecting spectroscopic and/or mass spectrometric data on individual components of the extract.

GC coupled with mass spectrometry (MS) is a robust chromatographic instrumental technique (Figure 5) with a high compound separation efficiency (peak widths of 2–5 s) that yields reproducible retention times. This feature allows the quick building of spectral libraries of reference analytes that boost the identification of a compound profile in the extract, with a high level of certainty in identification [49]. However, GC is only able for the analysis of thermally stable and volatile compounds (directly from the extract, or once the extract is derivatized to produce volatile products), such as carotenoid degradation products (Figure 4) or phytol (arising from chlorophyll degradation, Figure 3). Additionally, the availability of standards of carotenoid and chlorophyll volatile metabolites is still very limited, so the great advantages of the reproducibility of GC retention times and direct matching with mass spectral libraries are fully usable in metabolomic studies of primary metabolites, which have commercially available standard compounds [50–52], but not in pigmentomic studies. Despite these limitations, GC-MS is a suitable technique for the identification of the links between carotenoids, their putative signaling molecules (aroma profile), and the antioxidant potential during fruit ripening, as shown in melon [53], red pepper [54], citrus, and tomato [55], or during the processing of black tea [56] and *Mentha* species [57]. However, the applications remain scarce in the case of chlorophylls [58].

LC emerged from the principles of classic chromatography and the instrumental advances designed for GC, typically used for chlorophylls and carotenoids (Figure 5). The number of possible combinations for mobile phase composition, the increasing amount of packing materials for column building, and the high speed of the cumulative working pressure have definitively improved the efficiency and resolution of this technique, which could be easily combined with a wide range of detection systems in a single workflow [59–61]. Liquid chromatography in the classic high-pressure arrangement or the modern ultra-performance technology is typically coupled with different detectors based on optical detection (UV-visible, diode array, fluorescence, evaporative light-scattering, and differential refractive index detectors) applied to carotenoids and chlorophylls in foods [62,63] and biological samples [64,65], or in electrical detection (conductivity, electrochemical, and Corona-charged aerosol detectors), as was shown for the measurement of carotenoid bioavailability [66] and antioxidant potential [67] and in vitamin A equivalence studies [68] in humans. However, while the application of electrochemical detectors for chlorophyll analysis is rather limited [69], the holistic strategy that features metabolomics requires the application of further instrumental techniques to obtain as much information as possible from a single run, so the above-noted detection systems have begun to be combined with infrared, Raman, and NMR spectroscopies. This is the case of the metabolite profiling of microalgae species [70] and vegetable purees [71]. Soft-ionization techniques (electrospray ionization, ESI; atmospheric pressure chemical ionization, APCI) that yield protonated (positive mode) or de-protonated (negative mode) molecular ions are appropriate for the analysis of the most relevant groups of plant secondary metabolites [29,72,73], which are mainly separated with a reversed-phase column providing an efficient retention time and separation index, with a particular emphasis on the detection of isomeric compounds. Usually, APCI is used for carotenoids [74–76] and non-polar chlorophylls (chlorophylls and pheophytins) [77–79], and ESI is used in the analysis of polar chlorophylls (pheophorbide and chlorophyllide) and phyllobilins [80,81]. However, different configurations of both the ion source and mass analyzer have been implemented, including ion mobility [82] and MALDI [83,84].

To increase the reliability of data in the case of metabolite profiling/targeted metabolomics, where (tentative) identification of pigments is the aim, the acquisition of MS in a high-resolution mode, in combination with tandem MS, is almost a pre-requisite. This combination of working conditions and online experiments allows the analyst to obtain different pieces of information that are conveniently arranged in pairs of independent and orthogonal data of physicochemical properties, facilitating the implementation of workflow protocols for the characterization of chlorophyll and carotenoid metabolic profiles tailored to the selected strategy implemented in the study (targeted metabolomics, fingerprinting, or metabolite profiling) [85].

3.3. Application of Different Approaches to Pigment Metabolomics

Table 1 contains some representative work dealing with mass spectrometry in the analysis of chlorophylls and carotenoids that we examine in detail in this section. These studies may serve as the starting point to follow current strategies that successfully enhance the analysis of these plant pigments (Figure 5). The aims of these studies were diverse, so the difficulties and bottle-neck issues that were faced boosted the application of methodological approaches and solutions. Hegeman et al. [86] applied the stable isotope-assisted assignment of elemental composition to constrain the number of potential positive hits for a mass peaking procedure in the identification of chlorophyll derivatives. Similarly, Giavalisco et al. [87] provided a comprehensive multi-isotope labeling-based strategy in combination with a fractionated metabolite extraction protocol to perform unambiguous qualitative and quantitative metabolomics using *A. thaliana* leaf and root extracts.

The characteristic isotopic pattern of the copper chlorophyll derivatives is selected as a fast and specific procedure to charaterize precisely the presence of metallo-chlorophyll complexes applied to improve the green coloration of food products [88]. A novel strategy that combines UPLC coupled with traveling wave ion mobility (TWIN) and UV-visible detection is proposed to improve the characterization of chlorophylls and carotenoids analyzed in complex biological matrices [82]. A workflow strategy to perform targeted metabolomics of chlorophyll catabolites is applied to data analysis obtained by HPLC/ESI-hr-QTOF-MS from leaf and fruit senescent tissues [81,89]. Automated data analysis using multivariate curve resolution algorithms to study multi-component systems that follow additive bilinear models (pure spectrum and related time profile) is also an appropriate strategy for the analysis of pigment metabolites. With this method, Wehrens et al. [90,91] examined the metabolite profiles (carotenoids, tocopherols, and chlorophylls) of grapes (*Vitis vinifera*) and cassava (*Manihot escullenta*). Watanabe et al. [92] described a combination of analytic tools that can be used to obtain comprehensive metabolite profiles in the *A. thaliana* plant model. Another interesting approach in the metabolomic studies of chlorophylls is the determination of phytol, a direct metabolite produced by chlorophyll degradation [93] that is analyzed by GC-MS. The incorporation of Bayesian approaches to cluster accessions of *Brassica rapa* of different morphotypes and origins allows for the acquisition of association mapping between different markers and metabolites, including chlorophylls and carotenoids [94]. Authors correct for kinship and population structure with the main aim of reducing the rate of false-positive associations. The implementation of alternative separation procedures, such as supercritical fluid extraction/chromatography coupled with MS, which reduce the extraction time and analysis run time, is an increasingly applied option to achieve a reduction in inter-sample variability and the setting of batch-type applications [95,96].

Table 1. Brief description of representative works addressing mass spectrometry in the analysis of chlorophylls and carotenoids.

Raw Material	Extraction Solvent	Instrumental Techniques	Strategy for Metabolomic Study	Ref.
A. thaliana	MeOH:H$_2$O (8:2)	LC/ESI-TOF	Metabolite profiling based on isotope labeling-assisted elemental composition	[86]
A. thaliana	MeOH:MTBE:H$_2$O (1:3:1) and subsequent separation with MeOH:H$_2$O (1:3)	Multiplatform approach (UPLC-FT-MS and MS/MS, GC-MS, nUPLC-QTOF-MS, and MS/MS)	Metabolite profiling based on isotope labeling-assisted elemental composition	[87]
Olive oil, canned green vegetables	N,N-dimethylformamide	LC/APCI-ESI/hr-QTOF-MS	Metabolite profiling based on isotopic pattern	[88]
Microalgae	EtOH:hexane (2:1) and H$_2$O:hexane 1:2	UPLC-UV-TWIM-MS	Untargeted metabolomics	[82]
Lemon (Citrus lemon L.)	Acetone	LC/ESI/hr-QTOF-MS	Targeted metabolomics	[89]
A. thaliana	Ethanol	UPLC/TOF-MS	Targeted metabolomics	[92]
Wheat (Triticum aestivum)	Methanol:acetonitrile:water (4:4:2)	Multiplatform approach (GC-MS, GC-QTOF-MS, LC-MS, and LC-QTOF-MS)	Targeted and untargeted metabolomics	[93]
Tamarillo fruits (Solanum betaceum)	CO$_2$:MeOH (95:5 or 90:10)	SFE-SFC-MS	Untargeted metabolomics	[96]
Tomato (Solanum lycopersicum L.)	MeOH followed by hexane:acetone (1:1)	LC-APCI-QTOF-MS	Metabolite profiling	[97]
A. thaliana	Chloroform:MeOH:H$_2$O (2:6:2) and derivatization with methoxyamine hydrochloride and N-methyl-N-(trimethylsilyl) trifluoroacetamide	GC-TOF/MS	Metabolite profiling	[98]
S. lycopersicum L.	MeOH or MeOH:H$_2$O (75:25)	LC-QTOF-MS and LC-PDA-FD	Metabolite profiling	[99]
Zea mays	MeOH and dH$_2$O with ribitol; derivatization with methoxyamine, N,O bis(trimethylsilyl)trifluoroacetamide, and trimethylchlorosilane	GC-TOF-MS and spectrophotometry	Metabolite profiling	[100]
S. lycopersicum L.	MeOH and dH$_2$O with ribitol; derivatization with methoxyamine, N,O bis(trimethylsilyl)trifluoroacetamide, and trimethylchlorosilane	GC-TOF-MS and LC-PDA	Metabolite profiling	[101]
Cucumis melo L.	Hexane:acetone:ethanol (50:25:25)	LC-PDA	Metabolite profiling	[102]
Daucus carota, Brassica oleracea, S. lycopersicum L.	MeOH:chloroform:Tris-buffer (1.25:1:1.25, 50 mM, pH 7.5)	LC-PDA, LC-PDA-QTOF-MS, GC-MS, and ^{1}H-NMR	Targeted and untargeted metabolomics	[71]
Cuminum cyminum L.	N,N-dimethylformamide; trichloroacetic acid; chloroform:MeOH:phosphate buffer (1:2:0.9, pH 7.5)	Multiplatform approach (spectrophotometry, LC-PDA, LC-MS, and GC-MS)	Metabolite profiling	[103]
Potato (Solanum tuberosum)	MeOH:H$_2$O (87.5:12.5)	LC-ESI-QTOF-MS	Metabolite profiling	[104]

4. Metabolome-Based Studies of Chlorophylls and Carotenoids

During the initial development of metabolomics, compounds such as amino acids, organic acids, and carbohydrates were the focus of the studies. However, "pigmentomic analysis" is increasing exponentially, as researchers noticed the metabolic significance of chlorophylls and carotenoids in photosynthetic organisms. Indeed, they are valuable compounds for cells, with physiologic and economic implications. Next, we describe the main applicability areas where the metabolomics of chlorophyll and carotenoids contributes to deciphering a metabolic response. In some cases, the studies integrate metabolite and physiological data with transcriptional information to confirm both molecular and metabolic modifications. Figure 6 presents different pathways that might emerge during a metabolomics study related to chlorophylls and carotenoids.

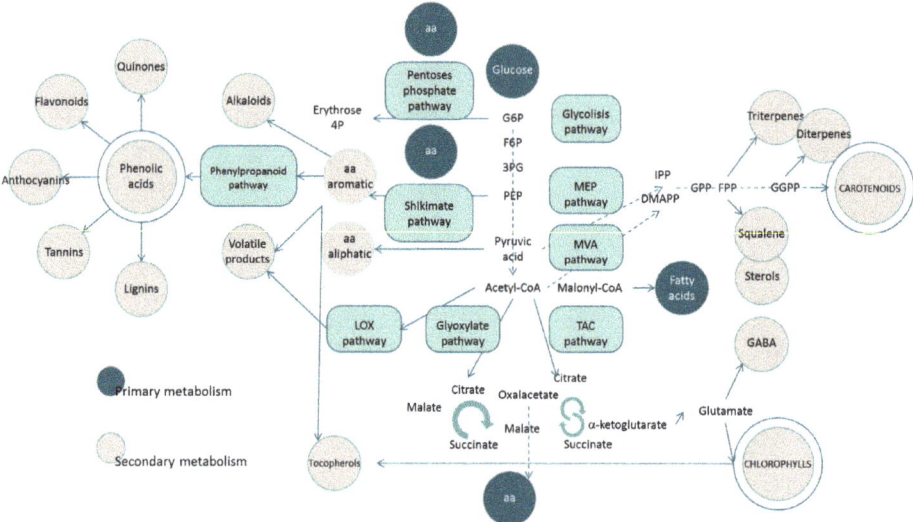

Figure 6. Brief description of some plant pathways and reaction processes that could be related to the biosynthesis and catabolism of chlorophyll and carotenoid pigments.

4.1. Application in Abiotic Factors Studies

Biological tolerance is a complex process that includes not only physio-biochemical modifications but also molecular changes. Such metabolic adjustments are required to respond to environmental signals. Consequently, metabolite profiling brings an opportunity to understand the fundamentals of tolerance by searching for modified or different signatures associated with tolerance ability. In this sense, the adaptations of the chlorophyll and carotenoid metabolism of cells exposed to different stresses have been investigated regarding cesium [105], nitric oxide [106], cadmium [107,108], graphene oxide [109], iron [110], nitrogen depletion [111–113], and extreme irradiation environments [114,115]. Another research field where the metabolomics of chlorophylls and carotenoids finds successful applications is the study of nanotoxicology, which aims to determine the toxicity of metals and micro- and nano-particles to environmental organisms and how the latter respond to the former. In this regard, it was studied how copper oxide nanoparticles, CuO microparticles, and copper ions perturb the metabolism of aquatics organisms [116] and even the effect of ZnO nanoparticles on the cultivation of terrestrial plants [117]. The most striking advance made in these studies is that the experimental design based in metabolic flux measurements might point out specific responses, which include chlorophyll breakdown and the tuning of carotenoids' metabolism. These responses reveal metabolomic-based strategies to allow acclimation of the organisms to the factor under study.

As chlorophylls and carotenoids are photosynthetic pigments, there are numerous metabolomics studies regarding the influence of light in the physiology of organisms. In this sense, the metabolomics approach has been applied to investigate the effect of light and dark cycles on the lipid metabolome [118], lineage-specific pathways [119], irradiation-induced stress [120], and photo-regulatory processes [121]. Moreover, metabolomics studies were developed to investigate the influence of LED light on the modulation of the fruit metabolome [122] or the porphyrin and chlorophyll metabolism itself [123]. These studies make use of state-of-the-art comprehensive omics analysis, together with a holistic effective treatment of data, although the translation of the results to the productive field requires further testing.

Another line of research is drought stress, considered to be one of the most important limiting environmental factors for agriculture and responsible for great losses of global food production. Once cells detect water stress, a cascade of signals activates multiple biochemical pathways (Figure 6): hormone induction, gene expression regulation, reactive oxygen species scavenging, carbohydrate and energy metabolism, nitrogen assimilation and amino acid metabolism, fatty acid metabolism, etc. Consequently, high-throughput "omics" techniques are essential to gain a holistic panoramic view of the plant response. In general, transcriptome and metabolite profiling reveals that plants respond to drought by modulating several secondary metabolic pathways and particularly by modifying the production of carotenoids or chlorophylls [124–128], including the extreme example of the adaptation to desiccation, as exhibited by resurrection plants [129].

Regarding salinity, plants have developed several mechanisms to adapt to this stress caused by osmoregulation, such as vacuolar H^+-ATPases, which are key in cytosol detoxification, as they create an electrochemical H^+ gradient across the membranes [130]. Transcriptome and metabolome analyses revealed the crucial biological pathways involved in the fast-adaptive response to salt stress, including carotenoid biosynthesis and the metabolism of porphyrin and chlorophyll [131–134]. An additional multi-omics analysis was used to unveil thermal adaptation strategies of extremophile bacteria [135] and plants [136,137], where the lipid or carotenoid metabolism seems to be implicated. The main effort that requires multi-omics analysis is to select complementary signals in the experimental design, so that the studied signals allow for a deeper understanding of the molecular adaptation of the organism to stress.

A completely different research area where pigment metabolomics was applied is the study of the environmental metabolome, which elucidates the relationship between living organisms and their ecosystem. Through the characterization of the metabolites obtained from the environment, paleometabolites (diagenetic products of chlorophylls and carotenoids derived from photosynthetic algae and bacteria) can be identified [138]. In addition, this technique can be used to determine the toxic effects of organophosphates on the species in freshwater ecosystems [139]. The above compilation is an example of the increasing research areas where chlorophyll and carotenoid metabolomics are involved. This growing trend is broadening our horizons in new, diverse disciplines with a variety of research focuses, such as the determination of metabolic turnover [140], the effects of biostimulants on the metabolome [141,142], and sustainable soil control [143]. All of them are examples of research areas where the metabolomics of chlorophylls and carotenoids has revealed as a successful approach to gather essential information.

4.2. Application in Biotic Factor Studies

The interrelation between organisms is a subject that has been scarcely studied, surely due to its complexity. However, it is in this subject where metabolomics could successfully contribute to advances in knowledge because of the inherent capacity of this approach of studying several physiological pathways, responses (Figure 6), and behaviors at the same time. Therefore, through metabolomic studies of chlorophylls and carotenoids, significant advances have been achieved with regard to the interplay between biotic stressors; the effect of single- vs. multiple-pest infestations on the biochemistry of plants [144]; the fluctuation of the leaf metabolome in response to arbuscular mycorrhizal fungi [145]; the microbial networks established during the assemblage of symbiotic microbials, such as in lichens [146]; the ecological interactions that occur at algal surfaces within microbial communities [147]; and the study of the evolutionary origin of symbiosis [148]. In parallel, metabonomic studies have also investigated the interplay between biotic and abiotic stresses, such as the effect of selenium treatments on the oxidative stress response of plants when infected [149] by the enhanced content of chlorophylls and carotenoids and related enzymatic activities. It should be highlighted that this application is very complex, such as the response to interrelation changes at the organ level, while triggering different biosynthetic pathways to down- or up-regulate them.

4.3. Application in Physiologic and Molecular Biology Studies

Metabolomics is a powerful tool that can be used not only to analyze the response of photosynthetic organisms to external abiotic or biotic stressors but also to conduct an in-depth investigation of their physiology in the widest meaning of the term. This promising research line with economic consequences is, nevertheless, a complicated area of study, taking into account the multiple variables that are accounted for. Examples of the potentiality of this implementation include the study of the mechanisms controlled by the circadian clock [98] and the effects of the auto-tetra-polyploidy on the balance between the primary and secondary metabolisms [150]. However, the main area of applicability is the behavioral patterns in the accumulation of metabolites (chlorophylls, carotenoids, etc.) paired with specific ripening stages, harvesting periods, cultivars, traceability, and plant tissue functions. Sometimes, such correlations are successfully established despite the genetic background or in a timeline fashion [101,151–156]. Additionally, it is possible to distinguish different genetic backgrounds with chemotaxonomic purposes [157], establishing species- and lineage-specific metabolites in marine microalgae [158], and differentiate chemotypes of selected accessions [159]. The metabolomics of chlorophylls and carotenoids could also be used to analyze the effects of postharvest treatments on the metabolism of edible plants [160–164] or for the identification of fast and unequivocal biochemical markers in breeding programs [165,166].

Another field of application is the utilization of metabolomics as a tool to investigate the biochemical pathways implied in the biosynthesis and degradation of these pigments (Figures 3 and 4), identifying pathway cascades [167,168] and revealing the effects of specific genes [169–173]. As a further step, the metabolomics study of pigments could be used as a platform for the development of strategies to engineer fluxes in complex biosynthetic networks [174–176]. A subset of pigment-targeted metabolomics is synthetic biology, which combines known molecular components and genes for the implementation of different molecular pathways displaying novel functions and dynamic behavior that do not occur naturally [177]. A workflow that combines gene expression and quantitative metabolomics with mathematical modeling to identify strategies in order to increase production yields of nutritionally significant pigments has even been proposed [178]. This overall approach, although highly informative and practical, could become difficult to apply as a routine method. Lastly, metabolomics is a common and useful approach for identification purposes [57,85,179,180] and the detection of food processing [71,181,182].

4.4. Application in Human Health (Health Status, Cancer, Hypertension, and Digestive Efficiency) Studies

In addition to all of these applicability areas, pigment metabolomics has also been applied to the investigation of human health. This is possible thanks to the fact that the concept of health status has moved from just "a state" to "the ability to adapt", which was denoted as phenotypic flexibility. In this context, metabolomics and proteomics were adopted to correlate micronutrients with the characteristics of metabolic parameters and, ultimately, to health-related processes [183]. A poorly scientifically explored research area is the potential bioactivity of metabolites yielded via the catabolism of chlorophylls and carotenoids. The wide array of catabolic products (Figures 3 and 4), including phyllobilins (bilin-type catabolites of chlorophylls) [184], volatile and non-volatile apocarotenoids arising from the asymmetric cleavage of carotenoids [185], and the carotenoid-derived hormones abscisic acid and strigolactones [186], deserve attention, because they perform antioxidant activities in their natural environment. Additionally, metabolomics was also proposed as a non-invasive and reliable screening technology as an alternative for cancer detection. Currently, diagnostic procedures are costly and invasive, and novel methodologies that could reduce such features of evaluation tests for patients are urgently required [187,188]. Alternative strategies to address the study of cancer are the identification of new compounds against the proliferation of selected cancer cells [189] and the review of the validity of established biomarkers of dietary intake and the identification of novel ones [190]. These studies are still in the hypothesis testing stage and although they embrace a great potential, the focus should be to establish the complex map of cancer-related activities before pointing out a direct link, either positive or negative, between carotenoids and chlorophyll metabolites and cancer effects.

5. Conclusions

Chlorophylls and carotenoids, known antioxidants, are often evaluated in metabolomics studies with regard to the matter under scrutiny (abiotic/biotic stress) and not as a marker of the metabolic status of the organism. Moreover, the evaluation of these plant pigments is performed with instrumental techniques that yield a global profile count rather than via an in-depth description of both the qualitative and quantitative aspects of the pigment catabolites. This review suggests that the assessment of processes for both primary and secondary metabolisms should consider chlorophylls and carotenoids as key contributors to metabolic study and not simply as "signaling" compounds to determine easily whether something is going wrong or well. Thus, the recently increasing number of published papers, summarized in this manuscript, addressing photosynthetic pigments and metabolomics is generating strong expectations for significant advances in our knowledge of metabolomics as a central piece of functional genomics. Indeed, the study of chlorophyll and carotenoid metabolites requires the development of a wide range of protocols, technical applications, and methodologies. This fact reflects the key role of photosynthetic pigments in the plant metabolism, chemotaxonomy, food technology, and animal health.

Author Contributions: Conceptualization, M.R. and A.P.-G.; investigation, M.R. and A.P.-G.; writing-original draft preparation, M.R. and A.P.-G.; writing-review and editing, M.R. and A.P.-G.; funding acquisition, M.R. and A.P.-G. All authors have read and agreed to the published version of the manuscript.

Funding: This research is part of the I+D+i projects RTI2018-095415-B-I00 and AGL2017-87884-R funded by MCIN/ AEI /10.13039/501100011033, and the European Regional Development Fund, FEDER "Una manera de hacer Europa".

Acknowledgments: This article is dedicated to Juan Garrido-Fernández, scientist, teacher, friend, on his retirement as *Ad Honorem Scientist* at the Spanish National Research Council (CSIC).

Conflicts of Interest: The authors declare no conflict of interest.

References

1. Hall, R.; Beale, M.; Fiehn, O.; Hardy, N.; Sumner, L.; Bino, R. Plant metabolomics: The missing link in functional genomics strategies. *Plant Cell* **2002**, *14*, 1437–1440. [CrossRef]
2. Sumner, L.W.; Lei, Z.; Nikolau, B.J.; Saito, K. Modern plant metabolomics: Advanced natural product gene discoveries, improved technologies, and future prospects. *Nat. Prod. Rep.* **2015**, *32*, 212–229. [CrossRef]
3. Pérez-Gálvez, A.; Viera, I.; Roca, M. Carotenoids and chlorophylls as antioxidants. *Antioxidants* **2020**, *9*, 505. [CrossRef]
4. Fiehn, O. Combining genomics, metabolome analysis, and biochemical modelling to understand metabolic networks. *Comp. Funct. Genom.* **2001**, *2*, 155–168. [CrossRef]
5. Nicholson, J.K.; Lindon, J.C.; Holmes, E. "Metabonomics": Understanding the metabolic responses of living systems to pathophysiological stimuli via multivariate statistical analysis of biological NMR spectroscopic data. *Xenobiotica* **1999**, *29*, 1181–1189. [CrossRef] [PubMed]

6. Brzezowski, P.; Richter, A.S.; Grimm, B. Regulation and function of tetrapyrrole biosynthesis in plants and algae. *Biochim. Biophys. Acta* **2015**, *1847*, 968–985. [CrossRef] [PubMed]
7. Kuai, B.; Chen, J.; Hörtensteiner, S. The biochemistry and molecular biology of chlorophyll breakdown. *J. Exp. Bot.* **2018**, *69*, 751–767. [CrossRef] [PubMed]
8. Grimm, B. *Metabolism, Structure and Function of Plant Tetrapyrroles: Control Mechanisms of Chlorophyll Biosynthesis and Analysis of Chlorophyll-Binding Proteins*; Academic Press: London, UK, 2019.
9. Tanaka, R.; Tanaka, A. Chlorophyll cycle regulates the construction and destruction of the light-harvesting complexes. *Biochim. Biophys. Acta* **2011**, *1807*, 968–976. [CrossRef] [PubMed]
10. Vom Dorp, K.; Hölz, G.; Plohmann, C.; Eisenhut, M.; Abraham, M.; Weber, A.P.M.; Hanson, A.D.; Dörmann, P. Remobilization of phytol from chlorophyll degradation is essential for tocopherol synthesis and growth of Arabidopsis. *Plant Cell* **2015**, *27*, 2846–2859. [CrossRef]
11. Pérez-Gálvez, A.; Roca, M. Phyllobilins: A new group of bioactive compounds. In *Studies of Natural Products Chemistry*; Atta-ur-Rahman, F.R.S., Ed.; Elsevier: Amsterdam, The Netherlands, 2017; pp. 159–191.
12. Christ, B.; Hauenstein, M.; Hörtensteiner, S. A liquid chromatography-mass spectrometry platform for the analysis of phyllobilins, the major degradation products of chlorophyll in *Arabidopsis thaliana*. *Plant J.* **2016**, *88*, 505–518. [CrossRef]
13. Busch, A.W.; Montgomery, B.L. Interdependence of tetrapyrrole metabolism, the generation of oxidative stress and the mitigative oxidative stress response. *Redox Biol.* **2015**, *4*, 260–271. [CrossRef] [PubMed]
14. Yabuzaki, J. Carotenoids Database: Structures, chemical fingerprints and distribution among organisms. *Database* **2017**, *2017*, bax004. [CrossRef]
15. Wurtzel, E.T. Changing form and function through carotenoids and synthetic biology. *Plant Physiol.* **2019**, *179*, 830–843. [CrossRef]
16. Al-Babili, S.; Bouwmeester, H.J. Strigolactones, a novel carotenoid-derived plant hormone. *Annu. Rev. Plant Biol.* **2015**, *66*, 161–186. [CrossRef] [PubMed]
17. Felemban, A.; Braguy, J.; Zurbriggen, M.D.; Al-Babili, S. Apocarotenoids involved in plant development and stress response. *Front. Plant Sci.* **2019**, *10*, 1168. [CrossRef] [PubMed]
18. Serra, S. Recent advances in the synthesis of carotenoid-derived flavours and fragrances. *Molecules* **2015**, *20*, 12817–12840. [CrossRef]
19. Águila Ruiz-Sola, M.; Rodríguez-Concepción, M. Carotenoid biosynthesis in Arabidopsis: A colorful pathway. *Arabidopsis Book* **2012**, *10*, e0158. [CrossRef]
20. Eggersdorfer, M.; Wyss, A. Carotenoids in human nutrition and health. *Arch. Biochem. Biophys.* **2018**, *652*, 18–26. [CrossRef]
21. Viera, I.; Pérez-Gálvez, A.; Roca, M. Bioaccessibility of marine carotenoids. *Mar. Drugs* **2018**, *16*, 397. [CrossRef]
22. Shumskaya, M.; Wurtzel, E.T. The carotenoid biosynthetic pathway: Thinking in all dimensions. *Plant Sci.* **2013**, *208*, 58–63. [CrossRef] [PubMed]
23. Llorente, B. Regulation of carotenoid biosynthesis in photosynthetic organs. In *Carotenoids in Nature. Subcellular Biochemistry*; Stange, C., Ed.; Springer: Cham, Switzerland; New York, NY, USA, 2016; Volume 79, pp. 141–160.
24. Lado, J.; Zacarías, L.; Rodrigo, M.J. Regulation of carotenoid biosynthesis during fruit development. In *Carotenoids in Nature. Subcellular Biochemistry*; Stange, C., Ed.; Springer: Cham, Switzerland; New York, NY, USA, 2016; Volume 79, pp. 161–198.
25. Sun, T.; Yuan, H.; Cao, H.; Yazdani, M.; Tadmor, Y.; Li, L. Carotenoid metabolism in plants: The role of plastids. *Mol. Plant* **2018**, *11*, 58–74. [CrossRef] [PubMed]
26. Britton, G.; Liaaen-Jensen, S.; Pfander, H. *Carotenoids Handbook*; Birkhäuser Verlag: Basel, Switzerland, 2004.
27. Rivera, S.M.; Canela-Garayoa, R. Analytical tools for the analysis of carotenoids in diverse materials. *J. Chromatogr. A* **2012**, *1224*, 1–10. [CrossRef]
28. Rivera, S.M.; Christou, P.; Canela-Garayoa, R. Identification of carotenoids using mass spectrometry. *Mass Spectrom Rev.* **2014**, *33*, 353–372. [CrossRef] [PubMed]
29. Pérez-Gálvez, A.; Roca, M. Recent developments in the analysis of carotenoids by mass spectrometry. In *Progress in Carotenoid Research*; Zepka, L., Ed.; IntechOpen: London, UK, 2018; pp. 17–44.
30. Fernandes, A.S.; Petry, F.C.; Mercadante, A.Z.; Jacob-Lopes, E.; Zepka, L.Q. HPLC-PDA-MS/MS as a strategy to characterize and quantify natural pigments from microalgae. *Curr. Res. Food Sci.* **2020**, *8*, 100–112. [CrossRef]
31. Fiehn, O.; Sumner, L.W.; Rhee, S.; Ward, J.; Dickerson, J.; Lange, B.M.; Lane, G.; Roessner, U.; Last, R.; Nikolau, B. Minimum reporting standards for plant biology context information in metabolomics studies. *Metabolomics* **2007**, *3*, 195–201. [CrossRef]
32. Fernie, A.R.; Morgan, J.A. Analysis of metabolic flux using dynamic labelling and metabolic modelling. *Plant Cell. Environ.* **2013**, *36*, 1738–1750. [CrossRef]
33. Rodriguez-Amaya, D.B. Update on natural food pigments—A mini-review on carotenoids, anthocyanins, and betalains. *Food Res. Int.* **2019**, *124*, 200–205. [CrossRef] [PubMed]
34. Silva Miranda, P.H.; Dos Santos, A.C.; De Freitas, B.C.B.; De Souza Martins, G.A.; De Barros Vilas Boas, E.V.; Damiani, C. A scientific approach to extraction methods and stability of pigments from Amazonian fruits. *Trends Food Sci. Technol.* **2021**, *113*, 335–345. [CrossRef]
35. Macías-Sánchez, M.D.; Mantell, C.; Rodríguez, M.; Martínez de la Ossa, E.; Lubián, L.M.; Montero, O. Comparison of supercritical fluid and ultrasound-assisted extraction of carotenoids and chlorophyll a from *Dunaliella salina*. *Talanta* **2009**, *77*, 948–952. [CrossRef]

36. Pereira, C.G.; Meireles, M.A.A. Supercritical Fluid Extraction of Bioactive Compounds: Fundamentals, Applications and Economic Perspectives. *Food Bioprocess. Technol.* **2010**, *3*, 340–372. [CrossRef]
37. Pasquet, V.; Chérouvrier, J.; Farhat, F.; Thiéry, V.; Piot, J.; Bérard, J.; Kaas, R.; Serive, B.; Patrice, T.; Cadoret, J.; et al. Study on the microalgal pigments extraction process: Performance of microwave assisted extraction. *Process. Biochem.* **2011**, *46*, 59–67. [CrossRef]
38. Kaufmann, B.; Christen, P. Recent extraction techniques for natural products: Microwave-assisted extraction and pressurised solvent extraction. *Phytochem. Anal.* **2002**, *13*, 105–113. [CrossRef] [PubMed]
39. Kumar, K.; Srivastav, S.; Sharanagat, V.S. Ultrasound assisted extraction (UAE) of bioactive compounds from fruit and vegetable processing by-products: A review. *Ultrason Sonochem.* **2021**, *70*, 105325. [CrossRef]
40. Martínez, J.M.; Schottroff, F.; Haas, K.; Fauster, T.; Sajfrtová, M.; Álvarez, I.; Raso, J.; Jaeger, H. Evaluation of pulsed electric fields technology for the improvement of subsequent carotenoid extraction from dried Rhodotorula glutinis yeast. *Food Chem.* **2020**, *323*, 126824. [CrossRef]
41. Leonhardt, L.; Käferböck, A.; Smetana, S.; de Vos, R.; Toepfl, S.; Parniakov, O. Bio-refinery of Chlorella sorokiniana with pulsed electric field pre-treatment. *Bioresour. Technol.* **2020**, *301*, 122743. [CrossRef] [PubMed]
42. Martins, M.; De Souza Mesquita, L.M.; Vaz, B.M.C.; Dias, A.C.R.V.; Torres-Acosta, M.A.; Quéguineur, B.; Coutinho, J.A.P.; Ventura, S.P.M. Extraction and fractionation of pigments from Saccharina latissima (Linnaeus, 2006) using an ionic liquid + oil + water System. *ACS Sustain. Chem. Eng.* **2021**, *9*, 6599–6612. [CrossRef]
43. Krishnan, P.; Kruger, N.J.; Ratcliffe, R.G. Metabolite fingerprinting and profiling in plants using NMR. *J. Exp. Bot.* **2005**, *56*, 255–265. [CrossRef]
44. Wei, F.; Furihata, K.; Koda, M.; Hu, F.; Kato, R.; Miyakawa, T.; Tanokura, M. ^{13}C NMR-based metabolomics for the classification of green coffee beans according to variety and origin. *J. Agric. Food Chem.* **2012**, *60*, 10118–10125. [CrossRef]
45. Ward, J.L.; Baker, J.M.; Miller, S.J.; Deborde, C.; Maucourt, M.; Biais, B.; Rolin, D.; Moing, A.; Moco, S.; Vervoort, J.; et al. An inter-laboratory comparison demonstrates that [H]-NMR metabolite fingerprinting is a robust technique for collaborative plant metabolomic data collection. *Metabolomics* **2010**, *6*, 263–273. [CrossRef]
46. Florentino-Ramos, E.; Villa-Ruano, N.; Hidalgo-Martínez, D.; Ramírez-Meraz, M.; Méndez-Aguilar, R.; Velásquez-Valle, R.; Zepeda-Vallejo, L.G.; Pérez-Hernández, N.; Becerra-Martínez, E. ^{1}H NMR-based fingerprinting of eleven Mexican Capsicum annuum cultivars. *Food Res. Int.* **2019**, *121*, 12–19. [CrossRef]
47. Djukovic, D.; Liu, S.; Henry, I.; Tobias, B.; Raftery, D. Signal enhancement in HPLC/microcoil NMR using automated column trapping. *Anal. Chem.* **2006**, *78*, 7154–7160. [CrossRef]
48. Kang, S.W.; Kim, C.Y.; Song, D.G.; Pan, C.H.; Cha, K.H.; Lee, D.U.; Um, B.H. Rapid identification of furanocoumarins in *Angelica dahurica* using the online LC-MMR-MS and their nitric oxide inhibitory activity in RAW 264.7 cells. *Phytochem. Anal.* **2010**, *21*, 322–327. [CrossRef]
49. Lisec, J.; Schauer, N.; Kopka, J.; Willmitzer, L.; Fernie, A.R. Gas chromatography mass spectrometry-based metabolite profiling in plants. *Nat. Protoc.* **2006**, *1*, 1–10. [CrossRef] [PubMed]
50. Fiehn, O.; Kopka, J.; Dörmann, P.; Altmann, T.; Trethewey, R.N.; Willmitzer, L. Metabolite profiling for plant functional genomics. *Nat. Biotechnol.* **2000**, *18*, 1157–1161. [CrossRef] [PubMed]
51. Fernie, A.R.; Trethewey, R.N.; Krotzky, A.J.; Willmitzer, L. Metabolite profiling: From diagnostics to systems biology. *Nat. Rev. Mol. Cell Biol.* **2004**, *5*, 763–769. [CrossRef]
52. Vorst, O.; de Vos, C.H.R.; Lommen, A.; Staps, R.V.; Visser, R.G.F.; Bino, R.J.; Hall, R.D. A non-directed approach to the differential analysis of multiple LC-MS-derived metabolic profiles. *Metabolomics* **2005**, *1*, 169–180. [CrossRef]
53. Nagashima, Y.; He, K.; Singh, J.; Metrani, R.; Crosby, K.M.; Jifon, J.; Jayaprakasha, G.K.; Patil, B.; Qian, X.; Koiwa, H. Transition of aromatic volatile and transcriptome profiles during melon fruit ripening. *Plant Sci.* **2021**, *304*, 110809. [CrossRef] [PubMed]
54. Kim, T.J.; Hyeon, H.; Park, N.I.; Yi, T.G.; Lim, S.H.; Park, S.Y.; Ha, S.H.; Kim, J.K. A high-throughput platform for interpretation of metabolite profile data from pepper (Capsicum) fruits of 13 phenotypes associated with different fruit maturity states. *Food Chem.* **2020**, *331*, 127286. [CrossRef] [PubMed]
55. Rambla, J.L.; Granell, A. Determination of plant volatile apocarotenoids. *Methods Mol. Biol.* **2020**, *2083*, 165–175.
56. Wu, H.; Huang, W.; Chen, Z.; Chen, Z.; Shi, J.; Kong, Q.; Sun, S.; Jiang, X.; Chen, D.; Yan, S. GC-MS-based metabolomic study reveals dynamic changes of chemical compositions during black tea processing. *Food Res. Int.* **2019**, *120*, 330–338. [CrossRef]
57. Park, Y.J.; Baek, S.-A.; Choi, Y.; Kim, J.K.; Park, S.-U. Metabolic profiling of nine *Mentha* species and prediction of their antioxidant properties using chemometrics. *Molecules* **2019**, *24*, 258. [CrossRef]
58. Rydberg, J.; Cooke, C.A.; Tolu, J.; Wolfe, A.P.; Vinebrooke, R.D. An assessment of chlorophyll preservation in lake sediments using multiple analytical techniques applied to the annually laminated lake sediments of Nylandssjön. *J. Paleolimnol.* **2020**, *64*, 379–388. [CrossRef]
59. Moco, S.; Bino, R.J.; Vorst, O.; Verhoeven, H.A.; de Groot, J.; van Beek, T.A.; Vervoort, J.; de Vos, C.H. A liquid chromatography-mass spectrometry-based metabolome database for tomato. *Plant Physiol.* **2006**, *141*, 1205–1218. [CrossRef] [PubMed]
60. Glauser, G.; Veyrat, N.; Rochat, B.; Wolfender, J.L.; Turlings, T.C. Ultra-high pressure liquid chromatography-mass spectrometry for plant metabolomics: A systematic comparison of high-resolution quadrupole-time-of-flight and single stage Orbitrap mass spectrometers. *J. Chromatogr. A* **2013**, *1292*, 151–159. [CrossRef] [PubMed]

61. Navarro-Reig, M.; Jaumot, J.; Baglai, A.; Vivó-Truyols, G.; Schoenmakers, P.J.; Tauler, R. Untargeted comprehensive two-dimensional liquid chromatography coupled with high-resolution mass spectrometry analysis of rice metabolome using multivariate curve resolution. *Anal. Chem.* **2017**, *89*, 7675–7683. [CrossRef]
62. Oliver, J.; Palou, A. Chromatographic determination of carotenoids in foods. *J. Chromatogr. A* **2000**, *881*, 543–555. [CrossRef]
63. Delpino-Rius, A.; Cosovanu, D.; Eras, J.; Vilaró, F.; Balcells, M.; Canela-Garayoa, R. A fast and reliable ultrahigh-performance liquid chromatography method to assess the fate of chlorophylls in teas and processed vegetable foodstuff. *J. Chromatogr. A* **2018**, *1568*, 69–79. [CrossRef] [PubMed]
64. Zapata, M.; Rodriguez, F.; Garrido, J.L. Separation of chlorophylls and carotenoids from marine phytoplankton: A new HPLC method using a reversed phase C-8 column and pyridine-containing mobile phases. *Mar. Ecol. Prog. Ser.* **2000**, *195*, 29–45. [CrossRef]
65. Su, Q.; Rowley, K.G.; Balazs, N.D. Carotenoids: Separation methods applicable to biological samples. *J. Chromatogr. B* **2002**, *781*, 393–418. [CrossRef]
66. Unlu, N.Z.; Bohn, T.; Francis, D.; Clinton, S.K.; Schwartz, S.J. Carotenoid absorption in humans consuming tomato sauces obtained from tangerine or high-beta-carotene varieties of tomatoes. *J. Agric. Food Chem.* **2007**, *55*, 1597–1603. [CrossRef]
67. Lee, B.L.; Ong, C.N. Comprehensive high-performance liquid chromatographic method for the measurements of lipophilic antioxidants in human plasma. *J. Chromatogr. A* **2009**, *1216*, 3131–3137. [CrossRef]
68. Li, S.; Nugroho, A.; Rocheford, T.; White, W.S. Vitamin A equivalence of the ß-carotene in ß-carotene-biofortified maize porridge consumed by women. *Am. J. Clin. Nutr.* **2010**, *92*, 1105–1112. [CrossRef] [PubMed]
69. Puspitasari-Nienaber, N.L.; Ferruzzi, M.G.; Schwartz, S.J. Simultaneous detection of tocopherols, carotenoids, and chlorophylls in vegetable oils by direct injection C30 RP-HPLC with coulometric electrochemical array detection. *J. Am. Oil Chem. Soc.* **2002**, *79*, 633–640. [CrossRef]
70. Bustamam, M.S.A.; Pantami, H.A.; Azizan, A.; Shaari, K.; Min, C.C.; Abas, F.; Nagao, N.; Maulidiani, M.; Banerjee, S.; Sulaiman, F.; et al. Complementary analytical platforms of NMR spectroscopy and LCMS analysis in the metabolite profiling of *Isochrysis galbana*. *Mar. Drugs* **2021**, *19*, 139. [CrossRef] [PubMed]
71. Lopez-Sanchez, P.; de Vos, R.C.H.; Jonker, H.H.; Mumm, R.; Hall, R.D.; Bialek, L.; Leenman, R.; Strassburg, K.; Vreeken, R.; Hankemeier, T.; et al. Comprehensive metabolomics to evaluate the impact of industrial processing on the phytochemical composition of vegetable purees. *Food Chem.* **2015**, *168*, 348–355. [CrossRef] [PubMed]
72. Viera, I.; Roca, M.; Pérez-Gálvez, A. Mass spectrometry of non-allomerized chlorophylls a and b derivatives from plants. *Curr. Org. Chem.* **2018**, *22*, 842–876. [CrossRef]
73. Lei, Z.; Sumner, B.W.; Bhatia, A.; Sarma, S.J.; Sumner, L.W. UHPLC-MS analyses of plant flavonoids. *Curr. Protoc. Plant Biol.* **2019**, *4*, e20085. [CrossRef]
74. Breithaupt, D.E.; Wirt, U.; Bamedi, A. Differentiation between lutein monoester regioisomers and detection of lutein diesters from marigold flowers (*Tagetes erecta* L.) and several fruits by liquid chromatography-mass spectrometry. *J. Agric. Food Chem.* **2002**, *50*, 66–70. [CrossRef]
75. Giuffrida, D.; Pintea, A.; Dugo, P.; Torre, G.; Pop, R.M.; Mondello, L. Determination of carotenoids and their esters in fruits of sea buckthorn (Hippophae rhamnoides L.) by HPLC-DAD-APCI-MS. *Phytochem. Anal.* **2012**, *23*, 267–273. [CrossRef]
76. Rodrigues, D.B.; Mariutti, L.R.B.; Mercadante, A.Z. Two-step cleanup procedure for the identification of carotenoid esters by liquid chromatography-atmospheric pressure chemical ionization-tandem mass spectrometry. *J. Chromatogr. A* **2016**, *1457*, 116–124. [CrossRef]
77. Chen, K.; Ríos, J.J.; Pérez-Gálvez, A.; Roca, M. Development of an accurate and high-throughput methodology for structural comprehension of chlorophylls derivatives. (I) Phytylated derivatives. *J. Chromatogr. A* **2015**, *1406*, 99–108. [CrossRef] [PubMed]
78. Fernandes, A.S.; Nascimento, T.C.; Pinheiro, P.N.; De Rosso, V.V.; De Menezes, C.R.; Jacob-Lopes, E.; Zepka, L.Q. Insights on the intestinal absorption of chlorophyll series from microalgae. *Food Res. Int.* **2021**, *140*, 110031. [CrossRef] [PubMed]
79. Murador, D.C.; De Souza Mesquita, L.M.; Neves, B.V.; Braga, A.R.C.; Martins, P.L.G.; Zepka, L.Q.; De Rosso, V.V. Bioaccessibility and cellular uptake by Caco-2 cells of carotenoids and chlorophylls from orange peels: A comparison between conventional and ionic liquid mediated extractions. *Food Chem.* **2021**, *339*, 127818. [CrossRef] [PubMed]
80. Chen, K.; Ríos, J.J.; Roca, M.; Pérez-Gálvez, A. Development of an accurate and high-throughput methodology for structural comprehension of chlorophylls derivatives. (II) Dephytylated derivatives. *J. Chromatogr. A* **2015**, *1412*, 90–99. [CrossRef]
81. Roca, M.; Ríos, J.J.; Pérez-Gálvez, A. Mass spectrometry: The indispensable tool for plant metabolomics of colourless chlorophyll catabolites. *Phytochem. Rev.* **2018**, *17*, 453–468. [CrossRef]
82. Pacini, T.; Fu, W.; Gudmundsson, S.; Chiaravalle, A.E.; Brynjolfson, S.; Palsson, B.O.; Astarita, G.; Paglia, G. Multidimensional analytical approach based on UHPLC-UV-ion mobility-MS for the screening of natural pigments. *Anal. Chem.* **2015**, *87*, 2593–2599. [CrossRef]
83. Fraser, P.D.; Enfisse, E.M.A.; Goodfellow, M.; Eguchi, T.; Bramley, P.M. Metabolite profiling of plant carotenoids using the matrix-assisted laser desorption ionization time-of-flight mass spectrometry. *Plant J.* **2007**, *49*, 552–564. [CrossRef]
84. Calvano, C.D.; Ventura, G.; Cataldi, T.R.; Palmisano, F. Improvement of chlorophyll identification in foodstuffs by MALDI ToF/ToF mass spectrometry using 1,5-diaminonaphthalene electron transfer secondary reaction matrix. *Anal. Bioanal. Chem.* **2015**, *407*, 6369–6379. [CrossRef]

85. Maroneze, M.M.; Caballero-Guerrero, B.; Zepka, L.Q.; Jacob-Lopes, E.; Pérez-Gálvez, A.; Roca, M. Accomplished high-resolution metabolomic and molecular studies identify new carotenoid biosynthetic reactions in *Cyanobacteria*. *J. Agric. Food Chem.* **2020**, *68*, 6212–6220. [CrossRef]
86. Hegeman, A.D.; Schulte, C.F.; Cui, Q.; Lewis, I.A.; Huttlin, E.L.; Eghbalnia, H.; Harms, A.C.; Ulrich, E.L.; Markley, J.L.; Sussman, M.R. Stable isotope assisted assignment of elemental compositions for metabolomics. *Anal. Chem.* **2007**, *79*, 6912–6921. [CrossRef]
87. Giavalisco, P.; Li, Y.; Matthes, A.; Eckhardt, A.; Hubberten, H.-M.; Hesse, H.; Segu, S.; Hummel, J.; Köhl, K.; Willmitzer, L. Elemental formula annotation of polar and lipophilic metabolites using ^{13}C, ^{15}N and ^{34}S isotope labelling, in combination with high-resolution mass spectrometry: Isotope labelling for unbiased plant metabolomics. *Plant J.* **2011**, *68*, 364–376. [CrossRef]
88. Pérez-Gálvez, A.; Viera, I.; Roca, M. Development of an accurate and direct method for the green food colorants detection. *Food Res. Int.* **2020**, *136*, 109484. [CrossRef]
89. Ríos, J.J.; Roca, M.; Pérez-Gálvez, A. Systematic HPLC/ESI-high resolution-qTOF-MS methodology for metabolomic studies in nonfluorescent chlorophyll catabolites pathway. *J. Anal. Methods Chem.* **2015**, *2015*, 1–10. [CrossRef]
90. Wehrens, R.; Carvalho, E.; Masuero, D.; de Juan, A.; Martens, S. High-throughput carotenoid profiling using multivariate curve resolution. *Anal. Bioanal. Chem.* **2013**, *405*, 5075–5086. [CrossRef]
91. Wehrens, R.; Bloemberg, T.G.; Eilers, P.H.C. Fast parametric time warping of peak lists. *Bioinformatics* **2015**, *31*, 3063–3065. [CrossRef] [PubMed]
92. Watanabe, M.; Tohge, T.; Balazadeh, S.; Erban, A.; Giavalisco, P.; Kopka, J.; Mueller-Roeber, B.; Fernie, A.R.; Hoefgen, R. Comprehensive metabolomics studies of plant developmental senescence. In *Plant Senescence*; Guo, Y., Ed.; Springer: New York, NY, USA, 2018; Volume 1744, pp. 339–358.
93. Herzog, M.; Fukao, T.; Winkel, A.; Konnerup, D.; Lamichhane, S.; Alpuerto, J.B.; Hasler-Sheetal, H.; Pedersen, O. Physiology, gene expression, and metabolome of two wheat cultivars with contrasting submergence tolerance: Submergence tolerance in two wheat cultivars. *Plant Cell Environ.* **2018**, *41*, 1632–1644. [CrossRef] [PubMed]
94. Pino Del Carpio, D.; Basnet, R.K.; De Vos, R.C.H.; Maliepaard, C.; Paulo, M.J.; Bonnema, G. Comparative methods for association studies: A case study on metabolite variation in a *Brassica rapa* core collection. *PLoS ONE* **2011**, *6*, e19624. [CrossRef] [PubMed]
95. Bamba, T.; Lee, J.W.; Matsubara, A.; Fukusaki, E. Metabolic profiling of lipids by supercritical fluid chromatography/mass spectrometry. *J. Chromatogr. A* **2012**, *1250*, 212–219. [CrossRef]
96. Giuffrida, D.; Zoccali, M.; Arigò, A.; Cacciola, F.; Roa, C.O.; Dugo, P.; Mondello, L. Comparison of different analytical techniques for the analysis of carotenoids in tamarillo (*Solanum betaceum* Cav.). *Arch. Biochem. Biophys.* **2018**, *646*, 161–167. [CrossRef] [PubMed]
97. Cichon, M.J.; Riedl, K.M.; Schwartz, S.J. A metabolomic evaluation of the phytochemical composition of tomato juices being used in human clinical trials. *Food Chem.* **2017**, *228*, 270–278. [CrossRef]
98. Fukushima, A.; Kusano, M.; Nakamichi, N.; Kobayashi, M.; Hayashi, N.; Sakakibara, H.; Mizuno, T.; Saito, K. Impact of clock-associated *Arabidopsis* pseudo-response regulators in metabolic coordination. *Proc. Natl. Acad. Sci. USA* **2009**, *106*, 7251–7256. [CrossRef] [PubMed]
99. Moco, S.; Capanoglu, E.; Tikunov, Y.; Bino, R.J.; Boyacioglu, D.; Hall, R.D.; Vervoort, J.; De Vos, R.C.H. Tissue specialization at the metabolite level is perceived during the development of tomato fruit. *J. Exp. Bot.* **2007**, *58*, 4131–4146. [CrossRef] [PubMed]
100. Witt, S.; Galicia, L.; Lisec, J.; Cairns, J.; Tiessen, A.; Araus, J.L.; Palacios-Rojas, N.; Fernie, A.R. Metabolic and phenotypic responses of greenhouse-grown maize hybrids to experimentally controlled drought stress. *Mol. Plant* **2012**, *5*, 401–417. [CrossRef] [PubMed]
101. Bemer, M.; Karlova, R.; Ballester, A.R.; Tikunov, Y.M.; Bovy, A.G.; Wolters-Arts, M.; de Barros Rossetto, P.; Angenent, G.C.; de Maagd, R.A. The tomato FRUITFULL homologs TDR4/FUL1 and MBP7/FUL2 regulate ethylene-independent aspects of fruit ripening. *Plant Cell* **2012**, *24*, 4437–4451. [CrossRef]
102. Harel-Beja, R.; Tzuri, G.; Portnoy, V.; Lotan-Pompan, M.; Lev, S.; Cohen, S.; Dai, N.; Yeselson, L.; Meir, A.; Libhaber, S.E.; et al. A genetic map of melon highly enriched with fruit quality QTLs and EST markers, including sugar and carotenoid metabolism genes. *Theor. Appl. Genet.* **2010**, *121*, 511–533. [CrossRef]
103. Pandey, S.; Patel, M.K.; Mishra, A.; Jha, B. Physio-Biochemical composition and untargeted metabolomics of cumin (*Cuminum cyminum* L.) make it promising functional food and help in mitigating salinity stress. *PLoS ONE* **2015**, *10*, e014446. [CrossRef]
104. Acharjee, A.; Kloosterman, B.; de Vos, R.C.H.; Werij, J.S.; Bachem, C.W.B.; Visser, R.G.F.; Maliepaard, C. Data integration and network reconstruction with omics data using Random Forest regression in potato. *Anal. Chim. Acta* **2011**, *705*, 56–63. [CrossRef]
105. Le Lay, P.; Isaure, M.P.; Sarry, J.E.; Kuhn, L.; Fayard, B.; Le Bail, J.L.; Bastien, O.; Garin, J.; Roby, C.; Bourguignon, J. Metabolomic, proteomic and biophysical analyses of *Arabidopsis thaliana* cells exposed to a caesium stress. Influence of potassium supply. *Biochimie* **2006**, *88*, 1533–1547. [CrossRef]
106. León, J.; Costa, A.; Castillo, M.C. Nitric oxide triggers a transient metabolic reprogramming in Arabidopsis. *Sci. Rep.* **2016**, *6*, 37945. [CrossRef]
107. Mwamba, T.M.; Islam, F.; Ali, B.; Lwalaba, J.L.W.; Gill, R.A.; Zhang, F.; Farooq, M.A.; Ali, S.; Ulhassan, Z.; Huang, Q.; et al. Comparative metabolomic responses of low- and high-cadmium accumulating genotypes reveal the cadmium adaptive mechanism in *Brassica napus*. *Chemosphere* **2020**, *250*, 126308. [CrossRef]
108. Hédiji, H.; Djebali, W.; Cabasson, C.; Maucourt, M.; Baldet, P.; Bertrand, A.; Boulila Zoghlami, L.; Deborde, C.; Moing, A.; Brouquisse, R.; et al. Effects of long-term cadmium exposure on growth and metabolomic profile of tomato plants. *Ecotox. Environ. Safety* **2010**, *73*, 1965–1974. [CrossRef]

109. Hu, X.; Gao, Y.; Fang, Z. Integrating metabolic analysis with biological endpoints provides insight into nanotoxicological mechanisms of graphene oxide: From effect onset to cessation. *Carbon* **2016**, *109*, 65–73. [CrossRef]
110. Allen, A.E.; LaRoche, J.; Maheswari, U.; Lommer, M.; Schauer, N.; Lopez, P.J.; Finazzi, G.; Fernie, A.R.; Bowler, C. Whole-cell response of the pennate diatom Phaeodactylum tricornutum to iron starvation. *Proc. Natl. Acad. Sci. USA* **2008**, *105*, 10438–10443. [CrossRef]
111. Matich, E.K.; Ghafari, M.; Camgoz, E.; Caliskan, E.; Pfeifer, B.A.; Haznedaroglu, B.Z.; Atilla-Gokcumen, G.E. Time-series lipidomic analysis of the oleaginous green microalga species Ettlia oleoabundans under nutrient stress. *Biotechnol. Biofuels* **2018**, *11*, 29. [CrossRef] [PubMed]
112. Alipanah, L.; Rohloff, J.; Winge, P.; Bones, A.M.; Brembu, T. Whole-cell response to nitrogen deprivation in the diatom *Phaeodactylum tricornutum*. *EXBOTJ* **2015**, *66*, 6281–6296. [CrossRef] [PubMed]
113. Lee, D.Y.; Park, J.-J.; Barupal, D.K.; Fiehn, O. System sesponse of metabolic networks in *Chlamydomonas reinhardtii* to total available ammonium. *Mol. Cell. Proteom.* **2012**, *11*, 973–988. [CrossRef] [PubMed]
114. Luan, H.; Meng, N.; Fu, J.; Chen, X.; Xu, X.; Feng, Q.; Jiang, H.; Dai, J.; Yuan, X.; Lu, Y.; et al. Genome-wide transcriptome and antioxidant analyses on gamma-irradiated phases of *Deinococcus radiodurans* R1. *PLoS ONE* **2014**, *9*, e85649. [CrossRef] [PubMed]
115. Hansler, A.; Chen, Q.; Ma, Y.; Gross, S.S. Untargeted metabolite profiling reveals that nitric oxide bioynthesis is an endogenous modulator of carotenoid biosynthesis in *Deinococcus radiodurans* and is required for extreme ionizing radiation resistance. *Arch. Biochem. Biophys.* **2016**, *589*, 38–52. [CrossRef]
116. Wang, L.; Huang, X.; Sun, W.; Too, H.Z.; Laserna, A.K.C.; Li, S.F.Y. A global metabolomic insight into the oxidative stress and membrane damage of copper oxide nanoparticles and microparticles on microalga *Chlorella vulgaris*. *Environ. Pollut.* **2020**, *258*, 113647. [CrossRef]
117. Reddy Pullagurala, V.L.; Adisa, I.O.; Rawat, S.; Kalagara, S.; Hernandez-Viezcas, J.A.; Peralta-Videa, J.R.; Gardea-Torresdey, J.L. ZnO nanoparticles increase photosynthetic pigments and decrease lipid peroxidation in soil grown cilantro (*Coriandrum sativum*). *Plant Physiol. Biochem.* **2018**, *132*, 120–127. [CrossRef]
118. Maldini, M.; Natella, F.; Baima, S.; Morelli, G.; Scaccini, C.; Langridge, J.; Astarita, G. Untargeted metabolomics reveals predominant alterations in lipid metabolism following light exposure in broccoli sprouts. *Int. J. Mol. Sci.* **2015**, *16*, 13678–13691. [CrossRef]
119. Will, S.E.; Henke, P.; Boedeker, C. Day and night: Metabolic profiles and evolutionary relationships of six axenic non-marine cyanobacteria. *Genome Biol. Evol.* **2019**, *11*, 270–294. [CrossRef]
120. Gong, G.; Liu, L.; Zhang, X. Multi-omics metabolism analysis on irradiation-induced oxidative stress to *Rhodotorula glutinis*. *Appl. Microbiol. Biotechnol.* **2019**, *103*, 361–374. [CrossRef] [PubMed]
121. Llewellyn, C.A.; Airs, R.L.; Farnham, G.; Greig, C. Synthesis, regulation and degradation of carotenoids under low level UV-B radiation in the filamentous cyanobacterium *Chlorogloeopsis fritschii* PCC 6912. *Front. Microbiol.* **2020**, *11*, 163. [CrossRef] [PubMed]
122. Ntagkas, N.; de Vos, R.C.H.; Woltering, E.J.; Nicole, C.C.S.; Labrie, C.; Marcelis, L.F.M. Modulation of the tomato fruit metabolome by LED light. *Metabolites* **2020**, *10*, 266. [CrossRef]
123. Yan, Z.; Zuo, J.; Zhou, F.; Shi, J.; Xu, D.; Hu, W.; Jiang, A.; Liu, Y.; Wang, Q. Integrated analysis of transcriptomic and metabolomic data reveals the mechanism by which LED light irradiation extends the postharvest quality of Pak-choi (*Brassica campestris* L. ssp. *chinensis* (L.) Makino var. *communis* Tsen et Lee). *Biomolecules* **2020**, *10*, 252.
124. Savoi, S.; Wong, D.C.J.; Arapitsas, P.; Miculan, M.; Bucchetti, B.; Peterlunger, E.; Fait, A.; Mattivi, F.; Castellarin, S.D. Transcriptome and metabolite profiling reveals that prolonged drought modulates the phenylpropanoid and terpenoid pathway in white grapes (*Vitis vinifera* L.). *BMC Plant Biol.* **2016**, *16*, 67. [CrossRef]
125. Uarrota, V.G.; Segatto, C.; Voytena, A.P.L.; Coelho, C.M.M.; Souza, C.A. Metabolic fingerprinting of water-stressed soybean cultivars by gas chromatography, near-infrared and UV-visible spectroscopy combined with chemometrics. *J. Agron. Crop Sci.* **2019**, *205*, 141–156. [CrossRef]
126. Lucini, L.; Miras-Moreno, B.; Busconi, M.; Marocco, A.; Gatti, M.; Poni, S. Molecular basis of rootstock-related tolerance to water deficit in *Vitis vinifera* L. cv. *Sangiovese*: A physiological and metabolomic combined approach. *Plant Sci.* **2020**, *299*, 110600. [PubMed]
127. Paul, K.; Sorrentino, M.; Lucini, L.; Rouphael, Y.; Cardarelli, M.; Bonini, P.; Miras Moreno, M.B.; Reynaud, H.; Canaguier, R.; Trtílek, M.; et al. A combined phenotypic and metabolomic approach for elucidating the biostimulant action of a plant-derived protein hydrolysate on tomato grown under limited water availability. *Front. Plant Sci.* **2019**, *10*, 493. [CrossRef]
128. Li, Z.; Cheng, B.; Yong, B.; Liu, T.; Peng, Y.; Zhang, X.; Ma, X.; Huang, L.; Liu, W.; Nie, G. Metabolomics and physiological analyses reveal β-sitosterol as an important plant growth regulator inducing tolerance to water stress in white clover. *Planta* **2019**, *250*, 2033–2046. [CrossRef] [PubMed]
129. Fernández-Marín, B.; Nadal, M.; Gago, J.; Fernie, A.R.; López-Pozo, M.; Artetxe, U.; García-Plazaola, J.I.; Verhoeven, A. Born to revive: Molecular and physiological mechanisms of double tolerance in a paleotropical and resurrection plant. *New Phytol.* **2020**, *226*, 741–759. [CrossRef]
130. Rouphael, Y.; Raimondi, G.; Lucini, L.; Carillo, P.; Kyriacou, M.C.; Colla, G.; Cirillo, V.; Pannico, A.; El-Nakhel, C.; De Pascale, S. Physiological and metabolic responses triggered by omeprazole improve tomato plant tolerance to NaCl stress. *Front. Plant Sci.* **2018**, *9*, 249. [CrossRef] [PubMed]

131. Zhang, Y.; Li, D.; Zhou, R.; Wang, X.; Dossa, K.; Wang, L.; Zhang, Y.; Yu, J.; Gong, H.; Zhang, X.; et al. Transcriptome and metabolome analyses of two contrasting sesame genotypes reveal the crucial biological pathways involved in rapid adaptive response to salt stress. *BMC Plant Biol.* **2019**, *19*, 66. [CrossRef] [PubMed]
132. Jia, X.; Zhu, Y.; Hu, Y.; Zhang, R.; Cheng, L.; Zhu, Z.; Zhao, T.; Zhang, X.; Wang, Y. Integrated physiologic, proteomic, and metabolomic analyses of *Malus halliana* adaptation to saline-alkali stress. *Hortic. Res.* **2019**, *6*, 91. [CrossRef]
133. Chu, F.L.; Pirastru, L.; Popovic, R.; Sleno, L. Carotenogenesis up-regulation in *Scenedesmus* sp. using a targeted metabolomics approach by liquid chromatography—high-resolution mass spectrometry. *J. Agric. Food Chem.* **2011**, *59*, 3004–3013. [CrossRef]
134. Van Meulebroek, L.; Vanden Bussche, J.; Steppe, K.; Vanhaecke, L. High-resolution Orbitrap mass spectrometry for the analysis of carotenoids in tomato fruit: Validation and comparative evaluation towards UV-VIS and tandem mass spectrometry. *Anal. Bioanal. Chem.* **2014**, *406*, 2613–2626. [CrossRef]
135. Mandelli, F.; Couger, M.B.; Paixão, D.A.A.; Machado, C.B.; Carnielli, C.M.; Aricetti, J.A.; Polikarpov, I.; Prade, R.; Caldana, C.; Paes Leme, A.F.; et al. Thermal adaptation strategies of the extremophile bacterium *Thermus filiformis* based on multi-omics analysis. *Extremophiles* **2017**, *21*, 775–788. [CrossRef]
136. Almeida, J.; Perez-Fons, L.; Fraser, P.D. A transcriptomic, metabolomic and cellular approach to the physiological adaptation of tomato fruit to high temperature. *Plant Cell Environ.* **2020**, *2020*, 1–19. [CrossRef]
137. Fogelman, E.; Oren-Shamir, M.; Hirschberg, J.; Mandolino, G.; Parisi, B.; Ovadia, R.; Tanami, Z.; Faigenboim, A.; Ginzberg, I. Nutritional value of potato (*Solanum tuberosum*) in hot climates: Anthocyanins, carotenoids, and steroidal glycoalkaloids. *Planta* **2019**, *249*, 1143–1155. [CrossRef]
138. Chou, L.; Kenig, F.; Murray, A.E.; Fritsen, C.H.; Doran, P.T. Effects of legacy metabolites from previous ecosystems on the environmental metabolomics of the brine of Lake Vida, East Antarctica. *Org. Geochem.* **2018**, *122*, 161–170. [CrossRef]
139. Wang, L.; Huang, X.; Lim, D.J.; Laserna, A.K.C.; Li, S.F.Y. Uptake and toxic effects of triphenyl phosphate on freshwater microalgae *Chlorella vulgaris* and *Scenedesmus obliquus*: Insights from untargeted metabolomics. *Sci. Total Environ.* **2019**, *650*, 1239–1249. [CrossRef] [PubMed]
140. Stopka, S.A.; Mansour, T.R.; Shrestha, B.; Maréchal, É.; Falconet, D.; Vertes, A. Turnover rates in microorganisms by laser ablation electrospray ionization mass spectrometry and pulse-chase analysis. *Anal. Chim. Acta* **2016**, *902*, 1–7. [CrossRef] [PubMed]
141. Silva, S.; Brown, P.; Ponchet, M. (Eds.) *Proceedings of the 1st World Congress on the Use of Biostimulants in Agriculture*; Acta Horticulturae Number 1009; ISHS: Leuven, Belgium, 2013; pp. 1–251.
142. Barrajón-Catalán, E.; Álvarez-Martínez, F.J.; Borrás, F.; Pérez, D.; Herrero, N.; Ruiz, J.J.; Micol, V. Metabolomic analysis of the effects of a commercial complex biostimulant on pepper crops. *Food Chem.* **2020**, *310*, 125818. [CrossRef]
143. Sofo, A.; Fausto, C.; Mininni, A.N.; Dichio, B.; Lucini, L. Soil management type differentially modulates the metabolomic profile of olive xylem sap. *Plant Physiol. Biochem.* **2019**, *139*, 707–714. [CrossRef]
144. Errard, A.; Ulrichs, C.; Kühne, S.; Mewis, I.; Drungowski, M.; Schreiner, M.; Baldermann, S. Single- versus multiple-pest infestation affects differently the biochemistry of tomato (*Solanum lycopersicum* 'Ailsa Craig'). *J. Agric. Food Chem.* **2015**, *63*, 10103–10111. [CrossRef]
145. Aliferis, K.A.; Chamoun, R.; Jabaji, S. Metabolic responses of willow (*Salix purpurea* L.) leaves to mycorrhization as revealed by mass spectrometry and 1H NMR spectroscopy metabolite profiling. *Front. Plant Sci.* **2015**, *6*, 344. [CrossRef]
146. Garg, N.; Zeng, Y.; Edlund, A.; Melnik, A.V.; Sanchez, L.M.; Mohimani, H.; Gurevich, A.; Miao, V.; Schiffler, S.; Lim, Y.W.; et al. Spatial molecular architecture of the microbial community of a *Peltigera* lichen. *mSystems* **2016**, *1*, e00139-16. [CrossRef]
147. Paix, B.; Carriot, N.; Barry-Martinet, R. A multi-omics analysis suggests links between the differentiated surface metabolome and epiphytic microbiota along the thallus of a mediterranean seaweed holobiont. *Front. Microbiol.* **2020**, *11*, 494. [CrossRef]
148. Sørensen, M.E.S.; Wood, A.J.; Minter, E.J.A.; Lowe, C.D.; Cameron, D.D.; Brockhurst, M.A. Comparison of independent evolutionary origins reveals both convergence and divergence in the metabolic mechanisms of symbiosis. *Curr. Biol.* **2020**, *30*, 328–334.e4. [CrossRef]
149. Cheng, Q.; Jia, W.; Hu, C.; Shi, G.; Yang, D.; Cai, M.; Zhan, T.; Tang, Y.; Zhou, Y.; Sun, X.; et al. Enhancement and improvement of selenium in soil to the resistance of rape stem against *Sclerotinia sclerotiorum* and the inhibition of dissolved organic matter derived from rape straw on mycelium. *Environ. Pollut.* **2020**, *265*, 114827. [CrossRef]
150. Tan, F.Q.; Zhang, M.; Xie, K.D.; Fan, Y.J.; Song, X.; Wang, R.; Wu, X.M.; Zhang, H.Y.; Guo, W.W. Polyploidy remodels fruit metabolism by modifying carbon source utilization and metabolic flux in Ponkan mandarin (*Citrus reticulata* Blanco). *Plant Sci.* **2019**, *289*, 110276. [CrossRef]
151. Maulidiani, M.; Mediani, A.; Abas, F.; Park, Y.S.; Park, Y.-K.; Kim, Y.M.; Gorinstein, S. ^{1}H NMR and antioxidant profiles of polar and non-polar extracts of persimmon (*Diospyros kaki* L.)—Metabolomics study based on cultivars and origins. *Talanta* **2018**, *184*, 277–286. [CrossRef]
152. Masetti, O.; Ciampa, A.; Nisini, L.; Sequi, P.; Dell'Abate, M.T. A multifactorial approach in characterizing geographical origin of Sicilian cherry tomatoes using ^{1}H-NMR profiling. *Food Res. Int.* **2017**, *100*, 623–630. [CrossRef] [PubMed]
153. Llano, S.M.; Muñoz-Jiménez, A.M.; Jiménez-Cartagena, C.; Londoño-Londoño, J.; Medina, S. Untargeted metabolomics reveals specific withanolides and fatty acyl glycoside as tentative metabolites to differentiate organic and conventional *Physalis peruviana* fruits. *Food Chem.* **2018**, *244*, 120–127. [CrossRef] [PubMed]
154. Li, L.; Zhao, J.; Zhao, Y.; Lu, X.; Zhou, Z.; Zhao, C.; Xu, G. Comprehensive investigation of tobacco leaves during natural early senescence via multi-platform metabolomics analyses. *Sci. Rep.* **2016**, *6*, 37976. [CrossRef] [PubMed]

155. Heavisides, E.; Rouger, C.; Reichel, A.; Ulrich, C.; Wenzel-Storjohann, A.; Sebens, S.; Tasdemir, D. Seasonal variations in the metabolome and bioactivity profile of *Fucus vesiculosus* extracted by an optimised, pressurised liquid extraction protocol. *Mar. Drugs* **2018**, *16*, 503. [CrossRef] [PubMed]
156. Masetti, O.; Ciampa, A.; Nisini, L.; Valentini, M.; Sequi, P.; Dell'Abate, M.T. Cherry tomatoes metabolic profile determined by ^{1}H-High Resolution-NMR spectroscopy as influenced by growing season. *Food Chem.* **2014**, *162*, 215–222. [CrossRef] [PubMed]
157. Juin, C.; Bonnet, A.; Nicolau, E.; Bérard, J.-B.; Devillers, R.; Thiéry, V.; Cadoret, J.-P.; Picot, L. UPLC-MSE Profiling of phytoplankton metabolites: Application to the identification of pigments and structural analysis of metabolites in *Porphyridium purpureum*. *Mar. Drugs* **2015**, *13*, 2541–2558. [CrossRef]
158. Marcellin-Gros, R.; Piganeau, G.; Stien, D. Metabolomic insights into marine phytoplankton diversity. *Mar. Drugs* **2020**, *18*, 78. [CrossRef]
159. Drapal, M.; Carvalho, E.; Ovalle Rivera, T.M.; Becerra Lopez-Lavalle, L.A.; Fraser, P.D. Capturing biochemical diversity in Cassava (*Manihot esculenta* Crantz) through the application of metabolite profiling. *J. Agric. Food Chem.* **2019**, *673*, 986–993. [CrossRef]
160. Ceccanti, C.; Landi, M.; Rocchetti, G.; Miras Moreno, M.B.; Lucini, L.; Incrocci, L.; Pardossi, A.; Guidi, L. Hydroponically grown *Sanguisorba minor* Scop.: Effects of cut and storage on fresh-cut produce. *Antioxidants* **2019**, *8*, 631. [CrossRef]
161. Yan, J.; Luo, Z.; Ban, Z.; Lu, H.; Li, D.; Yang, D.; Aghdam, M.S.; Li, L. The effect of the layer-by-layer (LBL) edible coating on strawberry quality and metabolites during storage. *Postharvest Biol. Technol.* **2019**, *147*, 29–38. [CrossRef]
162. Santin, M.; Lucini, L.; Castagna, A.; Chiodelli, G.; Hauser, M.-T.; Ranieri, A. Post-harvest UV-B radiation modulates metabolite profile in peach fruit. *Postharvest Biol. Technol.* **2018**, *139*, 127–134. [CrossRef]
163. Zhou, F.; Zuo, J.; Gao, L. An untargeted metabolomic approach reveals significant postharvest alterations in vitamin metabolism in response to LED irradiation in pak-choi (*Brassica campestris* L. ssp. chinensis (L.) Makino var. communis Tsen et Lee. *Metabolomics* **2019**, *15*, 155. [CrossRef] [PubMed]
164. Uarrota, V.G.; Moresco, R.; Coelho, B.; Nunes, E.d.C.; Peruch, L.A.M.; Neubert, E.d.O.; Rocha, M.; Maraschin, M. Metabolomics combined with chemometric tools (PCA, HCA, PLS-DA and SVM) for screening cassava (*Manihot esculenta* Crantz) roots during postharvest physiological deterioration. *Food Chem.* **2014**, *161*, 67–78. [CrossRef]
165. Fang, C.; Luo, J.; Wang, S. The diversity of nutritional metabolites: Origin, dissection, and application in crop breeding. *Front. Plant Sci.* **2019**, *10*, 01028. [CrossRef] [PubMed]
166. Sawada, Y.; Sato, M.; Okamoto, M. Metabolome-based discrimination of chrysanthemum cultivars for the efficient generation of flower color variations in mutation breeding. *Metabolomics* **2019**, *15*, 118. [CrossRef]
167. Hao, Z.; Liu, S.; Hu, L.; Shi, J.; Chen, J. Transcriptome analysis and metabolic profiling reveal the key role of carotenoids in the petal coloration of *Liriodendron tulipifera*. *Hortic. Res.* **2020**, *7*, 70. [CrossRef]
168. Heng, Z.; Sheng, O.; Huang, W.; Zhang, S.; Fernie, A.R.; Motorykin, I.; Kong, Q.; Yi, G.; Yan, S. Integrated proteomic and metabolomic analysis suggests high rates of glycolysis are likely required to support high carotenoid accumulation in banana pulp. *Food Chem.* **2019**, *297*, 125016. [CrossRef]
169. Enfissi, E.M.A.; Barneche, F.; Ahmed, I.; Lichtlé, C.; Gerrish, C.; McQuinn, R.P.; Giovannoni, J.J.; Lopez-Juez, E.; Bowler, C.; Bramley, P.M.; et al. Integrative transcript and metabolite analysis of nutritionally enhanced *DE-ETIOLATED1* downregulated tomato fruit. *Plant Cell.* **2010**, *22*, 1190–1215. [CrossRef] [PubMed]
170. Luo, W.; Gong, Z.; Li, N.; Zhao, Y.; Zhang, H.; Yang, X.; Liu, Y.; Rao, Z.; Yu, X. A negative regulator of carotenogenesis in *Blakeslea trispora*. *Appl. Environ. Microbiol.* **2020**, *86*, e02462-19. [CrossRef] [PubMed]
171. Farré, G.; Maiam Rivera, S.; Alves, R.; Vilaprinyo, E.; Sorribas, A.; Canela, R.; Naqvi, S.; Sandmann, G.; Capell, T.; Zhu, C.; et al. Targeted transcriptomic and metabolic profiling reveals temporal bottlenecks in the maize carotenoid pathway that may be addressed by multigene engineering. *Plant J.* **2013**, *75*, 441–455. [CrossRef] [PubMed]
172. Alós, E.; Roca, M.; Iglesias, D.J.; Mínguez-Mosquera, M.I.; Damasceno, C.M.B.; Thannhauser, T.W.; Rose, J.K.C.; Talón, M.; Cercós, M. An evaluation of the basis and consequences of a stay-green mutation in the *navel negra* citrus mutant using transcriptomic and proteomic profiling and metabolite analysis. *Plant Physiol.* **2008**, *147*, 1300–1315. [CrossRef] [PubMed]
173. Perez-Fons, L.; Wells, T.; Corol, D.I.; Ward, J.L.; Gerrish, C.; Beale, M.H.; Seymour, G.B.; Bramley, P.M.; Fraser, P.D. A genome-wide metabolomic resource for tomato fruit from *Solanum pennellii*. *Sci. Rep.* **2015**, *4*, 3859. [CrossRef] [PubMed]
174. Lee, J.J.L.; Chen, L.; Shi, J.; Trzcinski, A.; Chen, W.-N. Metabolomic profiling of *Rhodosporidium toruloides* grown on glycerol for carotenoid production during different growth phases. *J. Agric. Food Chem.* **2014**, *62*, 10203–10209. [CrossRef]
175. Farré, G.; Perez-Fons, L.; Decourcelle, M.; Breitenbach, J.; Hem, S.; Zhu, C.; Capell, T.; Christou, P.; Fraser, P.D.; Sandmann, G. Metabolic engineering of astaxanthin biosynthesis in maize endosperm and characterization of a prototype high oil hybrid. *Transgenic Res.* **2016**, *25*, 477–489. [CrossRef]
176. Saiman, M.Z.; Miettinen, K.; Mustafa, N.R.; Choi, Y.H.; Verpoorte, R.; Schulte, A.E. Metabolic alteration of *Catharanthus roseus* cell suspension cultures overexpressing geraniol synthase in the plastids or cytosol. *Plant Cell Tiss. Organ. Cult.* **2018**, *134*, 41–53. [CrossRef]
177. Hasunuma, T.; Takaki, A.; Matsuda, M.; Kato, Y.; Vavricka, C.J.; Kondo, A. Single-stage astaxanthin production enhances the nonmevalonate pathway and photosynthetic central metabolism in *Synechococcus* sp. PCC 7002. *Synth. Biol.* **2019**, *8*, 2701–2709.
178. Comas, J.; Benfeitas, R.; Vilaprinyo, E.; Sorribas, A.; Solsona, F.; Farré, G.; Berman, J.; Zorrilla, U.; Capell, T.; Sandmann, G.; et al. Identification of line-specific strategies for improving carotenoid production in synthetic maize through data-driven mathematical modeling. *Plant J.* **2016**, *87*, 455–471. [CrossRef]

179. Maroneze, M.M.; Zepka, L.Q.; Jacob-Lopes, E.; Pérez-Gálvez, A.; Roca, M. Chlorophyll oxidative metabolism during the phototrophic and heterotrophic growth of *Scenesdesmus obliquus*. *Antioxidants* **2019**, *8*, 600. [CrossRef] [PubMed]
180. Santos, P.M.; Batista, D.L.J.; Ribeiro, L.A.F.; Boffo, E.F.; de Cerqueira, M.D.; Martins, D.; de Castro, R.D.; de Souza-Neta, L.C.; Pinto, E.; Zambotti-Villela, L.; et al. Identification of antioxidant and antimicrobial compounds from the oilseed crop *Ricinus communis* using a multiplatform metabolite profiling approach. *Ind. Crop. Prod.* **2018**, *124*, 834–844. [CrossRef]
181. Capanoglu, E.; Beekwilder, J.; Boyacioglu, D.; Hall, R.; de Vos, R. Changes in antioxidant and metabolite profiles during production of tomato paste. *J. Agric. Food Chem.* **2008**, *56*, 964–973. [CrossRef] [PubMed]
182. Beleggia, R.; Platani, C.; Papa, R.; Di Chio, A.; Barros, E.; Mashaba, C.; Wirth, J.; Fammartino, A.; Sautter, C.; Conner, S.; et al. Metabolomics and food processing: From semolina to pasta. *J. Agric. Food Chem.* **2011**, *59*, 9366–9377. [CrossRef]
183. Van den Broek, T.J.; Kremer, B.H.; Rezende, M.M.; Hoevenaars, F.; Weber, P.; Hoeller, U.; van Ommen, B.; Wopereis, S. The impact of micronutrient status on health: Correlation network analysis to understand the role of micronutrients in metabolic-inflammatory processes regulating homeostasis and phenotypic flexibility. *Genes Nutr.* **2017**, *12*, 5. [CrossRef]
184. Moser, S.; Erhart, T.; Neuhauser, S.; Kräutler, B. Phyllobilins from senescence-associated chlorophyll breakdown in the leaves of Basil (Ocimum basilicum) show increased abundance upon herbivore attack. *J. Agric. Food Chem.* **2020**, *68*, 7132–7142. [CrossRef]
185. McQuinn, R.P.; Giovannoni, J.J.; Pogson, B.J. More than meets the eye: From carotenoid biosynthesis, to new insights into apocarotenoid signaling. *Curr. Opin. Plant Biol.* **2015**, *27*, 172–179. [CrossRef]
186. López-Ráez, J.A.; Kohlen, W.; Charnikhova, T.; Mulder, P.; Undas, A.K.; Sergeant, M.J.; Verstappen, F.; Bugg, T.D.H.; Thompson, A.J.; Ruyter-Spira, C.; et al. Does abscisic acid affect strigolactone biosynthesis? *New Phytol.* **2010**, *187*, 343–354. [CrossRef]
187. Huttanus, H.M.; Vu, T.; Guruli, G.; Tracey, A.; Carswell, W.; Said, N.; Du, P.; Parkinson, B.G.; Orlando, G.; Robertson, J.L.; et al. Raman chemometric urinalysis (Rametrix) as a screen for bladder cancer. *PLoS ONE* **2020**, *15*, e0237070. [CrossRef]
188. Nguyen, T.T.; Parat, M.-O.; Shaw, P.N.; Hewavitharana, A.K.; Hodson, M.P. Traditional aboriginal preparation alters the chemical profile of *Carica* papaya leaves and impacts on cytotoxicity towards human squamous cell carcinoma. *PLoS ONE* **2016**, *11*, e0147956. [CrossRef]
189. Arathi, B.P.; Sowmya, P.R.-R.; Kuriakose, G.C.; Vijay, K.; Baskaran, V.; Jayabaskaran, C.; Lakshminarayana, R. Enhanced cytotoxic and apoptosis inducing activity of lycopene oxidation products in different cancer cell lines. *Food Chem. Toxicol.* **2016**, *97*, 265–276. [CrossRef] [PubMed]
190. Rebholz, C.M.; Lichtenstein, A.H.; Zheng, Z.; Appel, L.J.; Coresh, J. Serum untargeted metabolomic profile of the dietary approaches to stop hypertension (DASH) dietary pattern. *Am. J. Clin. Nutr.* **2018**, *108*, 243–255. [CrossRef] [PubMed]

MDPI
St. Alban-Anlage 66
4052 Basel
Switzerland
Tel. +41 61 683 77 34
Fax +41 61 302 89 18
www.mdpi.com

Antioxidants Editorial Office
E-mail: antioxidants@mdpi.com
www.mdpi.com/journal/antioxidants

www.ingramcontent.com/pod-product-compliance
Lightning Source LLC
LaVergne TN
LVHW070452100526
838202LV00014B/1709